AF251982

Interference Analysis and Reduction for Wireless Systems

Interference Analysis and Reduction for Wireless Systems

Peter Stavroulakis

Artech House
Boston • London
www.artechhouse.com

Library of Congress Cataloging-in-Publication Data
Stavroulakis, Peter.
    Interference analysis and reduction for wireless systems / Peter Stavroulakis.
        p. cm. — (Artech House mobile communications series)
    Includes bibliographical references and index.
    ISBN 1-58053-316-7 (alk. paper)
        1. Radio—Interference.   2. Cellular telephone systems—Protection.   I. Title.
    TK6553 .S715   2003
    621.382'24—dc21   2002038273

British Library Cataloguing in Publication Data
Stavroulakis, Peter.
    Interference analysis and reduction for wireless systems. — (Artech House mobile
    communications series)
    1. Wireless communication systems   2. Electromagnetic interference
    I. Title
    621.3'845

    ISBN 1-58053-316-7

Cover design by Yekaterina Ratner

Figures 2.1, 2.8, 2.9, 2.10, 2.12, 4.10, 4.11, 4.12, 4.13, 4.19, 4.21, and 4.22 © 1999. Reprinted by permission of John Wiley & Sons, Inc., *Antennas and Propagation for Wireless Communication Systems*, by S. R. Saunders.

Figures 2.4, 2.6, and 2.7 © 2000. Reprinted by permission of John Wiley & Sons, Inc., *The Mobile Radio Propagation Channel*, by J. D. Parsons.

Figures 3.13, 3.16, 3.17, and 4.14–4.16, © 2000. Reprinted by permission of John Wiley & Sons, Inc., *Digital Communication over Fading Channels*, by M. K. Simon and M. S. Alouini.

Figure 5.17 © 2001. Reprinted by permission of John Wiley & Sons, Inc., *Wireless Local Loops, Theory and Applications*, by P. Stavroulakis.

Figures 6.13–6.15 © 2000. Reprinted by permission of John Wiley & Sons, Inc., *Advanced Digital Signal Processing and Noise Reduction*, by S. V. Vaseghi.

International Standard Book Number: 1-58053-316-7
Library of Congress Catalog Card Number: 2002038273

10 9 8 7 6 5 4 3 2 1

Contents

Preface

The subject of interference in communications systems is as old as communications itself. Agamemnon, the King of Mycenes, who captured Troy more than 800 miles away, to get back his niece, the beautiful Eleni, and wanted to notify his wife Clytaemistra about this happy event, used the most sophisticated communication techniques of that time to achieve his purpose. From that time until today, people have been aware of the importance of interference and the effect it can have on communications. Agamemnon used light sources at the peaks of mountains—by the motions of these sources, the information was coded and transmitted from mountain to mountain to arrive at Mycenes the same day.

If we analyze this communication system of Agamemnon, we find that he used three of the most important techniques still used today for interference suppression in communications. The first was the nature and form of the information signal (certain shape of flame), which corresponds to signal modulation techniques of today. The motion of the flames corresponds to modern coding techniques. The use of mountains corresponds to channel estimation techniques, which are used for the exploitation of favorable channel propagation characteristics or the avoidance of unfavorable characteristics through compensation of certain propagation parameters, fading, narrowband, or wideband characteristics.

Over the more than 3,000 years since Agamemnon, the necessary coexistence of information and interfering signals has been accommodated in the design of communication systems. Modern mathematical modeling and simulation techniques as new tools of study greatly facilitated this effort.

Of course, the subject of interference received special attention each time people concentrated on the usage of wireless systems on a large scale, as during the decade of 1970–1980 with the implementation of satellite systems and from 1995 until today with the large-scale applications of mobile systems. Most results of the worldwide efforts that had to do with interference analysis and design have been included in [1] and [2]. The purpose of this book is to present and analyze the techniques that are being used and can be used in the design of modern wireless systems in order to achieve an acceptable quality of service in an interference environment. Of course, many things have changed since Agamemnon and the early satellite implementations and special communication systems used during early space exploration. It is absolutely certain that the communications world is becoming digital, the wireless systems are converging into a universal standard, and the interference analysis and suppression techniques have become highly sophisticated because they have to be applicable to a universal communication system. As such, the material in this book becomes more and more comprehensive, from Chapter 4 on, and the reader—who can be an instructor, researcher, practicing engineer or a student—must have had a course in communication, signal processing or probability, and stochastic process in order to get the most out of it. The structure of this book is based on the methodology adapted by the author to present the subject matter of the book.

It is assumed that the reader will not have any difficulty proceeding along the steps that are formatted by the chapters that follow. Even though the interference signals (sources) and the general interference environment are discussed in Chapters 4 and 5, we shall briefly explain here in general terms the main theme of interference. As we shall see later, Chapters 4 through 7 introduce and analyze the subject of interference in detail.

For the readers who are not familiar with this subject, as far as this book is concerned, there exist two types of interfering signals no matter what their source is. One type is an additive signal, which enters the receiver and affects the detection process. Its source and nature can be a signal from a noiselike source, a signal from another friendly or nonfriendly system, or a signal produced by the nonlinearities of the system itself and its components (such as filters, which are exhibited as intermodulation signals and/or intersymbol interference). The other kind of interfering signals are the multiplicative types, which are mainly produced because of multipath phenomena in wireless systems, as we shall see in Chapter 4.

Before we embark on the main theme of this book, we consider it necessary to present an overview of the modern wireless systems in use and analyze the characteristics of the wireless channel and the transmission systems used. This background is necessary for the discussion that follows in the subsequent chapters. In Chapter 4, we analyze the techniques used for the study of wireless systems behavior when the interference environment is fading due to multipath interferers. In Chapter 5, we study the case where the interference environment is mainly characterized by additive interference effects. In Chapter 6, we review and use the results of Chapters 4 and 5 to develop interference suppression techniques and show how they can be used in real implementation. Finally, in Chapter 7, we present actual interference cancelers, which are used in real designs that utilize most of the techniques presented in previous chapters. The structure of the book also exhibits the methodology we propose for the analysis and design of wireless systems in an interference environment. We first need to quantify the parameters of the wireless systems that play a major role in the design, characterize the channel that will be used, and define the transmission system to be implemented. Subsequently, we must analyze and quantify the additive and/or multiplicative nature of the interfering signals, and finally we must utilize the appropriate technique to suppress or mitigate the effect of interference.

It is seen, therefore, that each chapter of this book has become an indispensable ring in the chain of steps necessary for a complete and integrated analysis and design of any wireless system in any interference environment as shown graphically in Figure P.1.

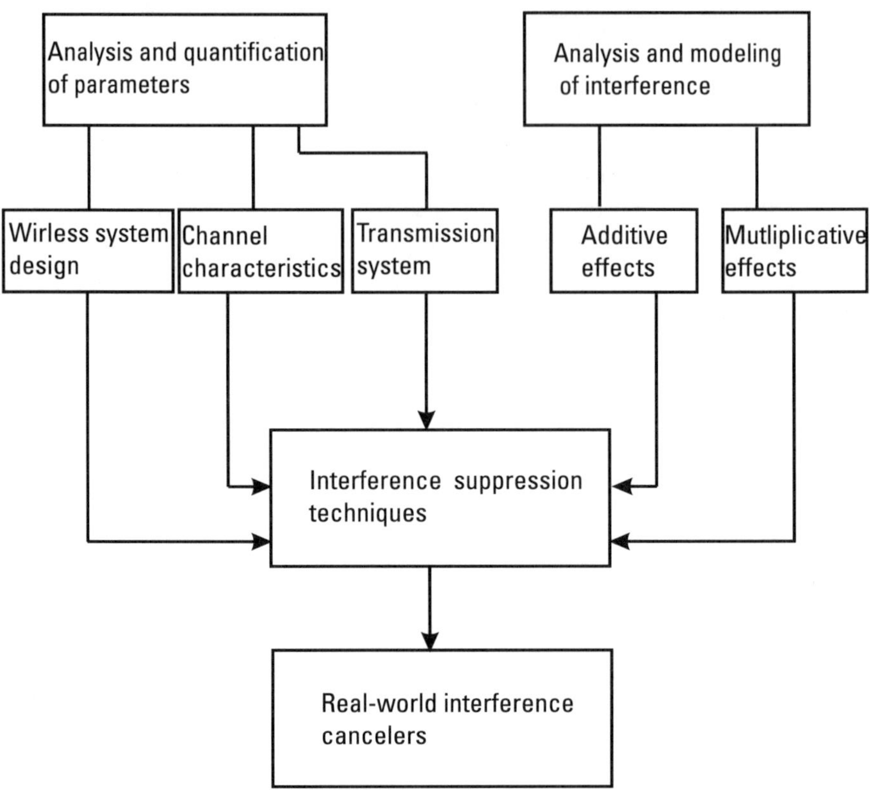

Figure P.1 Methodology of interference analysis and suppression.

References

[1]   Stavroulakis, P., *Interference Analysis of Communications Systems,* New York: IEEE Press, 1980.

[2]   Stavroulakis, P., *Wireless Local Loops,* New York: John Wiley, 2001.

Acknowledgments

This book required the work of many people during the various phases of its preparation. I feel indebted to my assistants, Miss Theano Lyrantonaki, Mr. Harris Kosmidis, my son, Peter, and Mr. Nick Farsaris, who worked endless hours to help me bring this important project to completion.

1

Overview of Wireless Information Systems

1.1 Introduction

We are all being exposed to a communications revolution that is taking us from a world where the dominant modes of electronic communications were standard telephone service and voiceband data communications carried over fixed telephone networks, packet-switched data networks, and high-speed local area networks (LANs) to one where a seamless and mobile communications environment has become a reality. Traditional wireless information networks, which include cordless and cellular telephones, paging systems, mobile data networks, and mobile satellite systems, have experienced enormous growth over the last decade and the new concepts of personal communication systems, wireless LANs (WLANs), and mobile computing have appeared in the industry [1–18].

In conjunction with this revolution, we are witnessing a transition in the infrastructure of our communication networks. After more than a century of reliance on analog-based technology for telecommunications, we now live in a mixed analog and digital world and are rapidly moving toward all-digital networks. In this chapter, we will briefly describe the various wireless systems in use and show that the wireless channel, which is the main vehicle of transmission of information, is not as predictable as the wired channel of the past.

In a wireless, all-digital world, we have to deal with many more forms of interference agents than in the wired world. The subject of interference over the years has triggered the interest of researchers in direct relation to the development of wireless communications. During the decade of 1970–1980, we have had a great upsurge in the study of this subject because of the development of satellite communications, as can be seen in [1]. We are now living the second decade of another revolution—mobile communications. The interest, therefore, in the subject of interference has increased, and the book at hand is a testament of that interest. We saw in the Preface that in order to study such a wide subject, we need to develop a methodology.

The structure of this book follows the methodology adapted and described in the preface. It is therefore important to start with the analysis of the wireless systems design parameters, which play a major role as these systems operate in an interference environment. In other words, in this chapter we pinpoint those wireless systems design characteristics that affect or can be affected by interference. This relationship and interdependence justifies the relevance of Chapter 1 in the structure of a book on interference. Moreover, it is in compliance with the methodology developed and adapted in the preface. This is very important because the interference environment of wireless systems is less controllable than that of wired systems.

In the wireless world, we encounter many more interference sources, which can be put into two categories. One category is the additive type of interference, which can be caused by cochannel, adjacent channel, intersystem, intermodulation, and intersymbol interfering signals. The second category includes multipath interference (i.e., the signals, which are produced by all sorts of reflections and diffractions from obstacles along the communication path, that interfere with the signal that bears the information). This type of interference affects the information signal in a multiplicative fashion, as we shall see in Chapter 4.

We shall, in this chapter, point out the vulnerable points of wireless systems in an interference- and distortion-based environment. In later chapters, we shall study the mechanisms of mitigating the effects of these distortions.

1.1.1 Wireless World Evolution

In Figure 1.1, we distinguish the various categories of the wireless networks and their evolution.

The evolution of the wireless world approaches convergence to a universal system, which, with the ability of pocket-sized personal stations, can access public and private communications networks through interoperable terrestrial and satellite media.

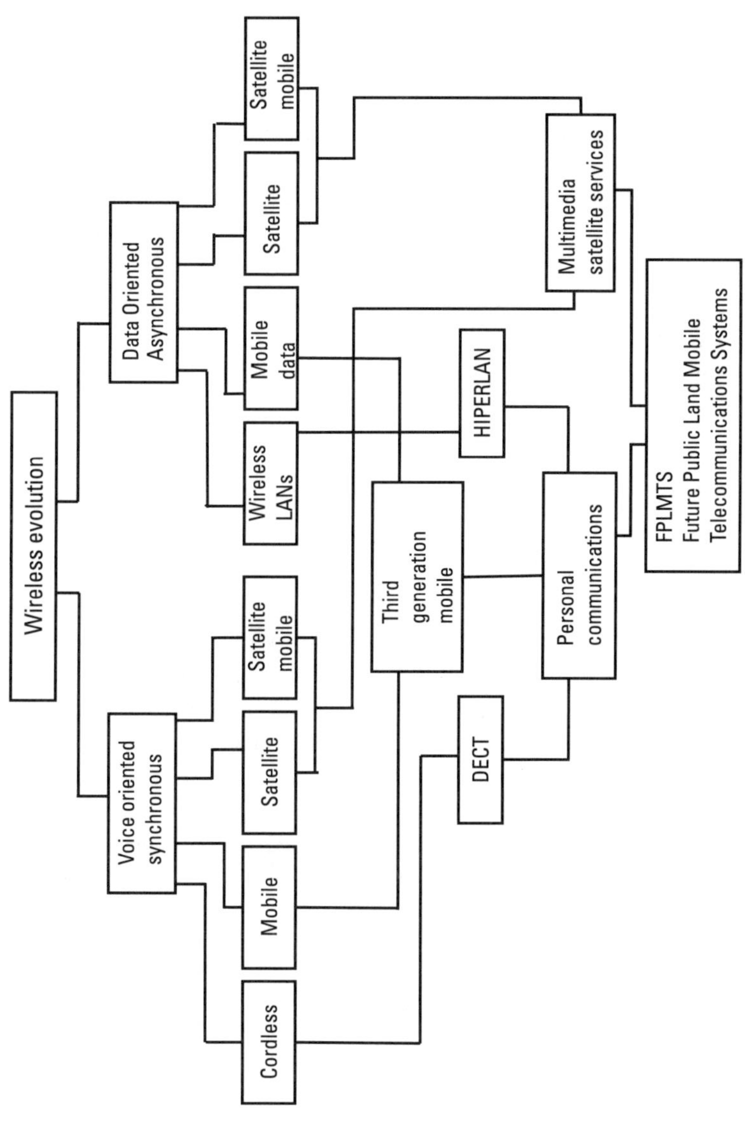

Figure 1.1 Evolution of wireless systems.

1.2 Historical Perspective

One hundred years ago, the notion of transmitting information in a wireless manner must have seemed like science fiction. Marchese Guglielmo Marconi made it possible. In 1896, the first patent for wireless communication was granted to him in the United Kingdom. He demonstrated the first wireless communication system in 1897 between a land-based station and a tugboat. Since then, important developments in the field of wireless communication have been taking place that shrink the world into a communication village. Such a system will provide communication services from one person to another in any place, at any time, in any form, and through any medium by using one pocket-sized unit at minimum cost, with acceptable quality and security through the use of a personal telecommunication reference number, or a PIN [1, 19, 20].

The wireless era in general can be divided into three periods: the pioneer era, the premobile era, and the mobile era, during which much of the fundamental research and development in the field of wireless communications took place [1].

1.3 First Generation Systems

The global communication village has been evolving since the birth of the first generation analog cellular system. Tables 1.1(a) and 1.1(b) [20] show a summary of analog cellular radio systems. Various standard systems were developed worldwide: advanced mobile phones service (AMPS) in the United States, Nordic mobile telephones (NMT) in Europe, total access communication systems (TACS) in the United Kingdom, Nippon Telephone and Telegraph (NTT) in Japan, and so on. The first AMPS cellular telephone service commenced operation in Chicago in 1983. In Norway, NMT-450 was launched in 1981 and later the NMT-900 was introduced. Similar systems were introduced in Germany, Portugal, Italy, and France. All the first generation systems used frequency modulation (FM) for speech and frequency shift keying (FSK) for signaling, and the access technique used was frequency division multiple access (FDMA).

1.4 Second Generation Systems

Advancements in digital technology gave birth to Pan-European digital cellular mobile systems, with general mobile systems (GSM) taking the acronym

Table 1.1(a)
Summary of Analog Cellular Radio Systems—AMPS, NMT-450, NMT-900, TACS, and ETACS

System	AMPS	NMT-450	NMT-900	TACS	ETACS
Frequency range (mobile Tx/base Tx) (MHz)	824–849/869–894	453–457.5/463–467.5	890–915/463–467.5	890–915/935–960	872–905/917–950
Channel spacing (kHz)	30	25	12.5*	25	25
Number of channels	832	180	1999	1,000	1,240
Region	The Americas, Australia, China, Southeast Asia	Europe	Europe, China, India, Africa	United Kingdom	Europe, Africa

*Frequency interleaving using overlapping channels; the channel spacing is half the nominal channel bandwidth.
(From: [20].)

Table 1.1(b)
Summary of Analog Cellular Radio Systems—C-450, RTMS, Radiocom-2000, JTACS/NTACS, and NTT

System	C-450	RTMS	Radiocom-2000	JTACS/NTACS	NTT
Frequency range (mobile Tx/base Tx) (MHz)	450–455.74/460–465.74	450–455/460–465	165.2–168.4/169.8–173 192.5–199.5/200.5–207.5 215.5–233.5/207.5–215.5 414.8–418/424.8–428	915–925/860–870 898–901/843–846 918.5–922/863.5–867	925–940/870–855 915–918.5/860–863.5 922–925/867–870
Channel spacing (kHz)	10*	25	12.5	25/12.5* 25/12.5* 12.5*	25/6.25* 6.25* 6.25*
Number of channels	573	200	256 560 640 256	400/800 120/240 280	600/2,400 560 480
Region	Germany, Portugal	Italy	France	Japan	Japan

*Frequency interleaving using overlapping channels; the channel spacing is half the nominal channel bandwidth.
(From: [20].)

from the French word; digital cordless systems (DCS)-1800, in Europe; personal digital cellular (PDC) systems in Japan; and interim standard (IS)-54/136 and IS-95 in North America, which are the second generation systems. A summary of digital cellular radio systems is shown in Table 1.2 [20].

We observe that time division multiple access (TDMA) is used as the access technique, except for IS-95, which is based in code division multiple access (CDMA). The second generation systems provide digital speech and short message services. More details will be given in Chapter 2. GSM has become deeply rooted in Europe and in more than 70 countries worldwide. DCS-1800 is also spreading outside Europe to East Asia and some South American countries. The development of new digital cordless technologies gave birth to the second-supplement generation systems—namely, personal handy phone systems (PHS, formerly PHP) in Japan, digital european cordless telephone (DECT) in Europe, and personal access communication services (PACS) in North America. Table 1.3 shows the features of the second-generation cordless systems [20].

In recent years DECT, PHS, and PACS/wireless access communications systems (WACS) have been introduced to provide cost-effective wireless connection in local loops (WLL) [3]. The term *local loop* stands for the medium that connects the equipment in the user's premises with telephone switching equipment. There are psychological and technical challenges that WLL faces in becoming acceptable to users. Because it replaces copper cable for connecting the user with the local exchange, the user may be apprehensive about reliability, privacy, and interference with wireless appliances like radio and television (manmade noise) and other WLL users. WLL must prove itself at least as good, if not better, than the services provided by physical cable. It should be able to carry and deliver voice, data, state-of-the-art multimedia services, and other modern services as efficiently as plain old telephone service (POTS) does [21, 22].

Although the second generation services and their supplements have covered local, national, and international areas, they still have one major drawback in terms of a universal service facility. In addition to the system discussed in this section, wireless data systems and WLANs are also very important in the field of wireless communications. Some features of wide area wireless packet data systems are shown in Table 1.4. Advanced radio data information service (ARDIS) and RAM mobile data (RMD) are the earliest and best-known systems in North America. Cellular digital packet data (CDPD) is a new wide area packet data network. The general packet radio service (GPRS) standard was developed to provide packet data service over the GSM infrastructure [23–25].

Table 1.2
Summary of Digital Cellular Radio Systems (From: [20].)

Systems	GSM/DCS-1800	IS-54	IS-95	PDC
Frequency range (base Rx/Tx, MHz)	GSM: Tx: 935–960 Rx: 890–915 DCS-1800: Tx: 1805–1880 Rx: 1710–1785	Tx: 869–894 Rx: 824–849	Tx: 869–894 Rx: 824–849	Tx: 810–826 Rx: 940–956 Tx: 1429–1453 Rx: 1477–1501
Channel spacing (kHz)	200	30	1,250	25
Number of channels	GSM: 124 DCS-1800: 375	832	20	1,600
Number of users per channel	GSM: 8 DCS-1800: 16	3	63	3
Multiple access	TDMA/FDMA	TDMA/FDMA	CDMA/FDMA	TDMA/FDMA
Duplex	FDD	FDD	FDD	FDD
Modulation	GMSK	$\pi/4$ DQPSK	BPSK/QPSK	$\pi/4$ DQPSK
Speech coding and its rate (Kbps)	RPE-LTP 13	VSELP 7.95	QCELP 8	VSELP 6.7
Channel coding	1/2 Convolutional	1/2 Convolutional	Uplink 1/3 Downlink 1/2 Convolutional	9/17 Convolutional
Region	Europe, China, Australia, Southeast Asia	North America, Indonesia	North America, Australia, Southeast Asia	Japan

Gaussian minimum shift keying (GMSK); regular pulse excited–long term prediction (RPE-LTP); vector excited linear predictor (VSELP); qualcomm code excited linear predictive coding (QCELP); Rx receiver (Rx); Tx Transmitter (Tx); and digital cellular system (DCS).

Table 1.3
Second Generation Cordless Systems

System	CT2/CT2+*	DECT	PHS	PACS
Frequency range (base Rx/Tx, MHz)	CT2: 864–868 CT2+: 944–948	1,880–1,990	1,895–1,918	TX: 1,850–1,910 Rx: 1,930–1,990
Channel spacing (kHz)	100	1,728	300	300
Number of channels	40	10	77	96
Number of users per channel	1	12	4	8
Multiple access	FDMA	TDMA/FDMA	TDMA/FDMA	TDMA/FDMA
Duplex	TDD	TDD	TDD	FDD
Modulation	GFSK	GFSK	$\pi/4$ DQPSK	$\pi/4$ DQPSK
Speech coding	ADPCM 32 32	ADPCM 32 32	ADPCM 32	ADPCM 32 32
Channel coding	None	CRC	CRC	CRC
Region	Europe, Canada, China, Southeast Asia	Europe	Japan, Hong Kong	United States

*CT2+ is the Canadian version of CT2. (From: [20].)

1.5 Third Generation Systems

The third generation systems are being employed via universal wireless personal communications (UWPC) systems, which will provide universal speech services and local multimedia services [2, 19, 20]. The third generation personal communication systems are in the process of implementation worldwide by the International Telecommunications Union (ITU) within the framework of the future public land mobile telecommunications systems (FPLMTS)/international mobile telecommunications-2000 (IMT-2000) activities and along the evolution path of Figure 1.1. In Europe, this is supported by the universal mobile telecommunications system (UMTS) program within the European community. Both the FPLMTS and UMTS

Table 1.4
Summary of Wide Area Wireless Packet Data Systems

System	CDPD	RAM Mobile (Mobitex)	ARDIS (KDT)	Metricom (MDN)
Data rate	19.2 Kbps	8 Kbps [19.2 Kbps]	4.8 Kbps [19.2 Kbps]	~76 Kbps
Modulation	GMSK BT = 0.5	GMSK	GMSK	GMSK
Frequency	~800 MHz	~900 MHz	~800 MHz	~915 MHz
Channel spacing	30 KHz	12.5 KHz	25 KHz	160 KHz
Status	1994 service	Full service	Full service	In service
Access means	Unused AMPS channels	Slotted Aloha CSMA		FH SS (ISM)
Transmit power			40W	1W

(From: [20].)

programs are tightly related and expected to lead to consistent and compatible systems.

A lot of research and development (R&D) activity is taking place worldwide in order to come to a consensus on issues such as frequency bands, multiple access protocols, interfacing, internetworking, and integration (asynchronous transfer mode [ATM], fiber, air, fixed, macrocells, microcells, picocells, and hypercells), system development (baseband, terminals, and antennas), multimedia communications, satellite (frequency allocation, channel characterization, radio access), and technology (low power, size, and cost). Figure 1.2 shows the evolution in time of services/systems in the wireless world.

1.5.1 UMTS Objectives and Challenges

UMTS is a third generation mobile communication system that provides seamless personal communication services anywhere and anytime. In particular, it provides mobile broadband multimedia services along the lines of the following objectives:

- User bit rates of 144 Kbps (wide area mobility and coverage) and up to 2 Mbps (local mobility and coverage);
- Provision of services via handheld, portable, vehicular-mounted, movable, and fixed terminals, in all radio environments based on single radio technology;

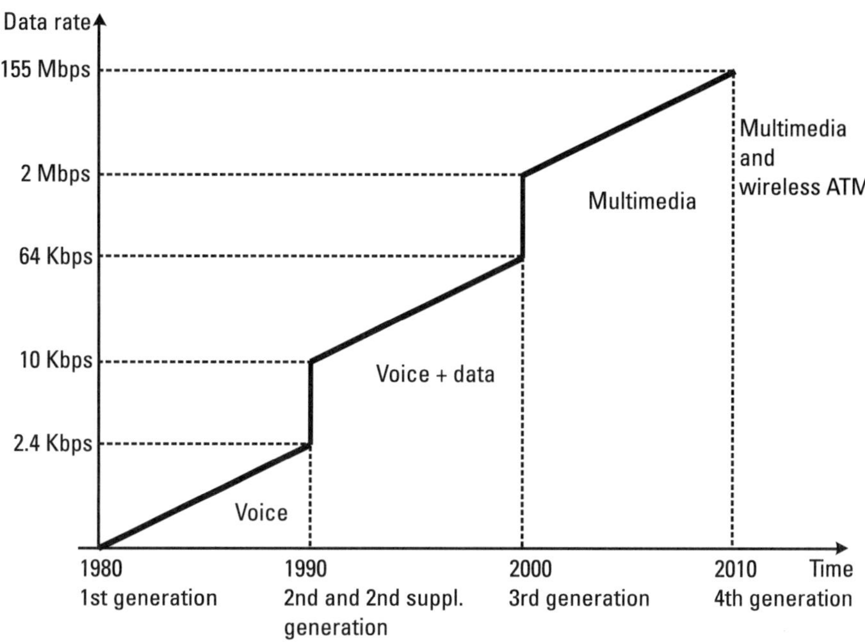

Figure 1.2 Worldwide service evolution with respect to time.

- High spectrum efficiency compared to the existing system;
- Speech and service quality at least comparable to current fixed network;
- Flexibility for the introduction of new services and technical capabilities;
- Radio resource flexibility to multiple networks and traffic types within a frequency band.

It is expected that the basis for the UMTS market will be the existing GSM/DCS market for speech, as well as low and medium bit rate data up to 100 Kbps. The GSM market will continue to grow even after UMTS introduction; thus, positioning UMTS toward GSM is important. Most likely, the first services offered by UMTS will complement the services offered by GSM/DCS. The bit rates of UMTS compared to existing and evolved second generation systems are shown in Figure 1.3 [20].

The goal of UMTS is to support a large variety of services, most of which are not yet known. UMTS air interface must be able to cope with variable and asymmetric bit rates, up to 2 Mbps, with different quality of

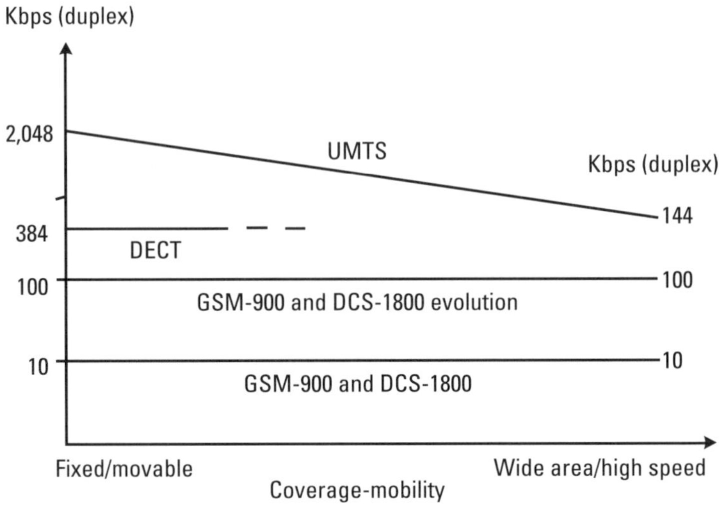

Figure 1.3 UMTS bit rates versus coverage and mobility. (From: [20].)

service requirements (bit error probability and delay), such as multimedia services with bandwidth on demand. Effective packet access protocol is also essential for the UMTS air interface to handle bursty real-time and nonreal-time data.

A UMTS objective is to cover all environments with a single interface. However, to use spectrum efficiently in different environments and for different services, the air interface has to be adaptable. Therefore, implementation of UMTS terminals has to take this adaptability into account by requiring configurable terminals. In addition, utilization of the existing infrastructure will require dual-mode terminals.

1.5.2 Standardization of UMTS

UMTS standards are now being developed. One of the main aspects of UMTS standardization is selection of the air interface. The strong support behind wideband CDMA (WCDMA) led to the selection of WCDMA as the UMTS terrestrial air interface scheme for frequency division duplex (FDD) bands by ETSI. The selection of WCDMA is also backed by the Asian and American operators. For time division duplex (TDD) bands, a time division CDMA (TDCDMA) concept has been selected. As far as access is concerned, UMTS will utilize a radio access network to be connected to several core networks. In ITU, the development of FPLMTS (also called

IMT-2000) is being carried out. FPLMTS can be seen as global interworking of mobile services and as the cornerstone of spectrum allocation for third generation mobile radio systems. UMTS standards are more detailed than FPLMTS recommendations, covering test specifications and focusing on the European market and existing systems.

R&D is also going on in Europe, Japan, and North America for the fourth generation mobile broadband systems (MBS) and wireless broadband multimedia communications systems (WBMCS). WBMCS is expected to provide its users with customer premises services with information rates exceeding 2 Mbps.

1.6 The Cellular Concept

In the beginning, mobile systems were developed much like radio or television broadcasting (i.e., a large area was covered by installing a single, high-power transmitter in a tower situated at the highest point in the area). A single high-power transmitter mobile radio system gave good coverage with a small number of simultaneous conversations depending on the number of channels N_c. The $(N_c + 1)$ caller was blocked. Those systems were also characterized by the lack of handoff.

To increase the number of simultaneous conversations, a large area can be divided into a large number of small areas, N_α. Each small area is called a *cell*. To cover a cell, a single low-power transmitter is required. If every cell uses the same frequency that is available for a large area, and its available bandwidth is divided into the number of channels, N_c, then instead of N_c simultaneous conversations for a large area, there would be N_c simultaneous conversations for each cell. Thus, now there can be $N_\alpha N_c$ simultaneous conversations in the entire large area as compared with only N_c [4–8].

The idea of using the same frequency in all the cells does not work because of the interference between mobile terminals operating on the same channel in adjacent cells. Therefore, the same frequency cannot be used in each cell, and it is necessary to skip a few cells before the same frequency is used. Cellular concept is illustrated in Figure 1.4.

The cellular concept, therefore, is a wireless system designed by dividing a large area into several small cells, replacing a single, high-power transmitter in a large area with a single, low-power transmitter in each cell, and reusing the frequency of a cell to another cell after skipping several cells. Thus, the limited bandwidth is reused in distant cells, causing a virtually infinite multiplication of the available frequency.

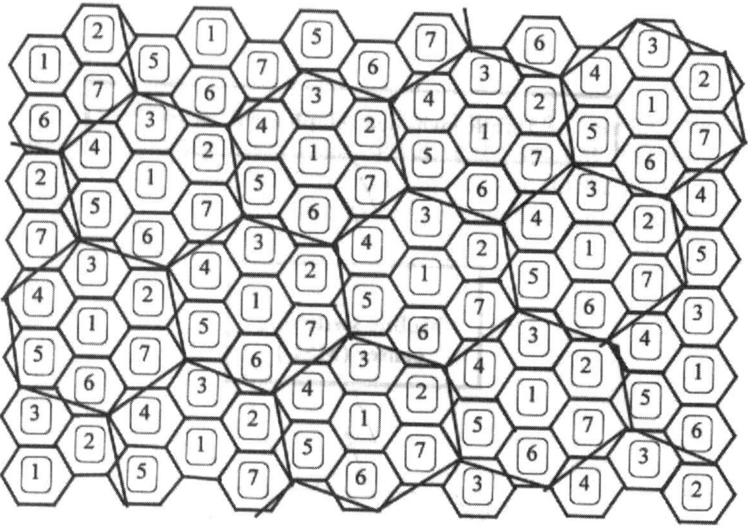

Figure 1.4 Cellular concept. (From: [20].)

Major design elements that are considered to efficiently utilize available frequency are frequency reuse, cochannel interference, carrier-to-interference ratio, handover/handoff mechanism, and cell splitting [20].

This section briefly introduces these basic elements, which will be important in the discussion about interference in later chapters. In addition, different types of cellular systems are reviewed along the same lines.

1.6.1 Frequency Reuse

The cellular structure was introduced due to capacity problems of mobile communication systems. In a cellular radio system, the area covered by the mobile radio system is divided into cells. In theory, the cells are considered hexagonal, but in practice they are less regular in shape. Each cell contains a base station, which is connected to the mobile switching center (MSC). This MSC is connected to the fixed telecommunication system—the public switched telephone network (PSTN). MSC serves as the central coordinator and controller for the cellular radio system and as the interface between mobile and PSTN. The cellular radio user in a car or train or in the street picks up a handset, dials a number, and immediately can talk to the person he or she called [20].

Each cell is assigned a part of the available frequency spectrum. Cellular radio systems offer the possibility of using the same part of the frequency spectrum more than once. This is called frequency reuse. Cells with identical channel frequencies (i.e., the same part of the frequency spectrum) are called cochannel cells. The cochannel cells have to be sufficiently separated to avoid interference. The distance between these cochannel cells is achieved by the creation of a cluster of cells. As explained earlier, cells with identical numbers make use of the same part of the frequency spectrum. The total number of channels N_{tc} in a cellular radio system is

$$N_{tc} = N_r N_c C \tag{1.1}$$

where N_r is the number of times a cluster is replicated within the system, N_c is the number of channels in a cell, and C is the cluster size (number of cells in a cluster).

It is not possible to choose an arbitrary value of the cluster size. The cluster size is determined by

$$C = i^2 + ij + j^2 \tag{1.2}$$

where i and j are nonnegative integers. So there can be only selected values of C.

$$C = 3, 4, 7, 9, 12, 13, \ldots \tag{1.3}$$

The cluster size can be chosen and it determines the amount of frequency reuse within a certain area. An important design parameter denoting the amount of frequency reuse in a certain area is called the normalized reuse distance. The normalized reuse distance, R_u, is defined as the ratio of the reuse distance, D, between the centers of the nearest cochannel cells and the cell radius, R, as shown in Figure 1.5. Hence,

$$R_u = \frac{D}{R} \tag{1.4}$$

Using Figure 1.5, C and D can be related

$$C = D^2 = i^2 + j^2 - 2ij \cos(120°) = i^2 + ij + j^2 \tag{1.5a}$$

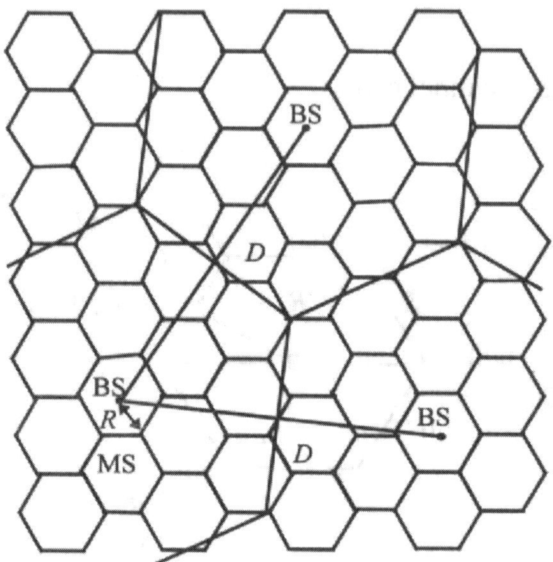

Figure 1.5 Normalized reuse distance. (Source: [20]. Reprinted with permission.)

From Figure 1.6, which shows how cells with identical numbers make use of the same part of the frequency spectrum, we obtain

$$R = \frac{1}{\sqrt{3}}$$
(1.5b)

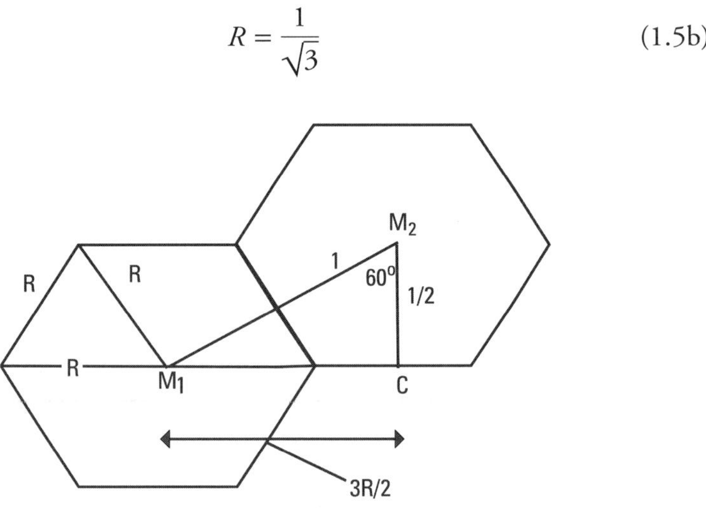

Figure 1.6 Cluster size length.

Thus, the relationship between R_u and C is obtained using (1.5a), (1.5b), and (1.4):

$$R_u = \sqrt{3C} \tag{1.6}$$

Therefore,

$$C = \frac{R_u^2}{3} \tag{1.7}$$

1.6.2 Handover/Handoff Mechanism

Handover, also known as handoff, is a process to switch an ongoing call from one cell to the adjacent cell as a mobile user approaches the cell boundary.

Figure 1.7 shows that as the user moves from cell 1 to cell 2, the channel frequencies will be automatically changed from the set f_1 to the set f_2. Handover is an automatic process, if the signal strength falls below a threshold level. It is not noticed by the user because it happens very quickly— within 200 to 300 ms [20].

The need for a handover may be caused by radio, operation and management (O&M), or by traffic. Radio causes the majority of handover requests. The parameters involved are low signal level or high error rate. This can be caused by a mobile moving out of a cell or signal blocking by objects.

O&M-generated handovers are rare. They evolve from the maintenance of equipment, equipment failure, and channel rearrangement. Handovers

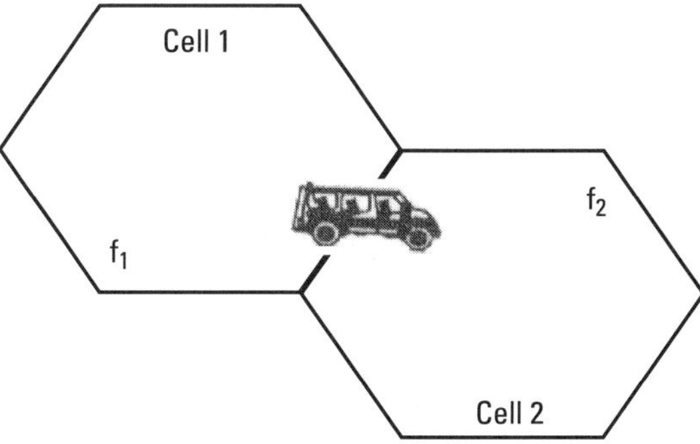

Figure 1.7 Handover/handoff mechanism. (After: [20].)

due to unevenly distributed traffic may cause some mobiles at the border of a cell to be handed over to an adjacent cell.

The performance metrics used to evaluate handover algorithms are handover blocking probability, call blocking probability, handover probability, call dropping probability, rate of handover, probability of an unnecessary handover, duration of interruption, and delay (distance).

A handover is performed in three stages. The mobile station (MS) continuously gathers information of the received signal level of the base station (BS) with which it is connected, and of all other BSs it can detect. This information is then averaged to filter out fast-fading effects. The averaged data is then passed on to the decision algorithm, which decides if it will request a handover to another station. When it decides to do so, handover is executed by both the old BS and the MS, resulting in a connection to the new BS.

As stated earlier, the received signal level suffers from fading effects. To prevent handover resulting from temporary fluctuations in the received signal level, the measurements must be averaged. An averaging window whose length determines the number of samples to be averaged is used. Longer averaging lengths give more reliable handover decisions, but also result in longer handover delays. Detailed studies were done to determine the averaging window shape—that is, to determine whether recent measurements should be treated as more reliable than older ones. The averaging window is used to trade off between handover rate and handover delay. More details are given in Chapters 2 and 4 [4, 8, 20, 26].

1.6.3 Cell Splitting

In principle, a cellular system can provide services for an unlimited number of users. However once a system is installed, it can only provide to a certain fixed number of users. As soon as the number of users increases and approaches the maximum that can be served, some technique must be developed to accommodate the increasing number of users. There are various techniques to enhance the capacity of a cellular system. One technique is cell splitting, a mechanism by which cells are split into smaller cells, each having the same number of channels as the original large cells, as shown in Figure 1.8.

1.6.4 Types of Cellular Networks

Based on the radius of the cells, there are three architectures of cellular networks:

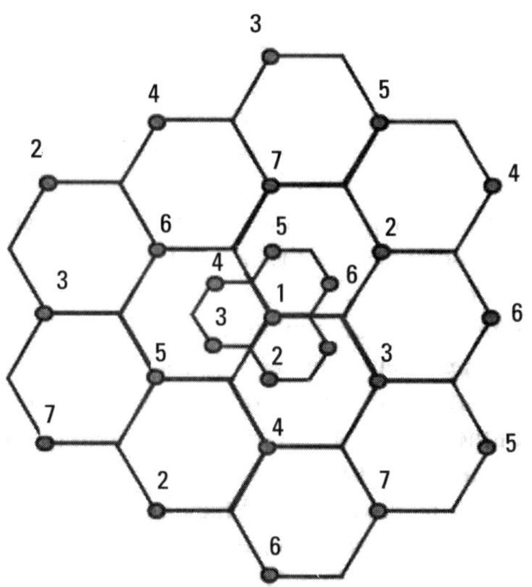

Figure 1.8 Cell splitting. (After: [20].)

1. Macrocells;
2. Microcells;
3. Picocells.

1.6.4.1 Macrocellular Radio Networks

Macrocells are mainly used to cover large areas with low traffic densities. These cells have radii between 1 and 10 km. A distinction between large macrocells and small macrocells should be made [4].

Large macrocells have radii between 5 and 10 km or even higher. They are used for rural areas. Small cells have radii between 1 and 5 km. These cells are used if the traffic density in large cells is so high that it will cause blocking of calls. They thus provide large cells with extra capacity (cell splitting). Planning small cells is more difficult because traffic predictions for relatively small areas cannot be easily done. The signals undergo multipath Rayleigh fading and lognormal shadowing. The standard deviation of lognormal shadowing signal lies between 4 and 12 dB. Typical root mean square (rms) delay spread is 8 μs. For more details, the reader is referred to Chapter 2 and to [4] and [27].

1.6.4.2 Microcellular Radio Networks

Microcellular radio networks are used in areas with high traffic density, like (sub)urban areas. The cells have radii between 200m and 1 km. For such small cells, it is hard to predict traffic densities and area coverage. Models for such parameters prove to be quite unreliable in practice. This is because the shape of the cell is time dynamic (i.e., the shape changes from time to time) due to propagation characteristics.

We can distinguish one- and two-dimensional microcells. One-dimensional microcells are placed in a chainlike manner along main highways with high traffic densities, whereas "two dimensional" refers to the case where an antenna transmits the main ray and two additional rays are reflected off buildings on both sides of the street. One-dimensional microcells usually cover one or two house blocks. Antennas are placed at street lamp elevations. Surrounding buildings block signals propagating to adjacent cochannel cells. This improves the ability to reuse frequencies, as cochannel interference is reduced drastically by the shadowing effect caused by the infrastructure. Microcells follow a dual path-loss law. Violation of this law depends on the type of environment and the position of the transmitting antenna. The signal undergoes Rician fading and lognormal shadowing. Typical rms delay spread is 2 μs.

1.6.4.3   Picocellular Radio Networks

Picocells or indoor cells have cell radii between 10 and 200m. For indoor applications, cells have three-dimensional structures. Fixed cluster sizes, fixed channel allocations, and prediction of traffic densities are difficult for indoor applications. Today, picocellular radio systems are used for wireless office communications. Various propagation characteristics of these types of networks are given in Table 1.5, and we shall further explain in Chapters 2 and 4. We will also see, later on, how these characteristics influence the interference aspects. Path loss exponent varies from 1.2 to 6.8. Signals in picocells are always Rician faded. The Rician parameter lies between 6.8 and 11 dB. Typical values of rms delay spread lie between 50 and 300 μs. For more details the reader is referred to the three chapters that follow.

1.7   Satellite Systems

Figure 1.9 sketches a simplified Earth-station-satellite connection. The transmission from the Earth station to the satellite is called uplink and from the satellite to the Earth station is called downlink. Transmitter power for Earth stations is generally provided by high-powered amplifiers, such as traveling wave tubes (TWTs) and klystrons. Because the amplifier and transmitting antenna are located on the ground, size and weight are not prime considera-

Table 1.5
Different Cell Characteristics

Cell Type	Size	Transmission Power	Antenna Height/ Location	Path Loss Exponent	Signal Characteristics	rms Delay Spread	Use
Macrocell	2–20 km diameter	0.6–10W	>30m, top of tall building	2–5	Rayleigh fading and lognormal shadowing	<8 μs	Large area coverage— reduce infrastructure cost
Microcell	0.4–2 km diameter	<20 mW	<10m street lamp elevation	Dual path-law	Rician fading and lognormal shadowing	<2 μs	Urban area coverage
Picocell	20–400m diameter	On the order of a few milliwatts	Ceiling/top of book shelf	1.2–6.8	Rician fading	50–300 ns	Mainly for indoor areas with high terminal density

(After: [20].)

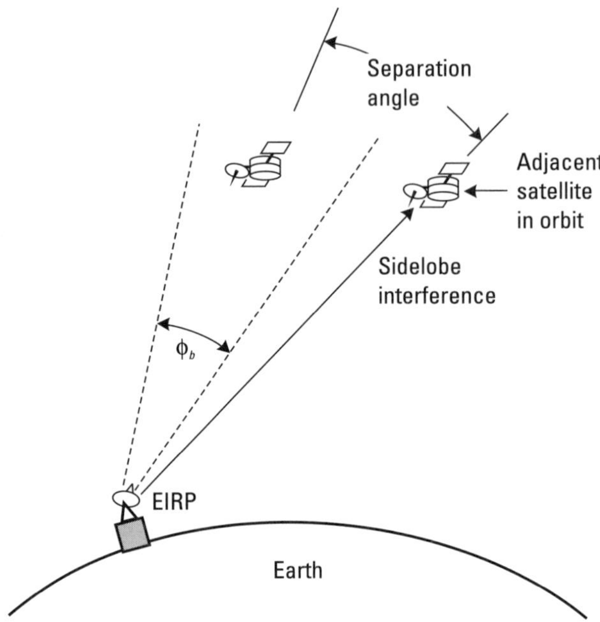

Figure 1.9 Satellite uplink (ϕ_b = halfpower beamwidth).

tions, and fairly high transmitter effective isotropic radiated power (EIRP) levels can be achieved. Earth-based power outputs of 40 to 60 dBw are readily available at frequency bands up through K-band, using a cavity-coupled traveling wave tube amplifier (TWTA) or klystrons. These power levels, together with the transmitting antenna gains, determine the available EIRP for uplink communications [9–18, 28, 29].

In the design of satellite uplinks, the beam pattern may often be of more concern than the actual uplink EIRP. Whereas the latter determines the power to the desired satellite, the shape of the pattern determines the amount of off-axis (sidelobe) interference power impinging on nearby satellites.

The beam pattern therefore establishes an acceptable satellite spacing, and thus the number of satellites that can simultaneously be placed in a given orbit with a specified amount of communication interference. The narrower the Earth-station beam, the closer an adjacent satellite can be placed without receiving significant interference. On the other hand, an extremely narrow beam may incur significant pointing losses due to uncertainties in exact satellite location. For example, if a satellite location is known only to within ±0.2°, a minimum Earth-station half-power beamwidth of about 0.6, is necessary. This sets the transmit antenna gain at about 55 dB. For parabolic ground antennas, this produces the off-axis gain curve shown in Figure 1.10.

For a 20-dB reduction in adjacent satellite interference, we see that the nearest satellite must be at least 3° away, as shown in Figure 1.10. That is, when observed from Earth, two satellites in the same orbit must be separated by about 3°. Thus, the uplink beamwidth is set by the pointing accuracy of the Earth station, whereas satellite orbit separation is determined by the acceptable sidelobe interference. If satellite pointing is improved, the uplink beamwidth can be narrowed, allowing closer satellite spacing in the same orbit. This would increase the total number of satellites placed in a common orbit, such as the synchronous orbit.

With the half-power beamwidth set, a higher carrier frequency will permit smaller Earth stations. Figure 1.11 shows the relation between Earth-station antenna diameter and frequency in producing a given uplink beamwidth and gain. Note that while increase of carrier frequency does not directly aid receiver power, we see that an advantage can be obtained in reducing Earth-station size and, possibly, in improving satellite trafficking (allowing more satellites in orbit).

With a 0.6° uplink beamwidth (gain ≈ 55 dB) Earth-station EIRP values of about 80 to 90 dBw are readily available. Table 1.6 lists an example of an uplink budget for computing the carrier-to-noise ratio (CNR) at the satellite in a 10-MHz bandwidth, showing the way in which individual budget elements affect the CNR.

Figure 1.12 generalizes this budget to show how uplink CNR will vary with Earth-station EIRP and satellite receiver characteristics (g/T_0).

Even with significant range losses (≈ 200 dB) and relatively low g/T_0 values, an acceptable uplink communication link can usually be established. A satellite downlink is constrained because the power amplifier and transmitting antenna must be spaceborne and add to the total payload considerably.

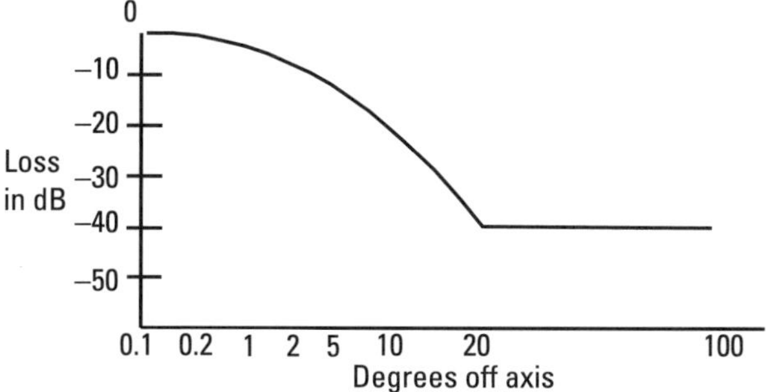

Figure 1.10 Uplink Earth-station antenna pattern. Angle measured from boresight.

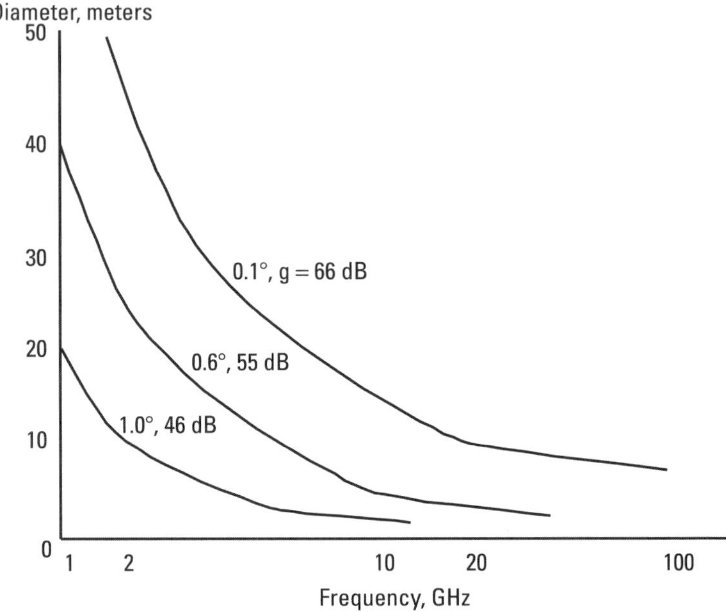

Figure 1.11 Earth-station antenna size versus frequency.

Table 1.6
C-Band Uplink Power Budget

Frequency, 6 GHz		
Transmitter power, $P_T = 30$W	15 dBW	
Transmitter antenna gain, g	55 dB	
EIRP		70 dBW
Path length, 23,000 miles		
Propagation loss, L_P		−199 dB
Atmospheric loss (rain)		−4 dB
Polarization loss		−1 dB
Pointing loss		−0.6 dB
Satellite receive antenna, 1.5 ft		
Beamwidth 8°		
Gain (efficiency = 55%)	26 dB	
Background temperature T = 100K		
Receiver noise figure $F = 7.86$ dB		
Receiver noise temperature $T_{eq} = 1{,}584$K	32 dB	
Receiver g/T_0		−6 dB
Boltzmann constant		228.6 dB
Bandwidth, 10 MHz		−70 dB
CNR		18 dB

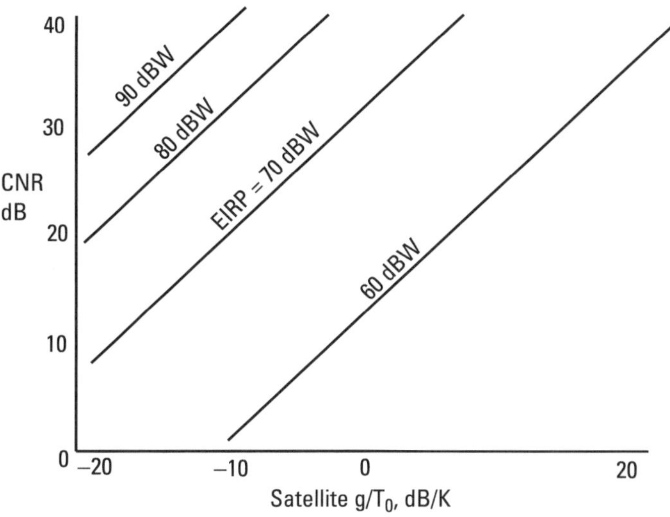

Figure 1.12 Uplink CNR versus satellite g/T_0 C-band link, bandwidth = 10 MHz.

A CNR plot in terms of satellite power and receiver g/T_0 is shown in Figure 1.13. It is evident that relatively large Earth-station g/T_0 is needed to overcome the smaller EIRP of the satellite. This means small Earth stations will be severely limited in their ability to receive wide bandwidth carriers.

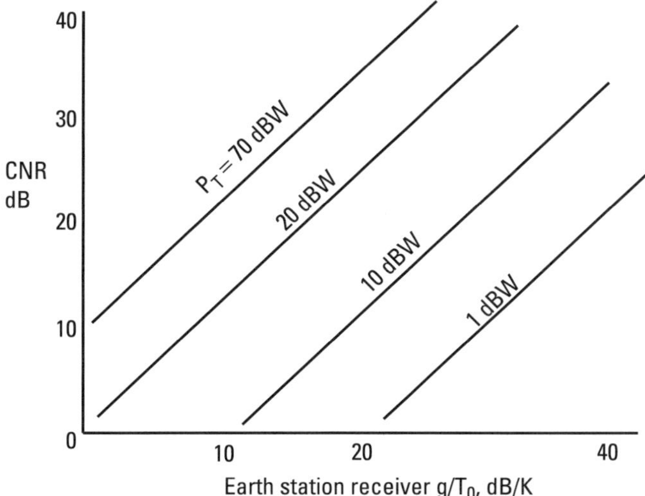

Figure 1.13 Downlink CNR versus receiver g/T_0. P_T = satellite power, global antenna, bandwidth = 10 MHz.

Figure 1.14 indicates that satellites and wireless LANs occupy opposite corners of a plane that displays two properties of a system—the bit rate available to terminals and the coverage area. A satellite system delivers low-bit-rate services (typically 64 Kbps or less) over continental or global coverage areas, while a wireless LAN operates at Mbps over a range measured in tens of meters.

1.7.1 Mobile Satellite Systems

A major trend in satellite communications in the 1990s has been toward smaller and smaller Earth terminals. This trend has produced a thriving direct broadcast satellite television industry, two-way communications between satellites and vehicles, two-way communications between satellites and ships, and the one-way global positioning system (GPS). With respect to personal communications, the main goal that can be served by satellites is ubiquitous coverage. Viewing a communications satellite as a base station, the cell dimensions are many orders of magnitude larger than those of terrestrial systems. Satellites can therefore provide communications services in areas where it is uneconomical to install the infrastructure of one of the systems described in this book [9, 16].

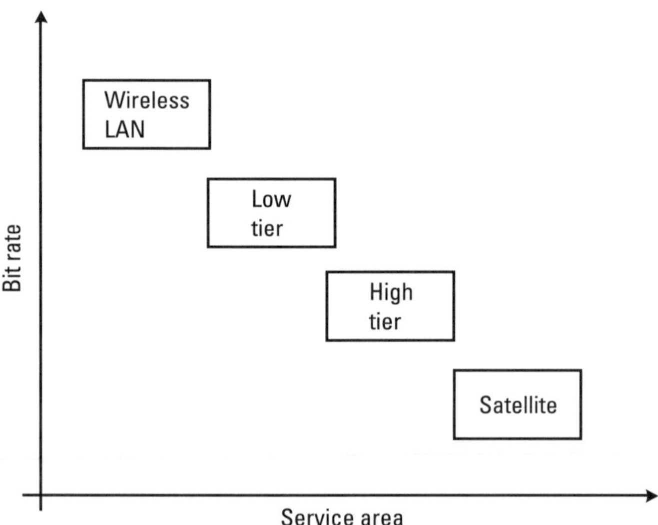

Figure 1.14 Coverage areas and bit rates of four types of wireless communications systems.

Satellite systems that provide mobile communications services fall into categories distinguished by the orbits of the satellites. The first mobile satellites are in geosynchronous orbits (GEO), at a distance of 35,800 km above the equator. Geosynchronous satellites have the advantage of a simple network configuration, as we saw in the previous section. The cell size of one satellite is approximately one-third of the Earth's surface. On the other hand, due to high transmission path attenuation, geosynchronous satellites require high-power transmitters in both the satellite and the mobile terminal. Other disadvantages are long propagation path delays and poor radio coverage at high latitudes. To overcome these disadvantages, mobile satellite systems planned to operate in low Earth orbits (LEOs), on the order of 500 to 2,000 km above the Earth, and others operating in medium Earth orbits (MEOs) at altitudes around 10,000 km will fill the void. LEO and MEO satellites have smaller coverage areas than GEO satellites, and they move with respect to the terminals they serve. Therefore, each system requires many satellites and network control that includes handoff from a satellite that moves out of range of a terminal to another satellite that moves into range. It is interesting to note that LEO and MEO satellite systems require handoff due to the mobility of base stations (in satellites) rather than the mobility of terminals.

By the nature of their architecture, these systems are diverse in their characteristics. They differ not only in orbit but also in the number of satellites per system (between 10 and 840), channel transmission rates (between 2,400 bps and 2 Mbps), and the sophistication of the communications tasks performed by each satellite. Some are *bent pipes,* which simply receive signals from one place on Earth and relay them to another place. Others perform sophisticated switching operations.

1.7.1.1　MEO and LEO Satellite Systems

In response to the growing needs for a global unified wireless communication system, many large telecommunications companies have proposed large satellite constellations. These projects are designed to provide hybrid cellular-satellite communication from virtually anywhere on the planet. In this chapter, the architecture of some satellite personal communication systems (SPCNs) is described, and special emphasis is given to the Iridium-type systems and their interference characteristics [3, 16].

Even though the Iridium system has been a commercial failure, it is believed that the technology on which it is based is being perfected as described in Figure 1.1, and these types of systems will be of wide use in the future. In other words, we believe that an Iridium-type system is a good model for the purposes of this book, as far as the Intersatellite links are concerned.

Intersatellite Links

The growth of the Internet Protocol (IP) traffic and the necessity to support real-time multimedia applications motivate the consideration of satellite communication networks from new perspectives. For a long time, satellites have been used for communication between distant areas. GEO satellites have been used extensively as simple repeaters for fixed communication links between two ground stations. The next generation of satellites support onboard processing, such as routing; however, for real mobile satellite systems, MEO and LEO systems are used.

In order to support multihop calls, the satellites are equipped with onboard switching technology and direct intersatellite links (ISLs). For these calls, it is no longer a simple task to find the optimal route between the source and sink nodes. Many constraints have to be taken into account. quality of service (QoS) connections such as multimedia and high-integrity traffic require special routing and call acceptance provisions.

The basic topology components we consider here for LEO satellite networks are given in Figure 1.15. Earth stations interconnect two distant areas through the satellite network. The customers of the network can be:

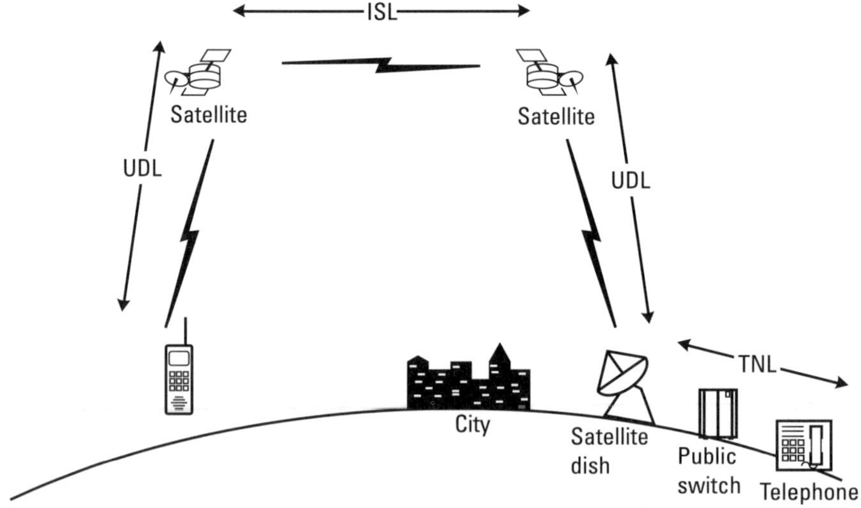

Figure 1.15 The Iridium end-to-end link.

- SPCN users (satellite telephony);
- Corporate intranets with branches in distant cities;
- Special users (military, emergency, media) that require survivable backup communications means.

The Earth gateway stations act as interfaces between satellite and terrestrial networks. Because the space segment of the network has routing capabilities, the number of such gateway stations can be reduced to three to four around the world. The gateway stations provide call setup, billing, user login, and other central control services. However, routing can be implemented exclusively in the space segment.

There are currently two design approaches for connectivity between satellites in the network. These approaches depend upon whether the satellites support real-time multimedia applications and act as communication repeaters. Satellites that serve as repeaters are used in a bent pipe architecture. A mobile user's transmitted signal is reflected off the satellite to a gateway in the same satellite footprint. The switch used to process the call is located at the gateway. This type of system requires a gateway in each satellite footprint in order to interface mobile users.

Satellites with onboard switching technology are able to use ISLs to route calls. A mobile user's transmitted signal is routed through several satellites and downlinked to either a regional gateway or another mobile user. This creates a network in the sky and allows the use of large regional gateways instead of gateways in each satellite footprint. Until recently, the technological complexity of utilizing ISLs to perform network routing was limited to military applications. The designers of the civilian networks have overcome these hurdles. Consequently, an Iridium-type network utilizes satellites with onboard switching technology and ISLs [16]. More details about the system design of satellite mobile systems and their operation in an interference environment are given in detail in [23–24].

1.8 Wireless Local Loops

The technical challenges of wireless local loops [3] are less stringent than those of communication systems that serve mobile terminals. Furthermore, operating companies have considerably less incentive to adopt systems that conform to published standards. As a consequence, the communications industry in 1997 offered a wide variety of communications systems to serve as wireless local loops. Some are adaptations of the standard systems described in this book. Others are proprietary systems, designed from the outset for this application. The interference characteristics of these systems will be studied in Chapters 5 and 6.

1.9 WLANs

WLANs [21, 22, 25–27] provide high throughput (Mbps) communications between stationary or slowly moving terminals in small coverage areas (on the order of tens of meters in diameter). Although the WLANs developed in the early 1990s use proprietary transmission protocols, it is likely that products produced at the end of the decade will conform to two published standards: IEEE 802.11, and HIPERLAN [21, 22, 25–27]. Both of these standards anticipate operation in unlicensed frequency bands, around 2.5 GHz, 5 GHz, and at higher frequencies. Another point of departure of wireless LANs from satellites and the systems studied in this book is that they allow terminals to communicate directly with one another, rather than through a network infrastructure containing base stations and switching equipment. Key issues in the design of WLANs derive from the distributed nature of the system architecture. Without the coordination of a base station, terminals contend for access to the same radio channel. Protocols are designed to promote fairness of access (equitable sharing of the channel among all terminals) and reliable operation even when some of the terminals are out of range of others (hidden-terminal problem).

1.9.1 The HIPERLAN System

Back in 1992, the Conference Europeene des Administration des Postes et de Telecommunications (CEPT) allocated the frequency bands in the 5.15- to 5.30-GHz and 17.1- to 17.3-GHz bands for the deployment of high-speed LANs. Neither frequency planning nor individual licensing is an obligation for these bands. Each user is responsible for possible interference with and/or protection from other users.

In an attempt to boost the wireless technology in the aforementioned bands, the European Telecommunication Standards Organization (ETSI) has developed a standard known as HIPERLAN type 1, which operates in the aforementioned band with bit rates of 23 Mbps. It is about a high-performance radio LAN, in which all nodes communicate a single shared communication channel. Its main properties are listed here:

   1. It provides an International Standards Organization (ISO) medium access control (MAC) compatible service.

2. It interconnects with other LANs according to the ISO MAC bridges specifications.

3. It is deployable in a prearranged or an ad hoc fashion.

4. It supports node mobility.

5. It may have coverage beyond the radio range limitation of a single node.

6. It supports both asynchronous and synchronous communications by means of a channel access mechanism with priorities.

7. It has the capability of rearranging the active receiver nodes for power conservation reasons.

The HIPERLAN is characterized by routing procedures that allow the dynamic establishment of routes between nodes that are not within each other's range. This avoids any range limitations and power-saving functions that allow terminals to switch on their circuits for small duty cycles. It uses an advanced channel access protocol whose contention is based on measurements on the state of the channel (busy/idle) with an adaptive receiving threshold, thus reducing the probability of collisions to a minimum. The modulation scheme used in HIPERLAN type 1 has two different bit rates:

1. The low bit rate (LBR) of roughly 1.5 Mbps is used for service information and indoor channels to be received without equalization.

2. The high bit rate (HBR) of 23 Mbps allows for the rapid transmission of the data. These receivers need equalizers for the HBR. The modulation format is FSK for the LBR and GMSK for the HBR.

The HIPERLAN 1 standard defines three layers: the MAC sublayer, the channel access control (CAC) sublayer, and the physical sublayer (PHY). These three layers correspond to the physical and the data link layers of the open system interconnection (OSI) reference model. HIPERLAN applications are considered as protocols belonging to a higher layer. Figure 1.16 shows the HIPERLAN reference model and its relationship with the OSI reference model.

1.9.1.1 Physical Layer

The main goals of the HIPERLAN physical layer can be summarized as follows:

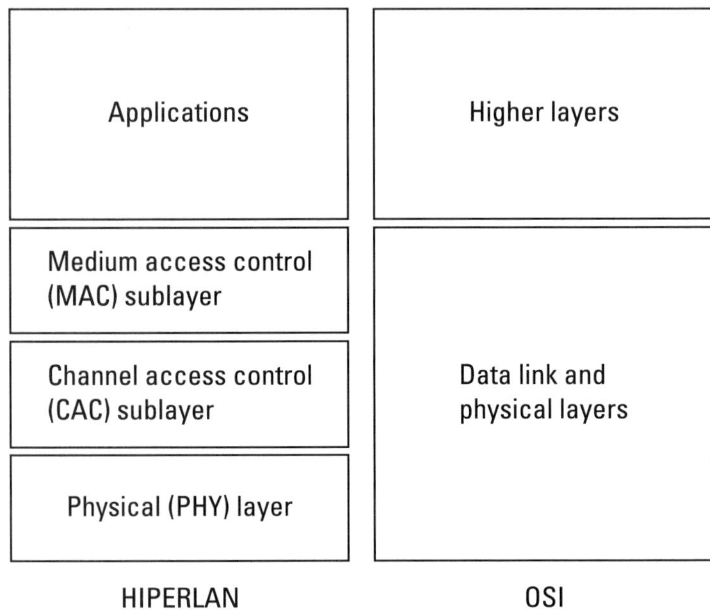

Figure 1.16 HIPERLAN reference model.

- To establish a physical link to deliver data from a transmitter to one or several receivers using the modulation formats described for the LBR and HBR transmission and the techniques for error correction as described in the standard;

- To assist the multiple access scheme by measuring the channel status according to predefined rules and maintaining an adaptive threshold to determine whether the channel is busy.

More details about the tasks of the PHY are given in [27].

1.9.1.2 Transmission Characteristics

There are five channels in the frequency band from 5,150 to 5,300 MHz. Table 1.7 lists the nominal values of carrier frequencies.

Channels 0, 1, and 2 are defined as default channels and are available in all countries under the CEPT regulations. The availability of channels 3 and 4 depends on national regulations. As HIPERLAN terminals can be taken to different countries, there are means of informing the terminals on the availability of these two channels. However, the whole problem of channel selection for a network is not addressed in this standard. It is left to the

Table 1.7
HIPERLAN 1 Carrier Frequencies

Carrier Number	Frequency
0	5,176.4680
1	5,199.9974
2	5,223.5268
3	5,247.0562
4	5,270.5856

higher network layers. All terminals belonging to an individual HIPERLAN must transmit and receive on the same channel. Transmitter frequency accuracy must be better than 10 pulses per minute (ppm) with respect to the nominal carrier frequency.

1.9.1.3 HIPERLAN Standards

The European standard HIPERLAN allows the implementation of WLANs based on standard equipment with highly advanced functionality. The main features of HIPERLAN that make it attractive when compared to other WLANs can be summarized as follows:

1. An HBR of 23 Mbps, comparable to or even higher than that of a WLAN;

2. Virtually unlimited coverage, as it does not depend on the radio range of individual nodes, thanks to the forwarding and routing functions and an easy interconnectability with other networks;

3. Dynamic configuration of the network, with a fast update upon making changes in the nodes;

4. High traffic capacity and the means to support time-bounded as well as asynchronous traffic;

5. Data encryption and power-saving functions.

The main disadvantages are related to the use of the 5.15- to 5.30-GHz band, instead of the lower band in the 2.5-GHz range, in which there exist readily available inexpensive products.

Bluetooth^{TM} is an initiative of five major manufacturers from the computer and cellular communications fields: Ericsson, Nokia, IBM, Toshiba, and Intel.

Bluetooth enables the easy and cheap wireless interconnection of electronic devices of any kind at the radio frequencies [27]. Communication is limited to the vicinity of the devices and need not be structured in networks with ad hoc configurations. On the contrary, Bluetooth networks, named piconets, are made up and canceled spontaneously according to the services needed. Some of the key applications include the connection of a laptop computer to a mobile phone (for example, to send data through a cellular network), the connection of a mouse to a personal computer, or the implementation of wireless headsets for mobile phones. The range of applications is not limited, nor is the profiles list closed. New profiles can be added according to newly identified needs. Table 1.8 lists the main characteristics of Bluetooth.

The Bluetooth specification provides low-cost connectivity by using a PHY specification with relaxed technical features compared to other systems. Also, as economies of scale are very important to reduce manufacturing costs, the first version of Bluetooth works in the ISM 2.4-GHz band (as does IEEE 802.11), which is available worldwide, although there are some restrictions in a number of countries.

The gross bit rate is 1 Mbps, with GFSK modulation. The hop sequences are selected on the basis of user identity and are not orthogonal but have a low probability of persistent interference. There is a TDMA/

Table 1.8
The Main Characteristics of Bluetooth

Parameter	Values
Band	2.45 ISM band Different channels according to the country (see Chapter 3)
Carrier's separation	1 MHz
Access	FH; TDD
Modulation	Gaussian frequency shift keying (GFSK) with $BT = 0.3$
Bit rate	1 Mbps
Voice channels	64 Kbps
Maximum transmit power	1 mW nominal 100 mW with closed loop power control
Receiver sensitivity	< -70 dBm (tentative)
Other features	Authentication, encryption, power-saving functions, interpiconet communications

(From: [27].)

TDD structure linked to the hopping pattern. The slot duration is 625 μs. The packet length is equal to one, three, or five slots. The carrier frequency is fixed within a packet but changes between subsequent packets.

High-quality audio at 64 Kbps can be provided through the use of a synchronous connection-oriented (SCO) link. Duplex slots (two continuous slots, one for each direction) are reserved at regular intervals. The remaining slots can be used by asynchronous connectionless (ACL) links, scheduled by the master.

With regard to the applications in WLANs, Bluetooth core includes the object (OBEX) protocol for interoperability with the proposal of Infrared Data Association (IrDa). There is also a LAN access profile. It defines how Bluetooth devices can access LAN services through a LAN access point (AP), with point to point protocol (PPP). Also, two Bluetooth devices can communicate using PPP as if they were part of a LAN. However, Bluetooth does not aim to establish a complete LAN.

Based on Bluetooth, the IEEE P802.15 working group is preparing a standard for wireless personal area networks. It is addressed to applications that need to communicate via devices that are around a person. These are the same applications that are the focus of Bluetooth.

Wireless Asynchronous Transfer Mode (WATM)

ATM is one of the leading technologies in fixed high-capacity networks. In most situations, ATM is implemented in optical fiber links, cables, or fixed microwave point-to-point links. The concept of WATM relates to the extension of ATM services to other scenarios through the use of wireless transmission and features mobility. It includes the wireless mobile ATM, which is the basis for providing services in the order of tens of megabits per second to mobile users, satellite ATM (where the large delays are significant), and WLANs.

With regard to WLANs, both HIPERLAN type 2 and the IEEE 802.11 extensions to HBRs can provide ATM services, as has already been mentioned. Also, the multimedia mobile access point (MMAC) project has the objective of delivering ultra-high-speed data rates to WLANs, for example, through the use of WATM.

Home Radio Frequency (RF)

Approximately 100 manufacturers from the computers, communications, and microelectronics fields make up the HomeRF^{TM} working group. The specification has been prepared to provide wireless voice and data networking in the home. The expected radio range is on the order of 50m indoor

(10–20m for low-power devices). The specified radio access is called shared wireless access protocol (SWAP). Some examples of applications of HomeRF are described as follows:

- Set up a wireless home network to share voice and data between PCs, peripherals, PC-enhanced cordless phones, and new devices such as portable remote display pads;
- Access the Internet from anywhere in and around the home from portable display devices;
- Share an Internet service provider (ISP) connection between PCs and other new devices;
- Share files/modems/printers in multiple-PC homes;
- Intelligently forward incoming telephone calls to multiple cordless handsets, fax machines, and voice mailboxes;
- Review incoming voice, fax, and e-mail messages from a small PC-enhanced cordless telephone handset;
- Activate other home electronic systems by simply speaking a command into a PC-enhanced cordless handset;
- Play multiplayer games and/or toys based on PC or Internet resources.

Table 1.9 lists the main technical features of SWAP.

The PHY specification is based on the IEEE 802.11 FH mode, with data rates of 0.8 and 1.6 Mbps and a hop time of 300 μs. The frame duration, equal to the hop time, is structured in two parts, called subframes:

- A TDMA/TDD subframe is intended for isochronous communications, mainly voice communications. A maximum of four simultaneous voice communications can be carried, with slots reserved for retransmissions. The voice link is based on DECT and uses 32 Kbps ADPCM high-quality voice coding.
- A CSMA/CA sub frame is based on IEEE 802.11 but with some of the more costly features removed. This is used for peer-to-peer asynchronous data communications, making up an effective WLAN.

A HomeRF connection point governs the net, and is connected to a PC (typically with Internet access) and to the PSTN. HomeRF asynchronous devices can connect to each other without the intervention of the control

Table 1.9
The Main Characteristics of SWAP

Parameter	Values
Band	2.45 ISM band; different channels according to the country (see Chapter 3)
Carrier's separation	1 MHz
Acess	FH; TDMA/TDD
Modulation	2-FSK, 4-FSK
Symbol rate	0.8 Ms/s
Bit rate	0.8 Mbps 1.6 Mbps
Voice channels	Four channels at 32 Kbps
Maximum transmit power	100–250 mW nominal 1–2.5 mW low power devices
Receiver sensitivity	−80 dBm (2-FSK)
Other features	USB connection to the PC; connection to the PSTN

(From: [27].)

point. The control point is necessary for voice communications, which can be streamed to the PSTN or be established between two users within the net.

Enhanced cordless telecommunications is provided by the presence of the PC, which can add features—for example, to route the incoming calls to a specific handset (based on the caller ID) or to store and generate voice messages.

The HomeRF-based WLAN cannot compete in the industrial and business fields with the 802.11 and HIPERLAN standards, because of its low range and reduced features. However, it is a good candidate for domestic LANs that only integrate a reduced number of devices (e.g., fixed PC, laptop, printer, and digital camera) in a limited physical range.

1.9.1.4 Future Broadband Radio Access Network Standards

ETSI is developing three new broadband radio access network (BRAN) standards to be allocated in these new bands. HIPERLAN type 2 (wireless IP, ATM, and UMTS short-range access) is designed to provide local wireless access to ATM, IP, and UMTS infrastructure networks for both moving and stationary terminals that interact with the access points connected to the infrastructure networks. It will be able to provide the same QoS that users would expect from a wired IP or ATM network. The typical operating environment is an indoor environment, with mobility restricted to the local service area. The data rate is on the order of 25 Mbps. The technical specification for the PHY was approved in October 1999. It is intended for the 5-GHz unlicensed band.

HIPERACCESS (wireless IP and ATM remote access), previously known as HIPERLAN 3, will provide, high-speed, outdoor (25-Mbps) wireless access to infrastructure networks. Unless it is type 2, HIPERACCESS is not designed to support mobility at high data rates, allowing the use of directional antennas with significant gains. Thus, the range can be increased to 5 km. This will allow the rapid deployment of broadband access WANs. It will operate in the licensed (3–60-GHz) as well as the unlicensed (5-GHz) bands. The HIPERACCESS specification was completed within 2000.

HIPERLINK (wireless broadband interconnect), previously known as HIPERLAN 4, is intended to provide point-to-point (up to 150m) very-high-speed wireless links. It will operate in the 17-GHz unlicensed band, with bit rates up to 155 Mbps. One of the envisaged applications is the interconnection of HIPERACCESS networks and/or HIPERLAN APs in a fully wireless network. Figure 1.17 presents an overview of the different BRAN standards and their allocation and baud rates. HIPERLAN type-2, HIPERLINK, and HIPERACCESS can be combined into an open wireless architecture that meets the needs of a very large user population.

ETSI has described two main environments for this kind of network: The domestic premises network (DPN) environment covers the home and its immediate vicinity. It typically includes a localized radio extension to a broadband network. It is characterized by individual cells, and supporting

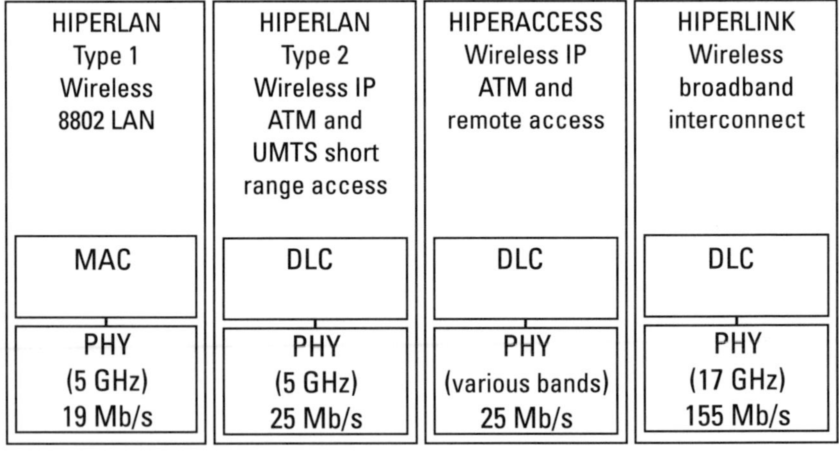

HIPERLAN Type 1 Wireless 8802 LAN	HIPERLAN Type 2 Wireless IP ATM and UMTS short range access	HIPERACCESS Wireless IP ATM and remote access	HIPERLINK Wireless broadband interconnect
MAC	DLC	DLC	DLC
PHY (5 GHz) 19 Mb/s	PHY (5 GHz) 25 Mb/s	PHY (various bands) 25 Mb/s	PHY (17 GHz) 155 Mb/s

Figure 1.17 Four different BRAN standards.

mobility beyond the coverage area is not required. The business premises network (BPN) environment covers a privately owned network over an extended area (such as university campuses or hospitals). It may offer access switching and management functions within an arbitrarily large coverage area serviced by multicellular wireless communications facilities. We can define some application scenarios within these environments for HIPERLAN 2, HIPERACCESS, and HIPERLINK, such as replacing infrastructure networks, enabling wireless access to infrastructure networks in the DPN, or interconnecting manufacturing devices in BPN. In BPN, delay and data losses are critical because of the need for supporting alarm data and other time-bound services.

1.10 Wireless Data Networks

Wide area wireless data systems [25] have been in operation since the early 1980s to serve specialized commercial needs, including transaction processing, such as credit card verification; broadcast services, such as road traffic advisories; and interactive services, such as wireless electronic mail. These systems provide two-way, low-speed, packet-switched data communications. Two of the early systems were Ardis, deployed throughout the United States, and Mobitex, used in public data networks in several European countries. Ardis [25] and Mobitex operate in *specialized mobile radio* frequency bands between 800 and 1,000 MHz. Channel bit rates range from 8 to 19.2 Kbps. The network architectures are similar to those of cellular systems. A newer system, introduced in the United States by Metricom, operates in an unlicensed frequency band at 900 MHz in the United States. The Metricom system relays packets through several radio transceivers between terminals and fixed-packet switches. The channel rate is 100 Kbps. In Europe, ETSI has adopted a standard for trans-European trunked radio (TETRA). It operates in 25 kHz channels between 380 MHz and 393 MHz, as well as in the 900-MHz band with a transmission rate of 36 Kbps.

TETRA is an all-digital, spectrum-efficient, trunked land mobile radio (LMR) radio system that uses a four-slot TDMA technology, which provides in a 25-kHz channel either four simultaneous voice and/or data paths or a *pipeline* mode that uses all four TDMA channels for high-speed data.

TETRA is designed for private mobile radio (PMR) and public access mobile radio (PAMR) use. TETRA provides voice and data communication with short data, circuit mode, and packet mode data services. A number of large TETRA users have generated specific-to-system specifications that further define the standard as is the Public Safety Radio Communication Project (PSRCP) emergency services project in the United Kingdom.

The TETRA technology provides one-to-one or one-to-many voice and/or data communications with "traditional" simplex or duplex, cellular-type operation. Trunked operation is the normal state but various managed and unmanaged direct modes (DMO) for direct subscriber-to-subscriber communication or, via gateways, subscriber-to-system communication. Very fast call set-up times are standard.

TETRA is an established European standard. It is an accepted standard in Russia, China, and in many Pacific Rim and South American countries. It is also within the standards acceptance procedure in the United States.

In general, the physical layout of TETRA consists of three interfaces: the air interface; the fixed network access point (FNAP), through which mobile users gain access to fixed users; and the mobile network access point (MNAP), which is a physical interface between the mobile terminating unit (MTU) and the data terminals. These terminals implement the X.25 protocol because the data port at the mobile station is a true X.25 interface. The characteristics and parameters of TETRA are listed below in Table 1.10.

In addition to these specialized wireless data networks, most of the systems described in this book have adopted standards for packet data transmission using the physical channels of the wireless personal communications system. These standards make it possible for service providers to offer wireless access to packet data networks, including the Internet. The first of these

Table 1.10
Characteristics and Parameters of TETRA

System	TETRA
Frequency band	
Base to mobile, (MHz)	(400 and 900 Bands)
Mobile to base, (MHz)	
RF channel spacing	25 kHz
Channel access/multiuser access	FDMA/DSMA & SAPR
Modulation method	$\pi/4$-QDPSK
Channel bit rate (Kbps)	36
Packet length	192 b (short), 384 b (long)
Open architecture	Yes
Private or public carrier	Public
Service coverage	European trunked radio
Type of coverage	Mobile

technologies to be deployed commercially is cellular digital packet data (CDPD), which uses analog mobile phone system (AMPS) radio channels. In GSM, the packet data technology has the designation generalized packet radio service (GPRS) [23].

1.11 Wireless Broadband Mobile Communication Systems

The theme of wireless broadband mobile communication systems (WBMCS) is to provide its users a means of radio access to broadband services supported on customer premises networks or offered directly by public fixed networks. WBMCS will provide a mobile/movable nonwired extension to wired networks for information rates exceeding 2 Mbps with applications foreseen in wireless LAN or mobile broadband systems. Thus, WBMCS will be a wireless extension to the broadband integrated services digital network (B-ISDN). It will be achieved with the transparent transmission of ATM cells. In short, WBMCS will provide novel multimedia and video mobile communication services, also related to wireless customer premises network (WCPN) and WLL [3].

The spectacular growth of video, voice, and data communication via the Internet and the equally rapid pervasion of mobile telephony justify great expectations for mobile multimedia [5]. Research and development in the field of mobile multimedia is taking place all over the world and is summarized next.

Within the European ACTS program, there are four European Union funded R&D projects: the magic wand (wireless ATM network demonstration), ATM wireless access communication system (AWACS), system for advanced mobile broadband applications (SAMBA), and wireless broadband CPN/LAN for professional and residential multimedia applications (MEDIAN). Table 1.11 summarizes the European projects.

In the United States, seamless wireless network (SWAN), broadband adaptive homing ATM architecture (BAHAMA), two major projects in Bell Laboratories, and wireless ATM network (WATMnet) are being developed in the computer and communication (C&C) research laboratories of Nippon Electric Company (NEC).

In Japan, Communication Research Laboratory (CRL) is busy developing several R&D projects, such as a broadband mobile communication system in the super-high-frequency (SHF) band (from 3 to 10 GHz) with

Table 1.11
Summary of European ACTS Projects

ACTS Project Parameter	WAND	AWACS	SAMBA	MEDIAN
Frequency	5 GHz	19 GHz	40 GHz	61.2 GHz
Data rate	20 Mbps	70 Mbps	2×41 Mbps	155 Mbps
Modulation	Orthogonal frequency division (OFDM) 16 carriers, 8-phase shift keying (PSK)	Offset quadrature PSK (OQPSK), coherent detection	OQPSK	OFDM, 512 carriers, differential QPSK (DQPSK)
Cell radius	20–50m	50–100m	6m × 200m 60m × 100m	10m
Radio access	TDMA/TDD	TDMA/TDD	TDMA/FDD	TDMA/TDD

(After: [20].)

a channel bit rate up to 10 Mbps and an indoor high-speed wireless LAN in SHF band with a target bit rate of up to 155 Mbps.

1.12 Millimeter Waves

During recent years, millimeter waves have gained increasing interest because of bandwidth scarcity, and therefore the study of millimeter wave communication systems has drawn the attention of many researchers. Within Europe, the Cooperation in the Field of Scientific and Technical Research (COST) group is investigating the promising features of the millimeter waves for communication applications in the COST 231 project. This project deals with the evolution of land mobile radio (including personal communications). The low millimeter-wave band from 20 to 60 GHz, which is nearly unused and allows for large bandwidth applications, combines the advantages of infrared (IR) (enough free bandwidth) and ultrahigh frequency (UHF) (good coverage). Systems operating particularly in the 60-GHz frequency band can have a small reuse distance because of oxygen absorption at the rate of 14 dB/km. However, the indoor radio channel shows adverse frequency selective multipath characteristics due to the highly reflective indoor environment, which results in severe signal dispersion and limits the maximum usable symbol rate. Another advantage is that this frequency region is not in use by any other communications medium, so every channel can be allocated a large bandwidth: 100-MHz channels can be used without any bandwidth problems. A third advantage of millimeter-wave technology is that antenna sizes are very small, so the equipment will be convenient. A fourth advantage is that the millimeter-wave spectrum has the potential to support broadband service access, which is especially relevant because of the advent of B-ISDN.

A major drawback of this frequency region is that the technology for transmitters and receivers has not yet been fully developed. As a consequence, the hardware will be expensive in the early stages. It is worth mentioning that no definitive evidence of any hazards has been shown to date to the general public arising from prolonged exposure in fields of less than 10 mW/cm^2 in millimeter waves, though the general population is still reluctant to accept it.

Within the European research program, the millimeter-wave spectrum has been selected for development of the mobile broadband systems (MBS). The MBS typically addresses services above 2 Mbps. The high data rates envisaged for MBS require operation at much higher frequencies, currently estimated to be in the 60-GHz bands. Study of MBS needs a lot of investigation in terms of propagation modeling, antenna diversity, and technology development.

1.13 Other Wireless Communications Systems

The systems described so far present what we believe will be the wireless systems that will eventually follow the evolution path of Figure 1.1. They are being implemented and are expected to meet a large majority of the world's demand for wireless personal communications. Meanwhile, many other wireless communications systems will be emerging to serve special needs that are not met well by the existing systems. These systems include in summary:

- Mobile communications satellites [9, 12, 16];
- WLANs [27];
- Wireless local loops [3];
- Wireless data networks [25].

The reader up to this point is expected to understand the architecture and system design constraints of the wireless systems in use and in the implementation stage. They will come up in the discussion in the later chapters, with regard to interference suppression and performance improvement as they operate in an interference- and distortion-based environment.

References

[1] Stavroulakis, P., *Third Generation Mobile Communication System,* UMTS and IMT-2000, Berlin: Springer, 2001.

[2] Tisal, J., *The GSM Network,* New York: John Wiley, 2001.

[3] Stavroulakis, P., *Wireless Local Loops, Theory and Applications,* New York: John Wiley, 2001.

[4] Lee, W. C. Y., *Mobile Communication Design Fundamentals,* second edition, New York: John Wiley, 1993.

[5] Hanzo, Lajos, P. J. Cherriman, and Jurgen Streit, *Wireless Video Communications,* New York: IEEE Press, 2001.

[6] Pandya, Raj, *Mobile and Personal Communication Services and Systems,* New York: IEEE Press, 2000.

[7]　Steele, R., and Lajos Hanzo, *Mobile Radio Communications: Second and Third Generation Cellular and WATM Systems,* second edition, New York: IEEE Press, 1999.

[8]　Gibson, J. D., *The Mobile Communication Handbook,* second edition, New York: IEEE Press, 1999.

[9]　Ohmori, S., and S. Wakana, *Mobile Satellite Communications,* London: Artech House, 1998.

[10]　Marol, J., and M. Bousquet, *Satellite Communication Systems,* third edition, New York: John Wiley, 1993.

[11]　Gordon, G. D., and Walter L. Morgan, *Principles of Communications Satelites,* New York: John Wiley, 1993.

[12]　Logsdon, T., *Mobile Communication Satellites, Theory and Application,* New York: McGraw-Hill, 1995.

[13]　Jamalipur, A., *Low Earth Orbital Satellite for Personal Communication Networks,* Norwood, MA: Artech House, 1998.

[14]　Ha T., Tri, *Digital Satellite Communications,* second edition, New York: McGraw-Hill, 1990.

[15]　Elbert, B. R., *Introduction to Satellite Communication,* Norwood, MA: Artech House, 1987.

[16]　Ananasso, F., and F. Vatalaro, *Mobile and Personal Satellite Communications,* Berlin, Germany: Springer, 1995.

[17]　Feher, K., *Digital Communications: Satellite/Earth Station Engineering,* Norcross, GA: Noble Publishing Company, 1997.

[18]　Gagliardi, R. M., *Satellite Communications,* second edition, New York: Van Nostrand Reinhold, 1991.

[19]　Holma, H., and Antti Toskala, *WCDMA for UMTS,* New York: John Wiley, 2001.

[20]　Prasad, R., *Universal Wireless Personal Communications,* Norwood, MA: Artech House, 1998.

[21]　Hernando, M. J., and F. Fontan-Perez, *Introduction to Mobile Communications Engineering,* Norwood, MA: Artech House, 1999.

[22]　Goodman, J. D., *Wireless Personal Communications Systems,* Boston: Addison-Wesley, 1997.

[23]　Pratt. S. R., et al., "An Operational and Performance Overview of the Iridium Satellite System," IEEE Communications surveys, second quarter, 1999.

[24]　Dimou, J., "Optimal Routing in a Satellite LEO/MEO Network using Algorithmic and Neural Techniques," Master's thesis, June 2000, Technical University of Crete, Greece.

[25]　Pahlavan, K., and H. A. Levesque, *Wireless Information Networks,* New York: John Wiley, 1995.

[26]　Clark, J. M., *Wireless Access Networks,* New York: John Wiley, 2000.

[27]　Asuncion, Santamaria, and J.F. Hernandez-Lopez, *Wireless LAN,* Norwood, MA: Artech House, 2001.

[28]   Jansky, M. D., and C. M. Jeruchim, *Communication Satellites in the Geostationary Orbit,* Norwood, MA: Artech House, 1987.

[29]   Stavroulakis, P., and S. C. Moorthy, "A Statistical Approach to the Interference Reduction of a Class of Satellite Transmissions," *National Telecommunications Conference (NTC 79),* Vol. 3, Washington, D.C., Dec. 1979, pp. 52.3.1–52.3.5.

2

Wireless Channel Characterization and Coding

2.1 Introduction

Following the methodology developed in the Preface and discussed in Chapter 1, we now proceed in Chapter 2 to explain the importance and relevance of wireless channel characteristics and coding in the study and development of interference suppression techniques, as is the main theme of this book. In Chapter 1, we introduced the design and architectural parameters of various wireless systems that have been developed over the years. We showed how these parameters could affect the behavior of a particular wireless system in an interference environment.

This approach can facilitate the development of comparative and quality measures for the operation of wireless systems used or planned to be used. In this chapter, we shall define the wireless channel in order to form the basis for the later chapters, which will analyze their interference characteristics. Besides the fact that wireless systems, especially mobile systems, exhibit very severe-path loss with respect to distance, which is much larger for wireless systems than path loss for fixed (LOS) wireless links, their BER performance is severely degraded by multipath fading. Both of these characteristics are analyzed in this chapter and related to the main theme of this book.

For the transmission of data, however, taking all known corrective measures is not sufficient to ensure acceptable link quality for data transmission. Given that the communications world is becoming digital and all types of information signals are converted to data, extra measures must be taken in a wider scale in order to satisfy the quality requirements of data transmission. These measures refer to error control techniques, which intend to reduce the probability of bit errors and thus achieve a high quality of data transmission in wireless communication channels. To detect or correct errors, we add some redundant bits to the source information by using an encoding rule that maximizes the error detection or error correction abilities. Such encoding is called channel encoding—thus its relevance to this chapter. There are two ways of achieving bit error improvement using coding: to detect unacceptable errors and request retransmission, or detect errors and try to correct them. The purpose of this chapter is to relate these three main aspects of the wireless channel (i.e., propagation, multipath fading, and channel coding) to the behavior of wireless systems in an interference environment.

2.2 The Wireless Communication Channel

The classic architecture of a generic communication system is illustrated in Figure 2.1. This was originally described by Claude Shannon of Bell Laboratories in his classic 1948 paper [1]. An information source attempts to send information to a destination. The source can be a person speaking, a video camera, or a computer sending data, for example, and the destination can be a person listening, a video monitor, or a computer receiving data. The data is converted into a signal suitable for transmission by the transmitter

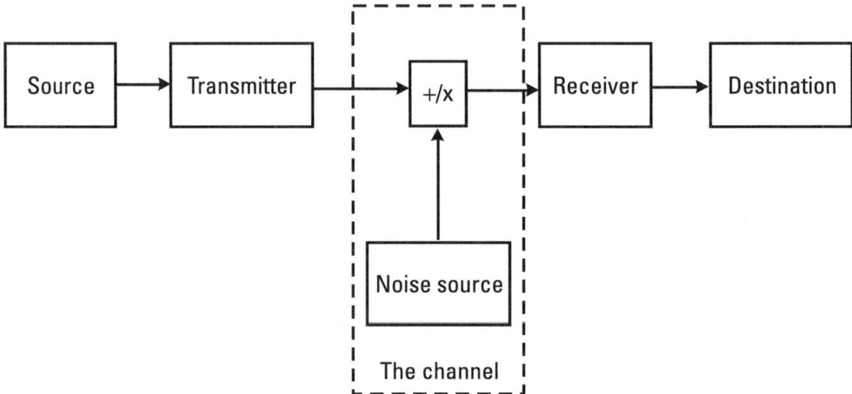

Figure 2.1 Architecture of a generic wireless channel (After: [2]. © 1999 John Wiley & Sons, Inc.)

and is then sent through the channel. The channel itself modifies the signal in ways that may be more or less unpredictable to the receiver because the wireless transmission path is more or less chaotic. The total effect is to produce a stochastic overall signal that must be treated as such. Appendix A provides the background of treating such signals. The receiver must be designed to automatically overcome these modifications and hence to deliver the information to its final destination with as few errors or distortions as possible.

This representation applies to all types of systems, whether wireless or otherwise. In the wireless channel, specifically, the noise sources or other interfering effects can be subdivided into multiplicative and additive effects, as shown in Figure 2.2. The additive noise arises from noise generated within the receiver itself, such as thermal and shot noise in passive and active devices, and also from external sources such as atmospheric effects, cosmic radiation, and interference from other transmitters and electrical appliances. Some of this interference may be intentionally introduced but carefully controlled, such as when channels are reused in order to maximize the capacity of a cellular radio system [3].

The multiplicative noise arises from the various communication processes encountered by transmitted waves on their way from the transmitter antenna to the receiver antenna, as described below.

- The directional characteristics of both the transmitter and receiver antennas;
- Reflection (from the smooth surfaces of walls and hill);
- Absorption (by walls, trees, and the atmosphere).

The way we design wireless systems to overcome these types of distortion agents (i.e., fading) will be examined in Chapter 4. In Chapter 5 and onward,

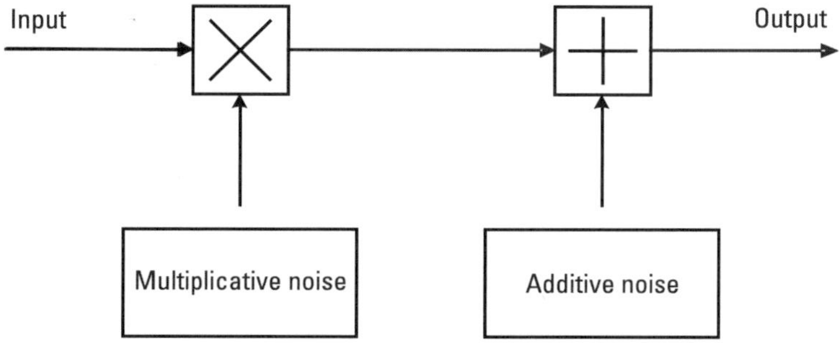

Figure 2.2 Two types of noise (distortions) in the wireless communication channel.

we will study the additive types of interfering agents as well as any other type of distortion that at the end acts and can be successfully overcome as an additive interferer, such as intermodulation effects and intersymbol interference. Nonetheless, we shall see that some of the techniques used in both cases are similar.

It is conventional to subdivide the multiplicative processes in the channel into three types of fading: path loss, shadowing (or slow fading), and fast fading (or multipath fading), which appear as time-varying processes between the antennas, as shown in Figure 2.3. All of these processes vary as the relative positions of the transmitter and receiver change [2, 4].

The path loss is an overall decrease in power, as the distance between the transmitter and the receiver increases. The physical processes that cause it are the outward spreading of waves from the transmit antenna and the obstructing effects of various intervening obstacles. A typical system may involve variations in path loss of around 150 decibels (dB) over the designed coverage area (satellite). Superimposed on the path loss is the shadowing, which changes more rapidly, with significant variations over distances of hundreds of meters and generally involving variations up to around 20 dB. Shadowing arises due to the varying nature of the particular obstructions between the transmitter and the receiver, such as particular tall buildings or dense woods. Fast fading involves variation on the scale of a half-wavelength (50 cm at 300 MHz, 17 cm at 900 MHz) and frequently introduces variations as large as 35 to 40 dB. It results from the constructive and destructive interference between multiple waves reaching the mobile from the base station.

In the following sections, we will discuss and analyze these three types of signal distortions and point out their characteristics, which will be used in Chapter 4 for developing corrective measures.

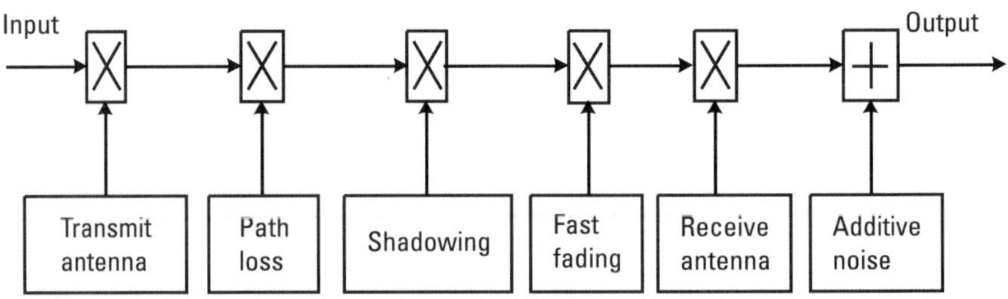

Figure 2.3 Contributions to noise in the wireless channel.

2.2.1 Path Loss

2.2.1.1 Outdoor Large-Zone Systems

When there are no obstacles around or between the base station (BS) and mobile station (MS), the propagation path characteristics are subject to free space propagation. In this case, the path loss is given by

$$L_{pf} \text{ (dB)} = 32.44 + 20 \log_{10} f_c + 20 \log_{10} d \qquad (2.1)$$

where

f_c = carrier frequency (megahertz);

d = distance between BS and MS (kilometers);

L_{pf} = path loss in decibels.

On the other hand, when there are many obstacles around or between the BS and MS, path loss is determined by many factors, such as irregular configuration of the natural terrain and irregularly arranged artificial structures [3].

Complicated propagation path loss characteristics based on a large amount of empirical data around Tokyo, Japan, were analyzed in [5]: First of all, they selected propagation path conditions and obtained the average path loss curves under flat urban areas as the standard propagation path conditions because most of the terminals are located in the urban areas. These curves are now called *Okumura curves*. Then, they obtained correction factors for the other propagation path conditions, such as:

- Antenna height and frequency;
- Suburban, quasi-open space, open space, or hilly terrain areas;
- Diffraction loss due to mountains;
- Sea or lake areas;
- Road slope.

Although Okumura curves are practical and effective when used to estimate coverage area for a system, it is not convenient to use them for the computational system designs, including system parameter optimization. To solve this problem, Hata derived empirical formulas for the median path loss that are fit to Okumura curves [6]. His contribution, which is *Hata's equation,* presents three models: typical urban, typical suburban, and rural area models. The results are given below as (a), (b), and (c) respectively:

 (a) Typical urban model

$$L_p \text{ (dB)} = 69.55 + 26.16 \log_{10} f_c + (44.9 - 6.55 \log_{10} h_b) \log_{10} d \quad (2.2)$$
$$- 13.82 \log_{10} h_b - a(h_m)$$

where

 f_c = carrier frequency (megahertz);

 d = distance between base and mobile stations (kilometers);

$d(h_m)$ is the correction factor for MS antenna height given by
(for large cities)

$$a(h_m) = \begin{cases} 8.29 \, [\log_{10}(1.54 h_m)]^2 - 1.1 & (f_c \leq 200 \text{ MHz}) \\ 3.2 \, [\log_{10}(11.75 h_m)]^2 - 4.97 & (f_c \geq 400 \text{ MHz}) \end{cases} \quad (2.3)$$

(for small and medium-size cities)

$$a(h_m) = [1.1 \, \log_{10}(f_c) - 0.7] h_m - [1.56 \, \log_{10}(f_c) - 0.8] \quad (2.4)$$

 (b) Suburban model

$$L_{ps} = L_p - 2\{\log_{10}(f_c/28)\}^2 - 5.4 \text{ [dB]} \quad (2.5)$$

where L_p is given by equation (2.2).

 (c) Rural area model

$$L_{po} = L_p - 4.78 (\log_{10} f_c)^2 + 18.33 \log_{10} f_c - 40.94 \text{ [dB]} \quad (2.6)$$

where L_p is given by equation (2.2).

2.2.1.2 Indoor Systems

Propagation path characteristics for indoor communication systems are very unique compared to outdoor systems. This is because there are so many obstacles that reflect, diffract, or shadow the transmitted radio waves, such as walls, ceilings, floors, and various office furniture. Various studies [7] have resulted in the conclusion that these situations can be effectively studied by categorizing the various situations into zone configurations as follows:

1. *Extra-large zone systems.* This configuration refers to the case when the BS is located outside the buildings, and it covers several buildings. In extra-large zone systems, the propagation path can be divided into the path outside a building and penetration into the building. As a result, path loss for extra-large zone systems [7–9] can be expressed as:

$$L_p(r) = L_r(r_0) \left(\frac{r}{r_0}\right)^{\alpha_1} \cdot L_B(r_0) \left(\frac{r}{r_0}\right)^{\alpha_2} A_F \qquad (2.7)$$

where

$L_r(r_0)$ = path loss attenuation due to propagation at a distance $r = r_0$;

$L_B(r_0)$ = attenuation due to building at $r = r_0$;

α_1 = attenuation factor of the propagation path loss with respect to distance;

α_2 = building attenuation factor;

A_F = building penetration loss.

and $L_r(r_0)$ and $L_B(r_0)$ are determined by the frequency and the density of obstacles nearby. The difference is that $L_r(r_0)$ increases with frequency, whereas $L_B(r_0)$ decreases. Therefore, the additional path loss at a higher frequency could be offset by lower building attenuation [7]. Also, α_1 is determined by the distribution of buildings between the base station and each terminal. In the case of LOS conditions, α_1 takes a value of around 2.0 and in the case of non-LOS conditions, α_1 takes a value in the range of $3 \le \alpha_1 \le 6$, the exact value being dependent on the obstacles around the building. On the other hand, α_2 depends less on the distance than α_1 and usually takes a value in the range of 0.5–1.5. Finally, A_F depends on the antenna height difference between the BS and each terminal as well as on the materials for windows. When the transmitter and receiver antennas are located at the same height, A_F is minimized.

On the other hand, when the antenna height difference is increased, A_F is also increased [8].

2. *Large-zone systems.* This configuration refers to the case when one BS is installed inside a building, and it covers the whole building. Large-zone systems cover all the terminals in a building by a BS in the building. Therefore, it is an extreme case of a middle-zone system, in which a BS covers several rooms in the building. The large-zone system may be effective for wireless private branch exchange (PBX) systems in a building with relatively low terminal density. Path loss for this system is given by:

$$L_p(r) = L_r(r_0)\,(r/r_0)^{\alpha_p} \qquad (2.8)$$

where α_p takes values 2–3 when the transmitter and the terminal are located on the same floor and it depends on its exact location, and takes values $\alpha_p \geq 3$ when they are located on the different floors [10–12].

3. *Middle-zone systems.* This configuration refers to the situation in which one BS covers several rooms. Middle-zone systems represent one of the most practical and widely applicable zone configurations for indoor systems. One of its path loss models is given by [12]:

$$L_p(r) = \left(\frac{4\pi f_c r}{c}\right)^2 F(r)^{k_1}\, W(r)^{k_2} R(r) \qquad (2.9)$$

where

$$c = \text{velocity of light;}$$

$$f_c = \text{carrier frequency;}$$

$$F(r) = \text{floor attenuation;}$$

$$W(r) = \text{wall attenuation;}$$

$$R(r) = \text{reflection loss;}$$

$$k_1 = \text{number of floors transversed;}$$

$$k_2 = \text{number of walls transversed.}$$

$F(r)$ is usually 20 to 40 dB and less dependent on r [13]. For the middle-zone systems, the coverage is restricted to within the same floor. For this purpose, larger $F(r)$ is preferable. Although floor attenuation in decibels linearly increases with the number of floors in most cases, it shows some nonlinearity with respect to the number of floors because of the power leakage through stairways or windows. $W(r)$ is a very important factor in determining the coverage. When a large coverage area is of interest, a moderate value of $W(r)$ is preferable. On the other hand, a larger value is necessary when we want to restrict zone radius. In [3], extensive studies were conducted on this attenuation factor. $R(r)$ is a reflection loss. In the case of indoor communications, however, $R(r)$ is sometimes a very small value, especially when the transmitted signal is propagated along the corridors. This is because the radiated wave outside a corridor is relatively small.

4. *Small-zone systems.* Small-zone systems are those systems for which one base station covers a single room. These types of systems achieve fewer service outages. This is especially true for those cases in which outages are more likely, as is the case of large buildings with high traffic density. Because only one BS covers one room, high wall attenuation as well as high floor attenuation are to be considered.

   Path loss for this system greatly depends on the number of obstacles between the BS and the terminal, and α_p can take values between 2 and 4. In the case of small-zone systems employing a higher frequency band, it is also a good strategy to improve system capacity to compensate for greater path loss [3, 14–16].

5. *Microzone systems.* Microzone systems, in which several BSs are installed in a room, are effective for covering a large business office with high terminal density. Path loss for this system is almost the same as that for middle-zone or small-zone systems, except that smaller α_p is more probable in the case of microzone systems.

2.2.2 Multipath Propagation

2.2.2.1 Shadowing (Slow Fading)

As the receiver moves, the received signal is not only the signal that was transmitted by the transmitter, but a combination of signals received at that point from different paths through reflecting differentials. The effect of the differential time delays of these will be to introduce relative phase shift between component waves, and superposition of these waves can lead to either constructive or destructive addition, as shown in Figure 2.4 [4].

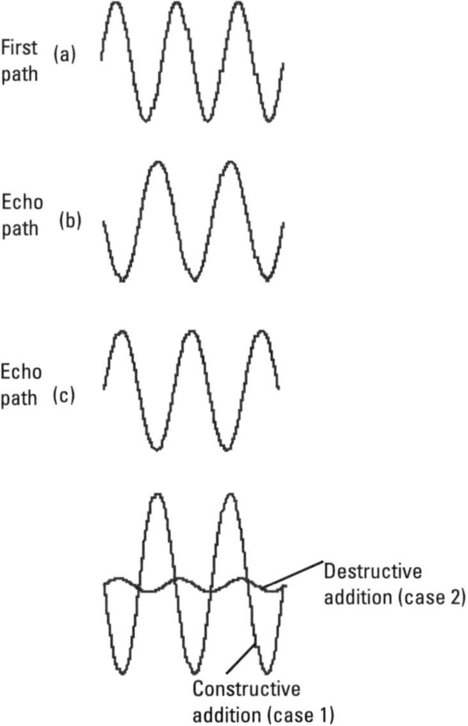

Figure 2.4 Constructive and destructive addition of two transmission paths. (After: [4]. © 2000 John Wiley & Sons, Inc.)

The distribution underlying signal powers is often lognormal. That is, the signal measured in decibels has a normal distribution. The process by which this distribution comes about is known as shadowing or slow fading.

The motion of the mobile in a particular path causes a Doppler effect that is exhibited as a Doppler frequency shift, as shown in Figure 2.5.

The incremental distance d traveled by the mobile receiver along path AA in Figure 2.5 is given by $d = v\Delta t$ where v is the velocity along the path. The phase change therefore [4]:

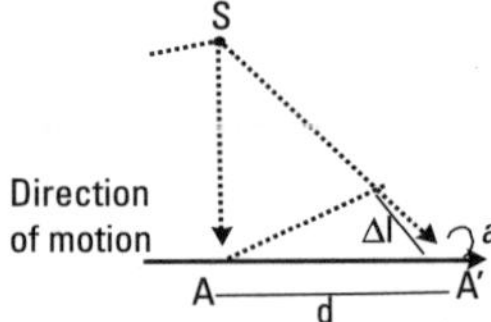

Figure 2.5 Doppler shift.

$$\Delta\phi = -\frac{2\pi}{\lambda}\,\Delta l = -\frac{2\pi v\Delta t}{\lambda}\,\cos a \qquad (2.10)$$

and the Doppler shift is given:

$$f = -\frac{1}{2\pi}\frac{\Delta\phi}{\Delta t} = -\frac{v}{\lambda}\,\cos a \qquad (2.11)$$

The maximum of the Doppler shift f_m is obtained when the waves arrive either directly from behind or ahead of the mobile giving

$$f_m = \pm v/\lambda \qquad (2.12)$$

In practical cases, the resultant signal envelope and phase will be random variables. To determine the behavior of the receiver to such a wave, we must be able to devise a mathematical model that will lead to results that are in accordance with the observed signal properties.

We shall assume that at a specific point the received signal is given mathematically by the equation

$$s(t) = I(t)\,\cos \omega_c t - Q(t)\,\sin \omega_c t \qquad (2.13)$$

where $I(t)$ and $Q(t)$ are stochastic processes.

The envelope of the signal $s(t)$ is given by

$$r(t) = \sqrt{I^2(t) + Q^2(t)} \qquad (2.14)$$

and the phase

$$\theta(t) = \tan^{-1}\frac{Q(t)}{I(t)} \qquad (2.15)$$

It can be shown [4] that the probability density function of $r(t)$ is given by

$$P_r(r) = \frac{r}{\sigma^2}\exp\left(-\frac{r^2}{2\sigma^2}\right) \qquad (2.16)$$

where σ^2 is the mean power and $\frac{r^2}{2}$ is the short term signal power.

This is the Rayleigh density function shown in Figure 2.6. The probability that the envelope does not exceed a specified value R is given by the cumulative distribution function

$$P_r(R) = P(r \le R) = \int_0^R p_r(r)\, dr = 1 - \exp\left(-\frac{R^2}{2\sigma^2}\right) \qquad (2.17)$$

This phase random variable has a probability density function given by [4]

$$P_\theta(\theta) = \frac{1}{2\pi} \qquad (2.18)$$

The mean value, the mean square value and variance of θ are given by π, $\dfrac{4\pi^2}{3}$ and $\dfrac{\pi^2}{3}$, respectively.

We expect that the signal composed of a number of components of random phase of the resultant signal would not have any bias. In wireless systems design, it is not very interesting nor meaningful to consider the absolute phase, we focus on the relative phase to another signal. The probability density function of the phase difference $\Delta\theta$ between points spatially separated by some distance can be determined and be able to show [4] that at spatial separations for which the envelope is uncorrelated, the phase difference is also uncorrelated.

Because the movement of the mobile produces a random change of phase with time equivalent to a random phase modulation, and because the

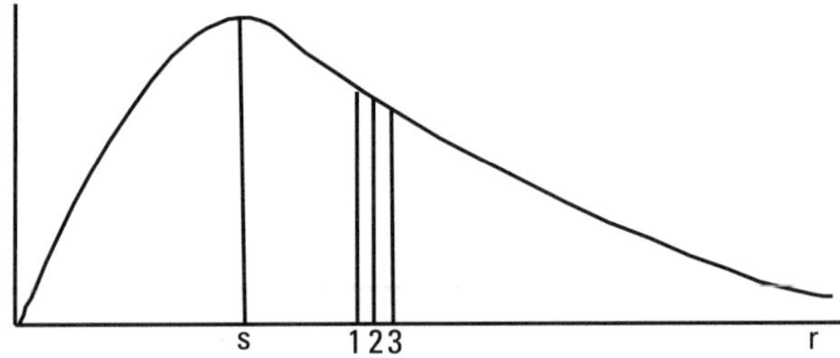

Figure 2.6 Probability density function (PDF) of the Rayleigh distribution: 1 = median (50%) value, 1.1774 σ: 2 = mean value, 1.2533 σ: 3 = rms value, 1.41 σ. (After: [4]. © 2000 John Wiley & Sons, Inc.)

time derivative of θ causes frequency modulation, which is detected by any phase detector, we have a phenomenon equivalent to frequency modulation (FM). This random FM spectrum is shown in Figure 2.7. It has been obtained as the Fourier transform of the expression of its correlation function in [17].

2.2.2.2 Correlated Shadowing (Slow Fading)

So far, we have considered the shadowing experienced on nearby paths as independent. In the practical situation depicted schematically in Figure 2.8, the four shadowing paths are not independent of each other because the four paths may include many of the same obstruction in the path profiles.

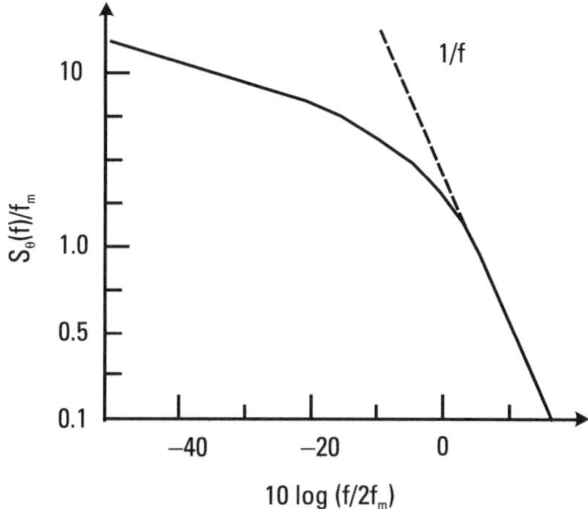

Figure 2.7 Power spectrum of random FM plotted as relative power on a normalized frequency scale (After: [4]. © 2000 John Wiley & Sons, Inc.)

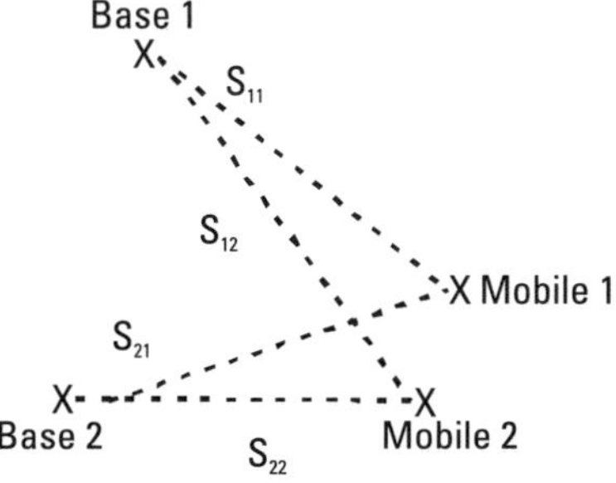

Figure 2.8 Definitions of shadowing correlations (After: [2]. © 1999 John Wiley & Sons, Inc.)

This correlation can take two forms. One form is the correlation at a single receiver by receiving signals from two different base stations; the other is two different mobiles receiving signals from the same base station.

- The first form is the correlation between S_{11} and S_{12} and between S_{21} and S_{22}. These are called serial correlations.
- The second form is the correlation between S_{11} and S_{21} or between S_{12} and S_{22} as shown in Figure 2.8.

The serial case affects the rate at which the total path loss experienced by a mobile varies in time as it moves around [2]. This has a particularly significant effect on power control processes, where the base station typically instructs the mobile to adjust its transmit power so as to keep the power received by the base station within prescribed limits. This process has to be particularly accurate in CDMA systems, where all mobiles must be received by the base station at essentially the same power in order to maximize system capacity. If the shadowing autocorrelation reduces very rapidly in time, the estimate of the received power that the base station makes will be very inaccurate by the time the mobile acts on the command, so the result will be unacceptable. If, on the other hand, too many power control commands are issued, the signaling overhead imposed on the system will be excessive.

The second case refers to site-to-site correlation, in which the two paths may be very widely separated and different in length. Because they may also involve rather different environments, the location variability associated with the paths may be different. The two base stations involved in the process may be on the same channel, in which case the mobile will experience some level of interference from the base station to which it is not currently connected. The system is usually designed to avoid this by providing sufficient separation between the base stations so that the interfering base station is considerably further away than the desired one, resulting in a relatively large signal-to-interference ratio (S/I). If the shadowing processes on the two links are closely correlated, the S/I will be maintained and the system quality and capacity is high. If, by contrast, low correlation is produced, the interference may frequently increase in level while the desired signal falls, significantly degrading the system performance. More details are given in Chapters 5 and 6.

It is therefore clear that the shadowing cross-correlation, which plays the role of interference, has a decisive effect upon system capacity. In addition, the use of realistic values is essential to allow accurate system capacity, and the use of realistic values is necessary for reliable system designs.

Here are some systems designs issues that may be affected by the shadowing cross-correlation [2]. The design parameters mentioned next were discussed in Chapter 1:

1. Optimum choice of antenna beamwidths when they are used for sectorization of cells.

2. Performance of soft hand-off and site diversity, including simulcast and quasi-synchronous operation, where multiple base sites may be involved in communication with a single mobile. Such schemes give maximum gain when the correlation is low in contrast to the conventional interference situation described earlier.

3. Design and performance of handover algorithms. In these algorithms, a decision to hand over to a new base station is usually made on the basis of the relative power levels of the current and the candidate base stations. In order to avoid *chatter,* where a large number of handovers occur within a short time, appropriate averaging of the power levels must be used. Proper optimization of this averaging window and of the handover process in general requires a knowledge of the dynamics of both serial and site-to-site correlations, particularly for fast-moving mobiles.

4. Optimum frequency planning for minimized interference and hence maximized capacity.

5. Adaptive antenna performance calculation.

As no well-agreed model exists for predicting the correlation, some approximate models have been proposed. These have some physical basis, but they require further testing against measurements. They include two main parameters:

1. The angle between the two paths between the base stations and the mobile;

2. The relative values of the two path lengths.

In conclusion, shadowing affects the dynamics of signal variation at the mobile, the percentage of locations that receive sufficient power, and the percentage that receive sufficient S/I.

The results of this section clearly show the importance of the methodology explained in the Preface. It is easy to see that the relationship between system design parameters and channel characteristics enable the comprehension and control of the behavior of wireless systems in an interference environment.

2.2.2.3 Narrowband Fast Fading

The inverse of Doppler spread has units of time, and it is called the coherence time. It is given by

$$T_c = \frac{1}{f_d} \tag{2.19}$$

The coherence time, T_c, of a channel measures the period of time over which the fading process is correlated (i.e., the channel response taken at the same frequency but different time instants is above a certain minimum threshold).

The fading is said to be slow if the symbol time duration, T_s, is smaller than the channel's coherence time, T_c ($T_s < T_c$). Otherwise, it is considered fast. If the transmitted signal bandwidth is much smaller than the channel coherence bandwidth, we refer to the case of narrowband systems. In other words, we refer to systems for which the transmitted signal bandwidth is much smaller than the channel's coherence bandwidth. The coherence bandwidth measures the frequency range over which the fading process is correlated. Also, the coherence bandwidth is the inverse of the maximum delay spread. If the spectral components of the transmitted signal are affected by different amplitude gains and phase shifts, the fading is called frequency selective, and it applies to wideband systems. In this section, we will discuss the narrowband case, whereas in Section 2.2.2.5, we will discuss the wideband case.

For the narrowband case, because the fading affects all frequencies in the modulated signal equally, one can model it as a single multiplicative process. It is called nonselective frequency fading. In such a case, there is no variation in the path loss over the signal bandwidth.

Hence the received signal at time t is given by

$$s(t) = A\alpha(t)u(t) + n(t) \tag{2.20}$$

where $u(t)$ is the modulated signal and $\alpha(t)$ is the complex fading coefficient at time t.

If we define

$$\gamma(t) = \text{SNR} = \frac{\text{Signal Power}}{\text{Noise Power}}$$

then

$$\gamma(t) = \frac{A^2|\alpha(t)|^2 E\{|u(t)|\}^2}{2P_N} = \frac{A^2|\alpha(t)|^2}{2P_N} \qquad (2.21)$$

where P_N noise power.
If we define by $\overline{\gamma} = E\{\gamma(t)\}$, then for a Rayleigh channel

$$\frac{E\{|\alpha(t)|^2\}}{2} = \sigma^2$$

and

$$P_R(r) = \frac{r}{\sigma^2} e^{-r^2/2\sigma^2}$$

where

$$r^2 = |\alpha(t)|^2$$

then

$$\gamma = \frac{A^2 r^2}{2P_N} \text{ and } \overline{\gamma} = \frac{A^2 E[r^2]}{2P_N} \qquad (2.22)$$

To find the distribution of γ given the distributions of r, we used the identity

$$p_\gamma(\gamma) = p_R(\gamma) \frac{dr}{d\gamma} = \frac{r}{\sigma^2} e^{-r^2/2\sigma^2} \frac{P_N}{A^2 r} \qquad (2.23)$$

Hence, we can write

$$p_\gamma(\gamma) = \frac{1}{\overline{\gamma}} e^{-\gamma/\overline{\gamma}} \qquad \text{for } \gamma > 0 \qquad (2.24)$$

where $\overline{\gamma} = \dfrac{A^2 \sigma^2}{2P_N}$, which is a Rayleigh distribution frequently used to model multipath fading with no direct line of sight (LOS) paths [18].

Narrowband Channel Simulations

As we will see in Chapter 5 and onward of this book, for the analysis and design of systems as immune to interference as possible, it is often necessary to emulate the channel under consideration. For accuracy purposes, any simulation must be consistent with at least the first- and second-order statistics of the mobile channel. An approach that seems to yield acceptable results is shown in Figure 2.9(a, b) [3].

A complex white Gaussian noise generator is used to represent the in-phase and quadrature signal components with unit power. These are passed

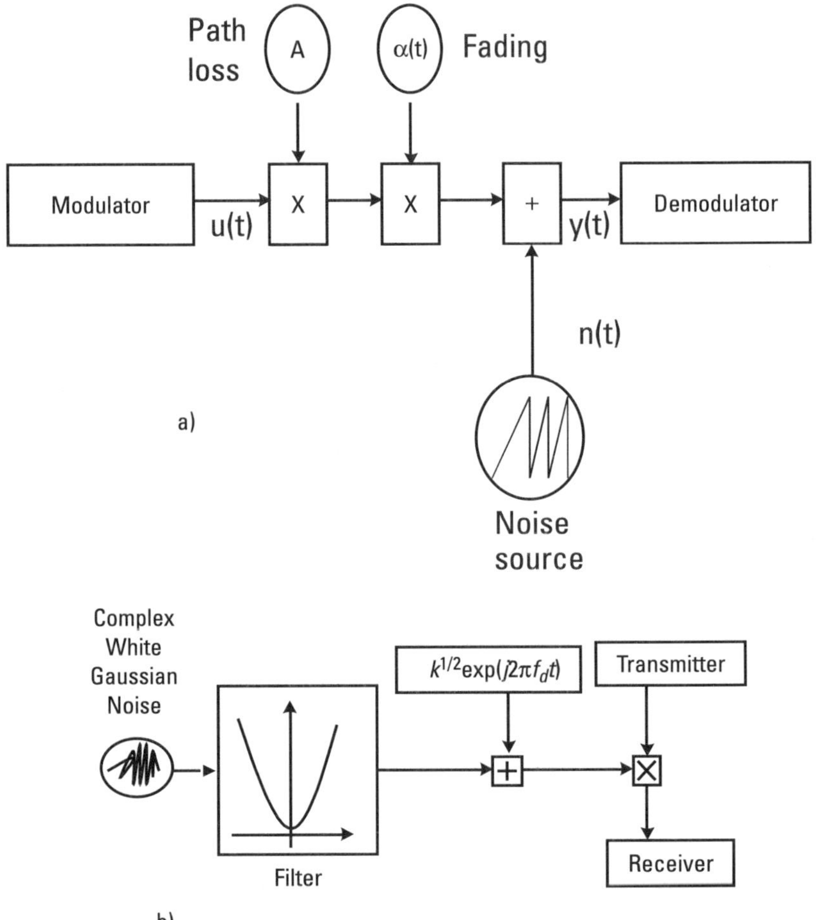

Figure 2.9 (a) The narrowband fading channel. (b) Baseband simulation of the Rice-fading channel as a narrowband channel. (Source: [2]. © 1999 John Wiley & Sons, Inc.)

to a filter, carefully designed to produce a close approximation to the classical Doppler spectrum at its output. The exact shape of the filter is not critical, as the average fade duration and level crossing rate will be correct, provided the variance of the noise spectrum at the filter output matches the variance of the desired classical spectrum. Other approaches to creating the signal at this stage are also available (e.g., a sum-of-sinusoids simulator [19]). A phasor of constant amplitude $\sqrt{k}$, where k is the desired Rice factor, is then added, representing the dominant coherent part of the channel. The phasor is usually given a nonzero frequency shift f_d, representing the Doppler shift associated with the LOS path. The final result can then be used to multiply the signal from any transmitter, either in a computer simulation or by creating a real-time implementation of the simulator in hardware. This permits real mobile radio equipment to be tested in laboratory conditions, which repeatedly emulate the practical mobile environment.

As a general rule, fading is described statistically by Rayleigh or Rice distributions with good accuracy, depending on where non-LOS or LOS conditions prevail, respectively. Both cases degrade the signal quality relative to the static case, where the channel can be described by simpler additive white Gaussian noise statistics. The rate or variation of the fading signal, due to the phenomenon of Doppler spread, is controlled by the carrier frequency, the speed of the mobile, and the angle-of-arrival distribution of waves at the mobile. The Doppler spread leads to characteristic fading behavior, which can be simulated using simple structures to provide a comparison of mobile equipment in realistic conditions.

Over the years, many approaches have been utilized for developing suitable simulation models in the analysis of fading channels. One approach described in [3] is to express the fading variation by the equivalent lowpass system using a multitone approximation. The resulting configuration, which can be tested in a laboratory, can acceptably emulate the fading effects on a transmitted signal. We observe that channel characterization results in the derivation of probability distribution of SNR, given the fading parameters that in turn will be used to derive quantitative measures of the effects of multipath fading in BER.

2.2.2.4 Rice and Other Distributions

In the LOS situation, the received signal is composed of a random multipath component whose amplitude is described by the Rayleigh distribution, plus a coherent LOS component that has essentially constant power. The theoretical distribution, which applies in this case, was derived and proved by Rice and it is called Rice distribution. It is given by:

$$P_R(r) = I_0\left(\frac{r}{\sigma^2}\right)\frac{r}{\sigma^2}\, e^{-\left(\frac{r^2 + k^2 r^2}{2\sigma^2}\right)} \tag{2.25}$$

where

σ^2 is the variance of the multipath part as we saw in (2.20) to (2.24);
k is the magnitude of the LOS component;
$I_0(0)$ is the Bessel function of the first kind and 0th order (see Appendix A).

We observe that if $k = 0$, which is called Rician constant, it reduces to Rayleigh distribution. If we use the procedure followed by (2.20) to (2.24), we obtain expressions $P_\gamma(\gamma)$ for various types of fading.

Other distributions have been developed over the years to model more complex situations, such as terrestrial and satellite land-mobile system, composite multipath/shadowing, and time-shared shadowed/unshadowed fading, as well as various forms of the Nakagami models. Table 2.1 tabulates various PDFs of the SNR per symbol γ for some common fading in various practical situations [20–22]. These results also give a justification for the relevance of this chapter to the main theme of the book.

2.2.2.5 Wideband Fast Fading

Mobile radio systems for voice and LBR data applications are designed with the consideration that the channel has purely narrowband characteristics, but the wideband mobile radio channel has assumed increasing importance in recent years as mobile data rates increase to support multimedia services. In nonmobile applications, such as television and fixed links, wideband channel characteristics have been important for a considerable period. If the relative delays are large compared to the basic unit of information transmitted on the channel (usually a symbol or a bit), the signal will then experience significant distortion, which varies across the channel bandwidth. The channel is then a wideband channel, and any models to be used for analysis of those channels must account for these effects [2].

The standard form of the model for wideband mobile channels is shown in Figure 2.10(a, b). The effects of scatterers in discrete channel delay ranges are lumped together into individual *taps* with the same delay. Each tap represents a single beam. The taps each have a gain, which varies in time according to the standard narrowband channel statistics. The taps are usually assumed to be uncorrelated from each other, as each arises from scatterers that are physically distinct and separated by many wavelengths. The channel is therefore a linear filter with a time-variant finite impulse

Table 2.1
PDF of SNR per Symbol γ for Some Common Fading

Type of Fading	Fading Parameter	PDF, $p_\gamma(\gamma)$
Rayleigh		$\dfrac{1}{\overline{\gamma}} \exp\left(-\dfrac{\gamma}{\overline{\gamma}}\right)$
Nakagami-q Hoyt	$0 \leq q \leq 1$	$\dfrac{(1 + q^2)}{2q\overline{\gamma}} \exp\left[-\dfrac{(1 + q^2)^2\gamma}{4q^2\overline{\gamma}}\right] \times I_0\left[\dfrac{(1 - q^4)\gamma}{4q^2\overline{\gamma}}\right]$
Nakagami-n (Rice)	$n \geq 0$ $n^2 = k$ $k =$ Rician constant	$\dfrac{(1 + n^2)\,e^{-n^2}}{\overline{\gamma}} \exp\left[-\dfrac{(1 + n^2)\gamma}{\overline{\gamma}}\right] \times I_0\left[2n\sqrt{\dfrac{(1 + n^2)\gamma}{\overline{\gamma}}}\right]$
Nakagami-m	$1/2 \leq m$	$\dfrac{m^m\gamma^{m-1}}{\overline{\gamma}^m\Gamma(m)} \exp\left(-\dfrac{m\gamma}{\overline{\gamma}}\right)$
Log-normal shadowing	$\mu,\ \sigma$ means and standard deviation of $10\log_{10}\gamma$	$\dfrac{4.34}{\sqrt{2\pi}\sigma\gamma} \exp\left[-\dfrac{(10\log_{10}\gamma - \mu)^2}{2\sigma^2}\right]$
Composite gamma/log-normal	Γ *Gamma function* $\Gamma(\cdot) = \displaystyle\int_0^\infty x^{(\cdot)-1}\,e^{-x}\,dx$ m and $0 \leq \sigma$ $\xi = \dfrac{10}{ln\,10}$	$\displaystyle\int_0^\infty \dfrac{m^m\gamma^{m-1}}{w^m\Gamma(m)} \exp\left(-\dfrac{m\gamma}{w}\right) \times \dfrac{\xi}{\sqrt{2\pi}\sigma w} \exp\left[-\dfrac{(10\log_{10}w - \mu)^2}{2\sigma^2}\right] dw$

(From: [20].)

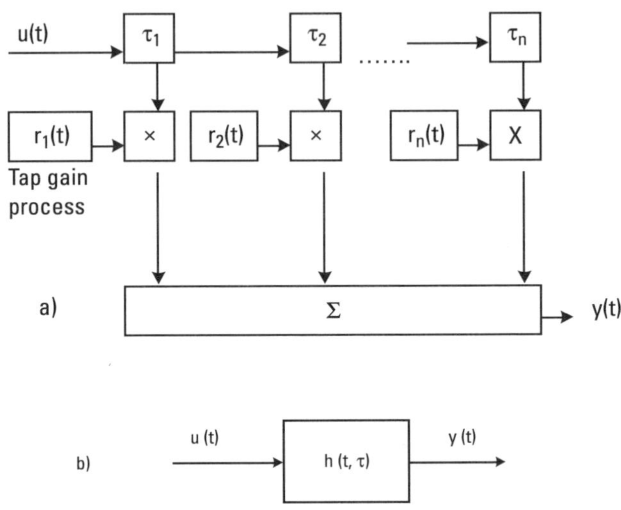

Figure 2.10 (a) Wideband fast fading and (b) impulse response. (Source: [2]. © 1999 John Wiley & Sons, Inc.)

response. It may be implemented for simulation purposes in digital or analog form. We next look at the parameters that characterize such models.

The basic function that characterizes the wideband channel is its time-variant impulse response. The output y at a time t can be found from the input u by convolving the inpute time series $u(t)$ with the impulse response $h(t, \tau)$ of the channel as it appears at time t, so

$$y(t) = u(t) * h(t, \tau) = \int_{-\infty}^{\infty} h(t, \tau) u(t - \tau) \, d\tau \qquad (2.26)$$

where * denotes convolution and τ is the delay variable. The time-variant impulse response is also known as the *input delay spread function*.

The channel is therefore a linear filter with a time-variant finite impulse response. In order for it to be shown graphically, we need three variables: relative power, relative delay, and time.

An interesting and useful parameter for such a model is the power delay profile (PDP), for the channel that represents the mean relative power of the taps. It is defined as the variation of mean power in the channel with delay:

$$P(\tau) = \frac{E\{|h(t, \tau)|^2\}}{2} \tag{2.27}$$

Each tap-gain may be either Rice or Rayleigh distributed. The PDP may be characterized by various parameters:

1. *Excess delay.* The delay of any tap relative to the first arriving tap.
2. *Total excess delay.* The difference between the delay of the first and last arriving tap; this is the amount by which the duration of a transmitted symbol is extended by the channel.
3. *Mean delay.* The delay corresponding to the *center of gravity* of the profile defined by

$$\tau_0 = \frac{1}{P_T} \sum_{i=1}^{n} P_i \tau_i \tag{2.28}$$

where the total power in the channel is

$$P_T = \sum_{i=l}^{n} P_i \tag{2.29}$$

4. *rms delay spread* (τ_{rms}). (τ_{rms}) represents the second moment, or spread, of the taps. This takes into account the relative powers of the taps as well as their delays, making it a better indicator of system performance than the other parameters. rms delay spread, τ_{rms}, is defined by

$$\tau_{rms} = \sqrt{\frac{1}{P_T} \sum_{i=l}^{n} P_i \tau_i^2 - \tau_0^2} \tag{2.30}$$

It is independent of the mean delay and hence of the actual path length, which is defined only by the relative path delays. The rms delay spread is a good indicator of the system error rate performance for moderate delay spreads (within one symbol duration). If the rms delay spread is very much less than the symbol duration, no significant intersymbol interference (ISI) is encountered, and the channel may be assumed as narrowband. Note that the effect of delayed taps in (2.30) is weighted by the square of the delay. This tends to overestimate the effect of taps with large delays but very small power, so that rms delay spread is not an unambiguous performance indicator. It nevertheless serves as a convenient way of comparing different wideband channels. The reader is referred to Chapters 3 to 5 for more details regarding various measures, which determine the performance of a wideband system operating in a fading environment.

2.2.2.6 Frequency Domain Model (Bello Functions)

In the frequency domain we must take the Fourier transform of $h(t, \tau)$ and this defines a time-variant transfer function $T(f, t)$, as shown in Figure 2.11 [2].

In practical channels, $T(f, t)$ is not known, and the only quantity measurable is the output in the frequency domain $Y(f, t)$. A standard method used to get a hold of the transfer function of a system where only the output is known is to determine correlation components of the output. In our case, the correlation between two components $\rho(t, t_2)$ of the channel transfer function with frequency separation Δt is defined by

$$\rho(\Delta f, \Delta t) = \frac{E\left[T(t, f)T^* (f + \Delta f, t + \Delta t)\right]}{\sqrt{E\left[|T(t, f)|\right]^2 \, E\left\{|T(f + \Delta f, t + \Delta t)|\right\}^2}} \quad (2.31)$$

we shall assume that the correlation $\rho(\Delta f, \Delta t)$ is independent of the particular time t and frequency f at which is proven in practice realistic along with the assumption that the scatterers are independent. This type of scattering is known as wide-sense stationary, uncorrelated scattering. For wideband systems and for signals with $\Delta t = 0$, it is shown that the bandwidth is inversely proportional to rms delay [23]. For an idealized model of a channel, the PDP can be modeled by the exponential function [23].

$$P(\tau) = \frac{1}{2\pi\tau_{\mathrm{rms}}} \, e^{-\tau/\tau_{\mathrm{rms}}} \quad (2.32)$$

Analyzing the phenomenon of time delay spread and Doppler delay spread by applying a system approach, we can follow the methodology of Bello, by using the Bello functions as shown in Figure 2.12 [19, 24, 25].

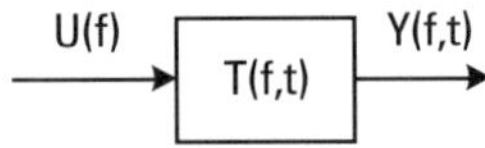

Figure 2.11 Time-variant transfer function.

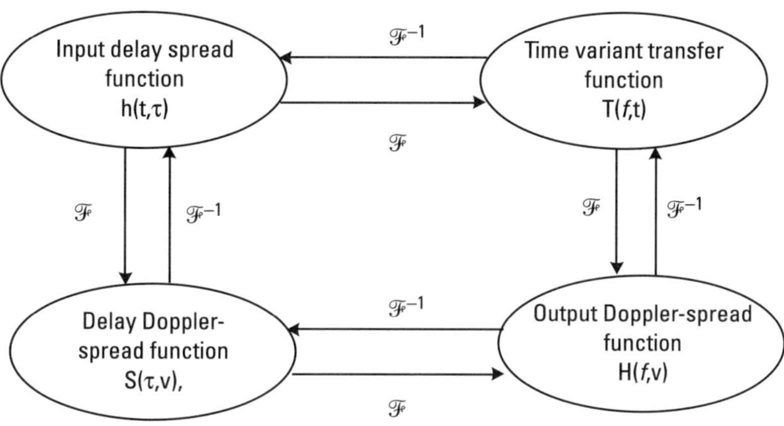

Figure 2.12 The Bello functions. (After: [2]. © 1999 John Wiley & Sons, Inc.)

We observe that these functions are obtained by taking the Fourier or the inverse Fourier transform of each pair, depending on what channel characteristics need to be studied. From the discussion so far it is clear that for the design of wideband channels, as is the case of mobile channels, we need both the channel characteristics. This, in turn, will determine the received signal distortion. This observation justifies both the relevance and inclusion of Chapter 1 and 2 in a book that deals with interference reduction techniques in wireless systems, according to the methodology developed in the Preface. Among the various ways that have been developed over the years to mitigate the effects of their distortions, we summarize the most important ones, which are discussed in detail and are the subject of Chapter 4.

1. *Directional antennas.* These allow the energy transmitted towards the significant scatterers to be reduced, thereby reducing far-out echoes.

2. *Small cells.* The maximum differential delay is reduced by limiting the coverage of a cell.

3. *Diversity.* This does not cancel the multipath energy directly, but instead makes better use of the signal energy by reducing the level of the deep fades. In this way, the SNR for a given BER can be reduced and error levels are lowered, although not removed. Various types of diversities exist, such as space, time, path, polarization, and frequency.

4. *Equalizers.* These work to transform the wideband channel back into a narrowband one. By applying an adaptive filter to flatten the channel frequency response or by making constructive energy use, the wideband channel performance can actually be better than the narrowband (flat fading) performance.

5. *Data rate.* One simple way of avoiding the effects of delay spread is simply to reduce the modulated data rate. By transmitting the required data simultaneously on a large number of carriers, each with a narrow bandwidth, the data throughput can be maintained. This is the OFDM concept, as used in digital broadcasting, and it can be combined with channel coding with CDMA to give very robust performance as we see in the following section.

2.3 Channel Coding

We have seen that the wireless channel introduces significant information signal deterioration by itself, before the nonlinearities of the components and filtering used as well as other interfering sources start contributing to this alteration. In order to prevent significant deterioration at the source level, at least, error control and correction techniques have been used via coding. A technique called interleaving has been used successfully [25].

2.3.1 Interleaving

Interleaving is used to obtain spreading of the source bits over time, which technically is called time diversity, without adding any overhead. As source signal digitization proliferates, interleaving has taken analogous prominence, at least in all second generation digital cellular systems. This is especially true in environments in which errors occur in bursts, such as in multipath fading situations.

The interleaver can be one of two forms—a block structure or a convolutional structure [25]. A block interleaver formats the encoded data into a rectangular array of m row and n columns. Usually, each row constitutes a codeword of length n. An interleaver of degree m consists of m rows. As seen, coded bits are read in as the row of the block but are read out columnwise and transmitted over the channel. At the receiver, the deinterleaver stores the data in the same rectangular format, but reads out the data rowwise, one codeword at a time. As a result of this reordering, a burst of errors of length $l = mb$ is broken up into m bursts of length b. Thus an (n, k) code that can handle burst errors of length $b < (n - k)/2$ can be combined with an interleaver of degree m to create an interleaved (mn, mk) block code that can handle bursts of length mb. An interleaver of degree 4 with codewords of length 7 is shown in Figure 2.13.

Convolutional interleavers can be used in place of block interleavers in much the same fashion. Convolutional interleavers are better matched for use with convolutional codes, as explained in Section 2.3.3.2. With all interleavers, there is an inherent processing delay that impacts the quality and perceived continuity of the transmission for the end user. For example, the interleaver length is practically limited to the delay that can be tolerated between bursts of human speech, which is of the order of 40 ms or less.

2.3.2 Channel Coding Fundamentals

In 1948, Shannon had demonstrated that by proper encoding of the information, errors induced by a noisy channel can be reduced to any desired level without sacrificing the rate of information transfer. One of Shannon's channel capacity formulas is applicable to the Additive white Gaussian noise (AWGN) channel and is given by

$$C = B \log_2 \left(1 + \frac{P}{N_0 B} \right) \tag{2.33}$$

where

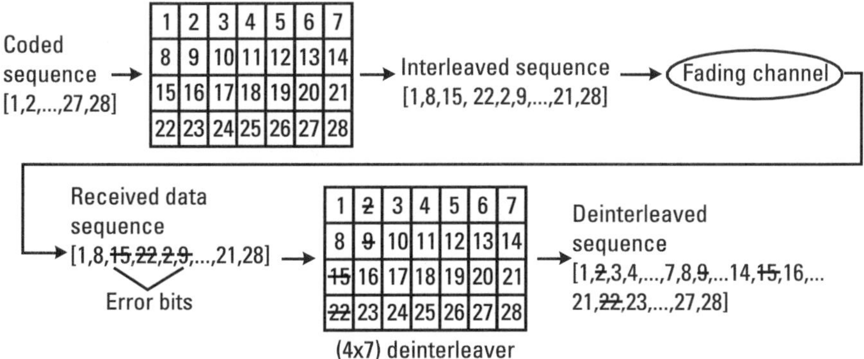

Figure 2.13 Block interleaver.

C is the channel capacity (bits per second);
B is the transmission bandwidth (hertz);
P is the received signal power (watts);
N_0 is the single-sided noise power density (watts/hertz).

The power received at the receiver is given as

$$P = E_b R_b \tag{2.34}$$

where

E_b is the average bit energy;
R_b is the transmission bit rate.

Equation (2.34) normalized by the transmission bandwidth represents the bandwidth efficiency and is given by

$$\frac{C}{B} = \log_2\left(1 + \frac{E_b}{N_0}\frac{R_b}{B}\right) \tag{2.35}$$

The basic purpose of error detection and error correction techniques is to introduce redundancies to the data and transmit codewords that have known properties. The introduction of redundant bits increases the data rate on the link, hence increasing the bandwidth requirement. However, it is well known that the use of orthogonal signaling enables us to make the probability of error arbitrarily small by expanding the signal set—that is, by making the number of waveforms $M \rightarrow \infty$, provided the SNR per bit $\text{SNR}_b \geq -1.6$ dB [21]. We can thus operate at the capacity of the AWGN channel in the limit as the bandwidth expansion factor $B \rightarrow \infty$ as can be seen by (2.35) [25]. This, however, results in such a large bandwidth for the resulting signal that it makes the communication impractical. The bandwidth expansion here grows exponentially with the block length k. Error control coding waveforms, on the other hand, have bandwidth expansion factors that grow linearly with the block length. Error correction coding thus offers advantages not only in bandwidth-limited applications but also in power-limited applications [2, 21, 25-31].

2.3.3 Types of Codes

The encoder maps the input information sequence into a code sequence for transmission over the channel. The purpose of this mapping is to improve communication efficiency by enabling the system to correct some transmission errors. In general, there exist forward error correction (FEC) techniques that are used to improve bit error rate performance and automatic repeat request (ARQ) techniques that are used to correct errors by requesting retransmission of corrupted data packets. There are two basic types of error correction and detection codes: *block codes* and *convolutional codes*.

2.3.3.1 Block Codes

Interleavers can combine with block codes to further improve digital information signal transmission. Block codes are FEC codes that enable a limited number of errors to be detected and corrected without retransmission, as explained in the examples that follow. Block codes can be used when other means of improvement, such as increasing transmitter power or using a more sophisticated demodulator, are impractical.

A simple example will show the advantage of the interleaving and block coding combination. An interleaver, for example, can have m rows and $n\text{-}bit$ words. Each word is made up of k source bits and $(n\text{-}k)$ bits from a block code. The combination will break up a burst of errors of length $l = mb$ into m bursts of length b. An (n, k) code that can handle burst errors of length $b < \dfrac{n-2}{2}$ can now handle with an interleaver of degree m bursts of length mb, as it can create interleaved block codes (mn, mk).

In block codes, data is segmented into known, fixed blocks of data. Parity bits are added to blocks of message bits to make codewords or code blocks of fixed length. In a block encoder, information bits are encoded into n code bits. A total of $n - k$ redundant bits are added to the k information bits for the purpose of detecting and correcting errors [5]. From a total possible set of 2^n codewords, we would select codewords to form the set of desired codewords. The block code is referred to as an (n, k) code and the rate of the code is defined as $R_c = k/n$ and is equal to the rate of information transmitted per channel user. The error correction capability of the code is due to the fact that not all possible code vectors are used by the code. Besides the rate, other important parameters are the distance and the weight of a code.

Examples of Block Codes

Hamming Codes   These were among the first of the nontrivial error correction codes [26]. These codes and their variations have been used for error control in digital communication systems. There are both binary and nonbinary Hamming codes. A binary Hamming code has the property that

$$(n, k) = (2^m - 1, 2^m - 1 - m) \tag{2.36}$$

where

k is the number of information bits used to form a n bit codeword;
m is any positive integer.

The number of parity symbols are $n - k = m$. Another parameter is the code distance, which is defined as the number of elements in which two codewords differ. The minimum Hamming distance is three, and the number of errors that they can correct is one. They can detect all combinations of double errors.

Hadamard Codes   These codes are obtained by selecting as codewords the rows of a Hadamard matrix. A Hadamard matrix A is a $N \times N$ matrix of ones (1) and zeroes (0) such that each row differs from any other row in exactly $N/2$ locations. One row contains all zeros, with the remainder containing $N/2$ zeros and $N/2$ ones. The minimum distance for these codes is $N/2$.
For example, for $N = 2$, the Hadamard matrix A is

$$A = \begin{bmatrix} 0 & 0 \\ 0 & 1 \end{bmatrix}$$

In addition to the special case considered when $N = 2^m$ (m being a positive integer), Hadamard codes of other block lengths are possible, but the codes are not linear.

Golay Codes   This is a linear binary code with a minimum distance of 7 and an error correction capability of 3 bits [27]. This is a special kind of code in that this is the only nontrivial example of a perfect code. Every codeword lies within a distance of three of any codeword, thus making possible maximum likelihood decoding of these codes.

Cyclic Codes   A cyclic code can be generated by using a generator polynomial $g(p)$ of degree $(n - k)$. The generator polynomial of an (n, k) cyclic code is a factor of $p^n + 1$ and has the general form

$$g(p) = p^{n-k} + g_{n-k-1}p^{n-k-1} + \ldots + g_1 p + 1$$

A message polynomial $x(p)$ can also be defined as

$$x(p) = x_{k-1}p^{k-1} + \ldots + x_1 p + x_0$$

where $(x_{k-1}, \ldots, x_0)$ represents the k information bits. The resultant codeword $c(p)$ can be written as

$$c(p) = x(p)g(p)$$

$c(p)$ is a polynomial of degree less than n.

Encoding for a cyclic code is usually performed by a linear feedback shift register, based on either the generator or parity polynomial. We will see more of these in Chapter 3.

Base-Chaudhuri-Hocquenghem (BCH) Codes These cyclic codes are among the most important block codes because they exist for a wide range of rates, achieve significant coding gains, and are implementable even at high speeds [28–31]. The block length of the codes is $n = 2^m - 1$ for $m \geq 3$, and the number of errors that they can correct is bounded by $t < (2^m - 1)/2$. The binary BCH codes can be generalized to create classes of nonbinary codes that use m bits per code symbol. The most important and common class of nonbinary BCH codes is the family of codes known as Reed-Solomon (RS) codes. The (63,47) RS code used in U.S. *cellular digital packet data* (CDPD) uses $m = 6$ bits per code symbol.

RS Codes These are nonbinary codes capable of correcting errors that appear in bursts and are commonly used in concatenated coding systems [28]. The block length of these codes is $n = 2^m - 1$. These can be extended to 2^m or $2^m + 1$. The number of parity symbols that must be used to correct e errors is $n - k = 2^m - 1 - 2e$. The minimum distance $d_{min} = 2e + 1$ RS codes achieve the largest possible d_{min} of any linear code. RS codes have extremely good error correcting properties for bursty channels, such as fading mobile radio channels, and thus are very popular for wireless communications.

2.3.3.2 Convolutional Codes

Convolutional codes are fundamentally different from block codes in that information sequences are not grouped into distinct blocks and encoded [29]. Instead, a continuous sequence of information bits is mapped into a continuous sequence of encoder output bits. This mapping is highly structured, enabling a decoding method considerably different from that of block codes to be employed. It can be argued that convolutional coding can achieve a larger coding gain than can be achieved using a block code with the same complexity.

A convolutional code is generated by passing the information sequence through a finite state shift register. In general, the shift register contains N (k *bit*) stages and m linear algebraic function generators based on the generator polynomials, as shown in Figure 2.14.

The input data is shifted into and along the shift register k bits at a time. The number of output bits for each k bit input data sequence is n bits. The code rate is $R_C = k/n$. The parameter N is called the constraint length and indicates the number of input data bits upon which the current output is dependent. It determines how powerful and complex the code is.

Because a convolutional coder uses shift registers and a continuous input data stream, there is an inherent delay when the input data stream is terminated. In order to zero out the convolutional decoder, a string of tail bits must be sent at the end of the data stream. This also allows the convolutional coder at the transmitter to reset.

There are different ways of representing convolutional codes. These include: generator matrix, generator polynomials, logic table, state diagrams, tree diagrams, and trellis diagrams.

Generator Matrix

The generator matrix for a convolutional code is semi-infinite because the input is semi-infinite in length. Hence, this may not be a convenient way of representing a convolutional code.

Generator Polynomials

Here, we specify a set of n vectors, one for each of the n modulo-2 adders used. Each vector of dimension $2k$ indicates the connection of the encoder to that modulo-2 adder. A one (1) in the lth position of the vector indicates

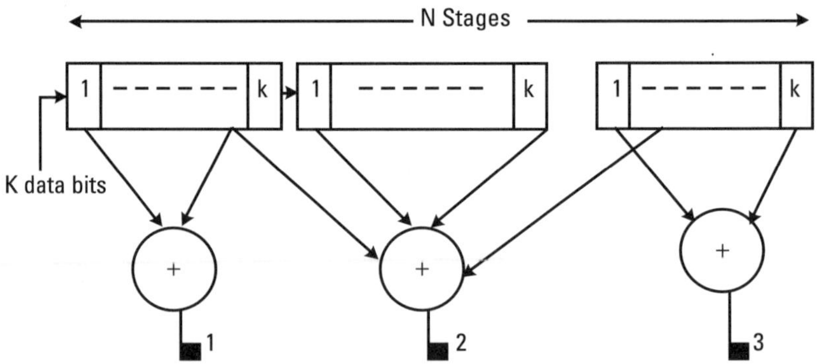

Figure 2.14 General block diagram of convolutional encoder. (After: [13].)

that the corresponding shift register stage is connected, and a 0 indicates no connection.

Logic Table

A logic table can be built showing the outputs of the convolutional encoder and the state of the encoder for the input sequence present in the shift register.

State Diagram

Because the output of the encoder is determined by the input and the current state of the encoder, a state diagram can be used to represent the encoding process. The state diagram is simply a graph of the possible states of the encoder and the possible transitions from one state to another.

Tree Diagram

The tree diagram shows the structure of the encoder in the form of a tree with the branches representing the various states and the outputs of the coder.

Trellis Diagram

Close observation of the tree shows that the structure repeats itself once the number of stages is greater than the constant length. It is observed that all branches emanating from two nodes having the same state are identical in the sense that they generate identical output sequences. This means that the two nodes having the same label can be merged. By doing this throughout the tree diagram, we can obtain another diagram called a trellis diagram, which is a more compact representation.

Decoding of Convolutional Codes

The function of the decoder is to estimate the encoded input information using a rule or method that results in the minimum possible number of errors. There is a one-to-one correspondence between the information sequence and the code sequence.

Table 2.2 summarizes the relationships between the input data, old and new states, and the output codeword.

There are two schematic diagrams that represent the relationship in Table 2.2, the state and trellis diagrams. Figure 2.15 shows the state diagram representation of Table 2.2. Solid lines indicate the logical *1*, and dashed lines indicate logical 0. Moreover, the 2-bit data written in each line represents

Table 2.2
State Transitions of the Encoder

Old State σ_n	Input a_{n+1}	New State σ_{n+1}	Codeword b_{2n+1}, b_{2n+2}
00	0	00	00
	1	10	11
01	0	00	11
	1	10	00
10	0	01	10
	1	11	01
11	0	01	01
	1	11	10

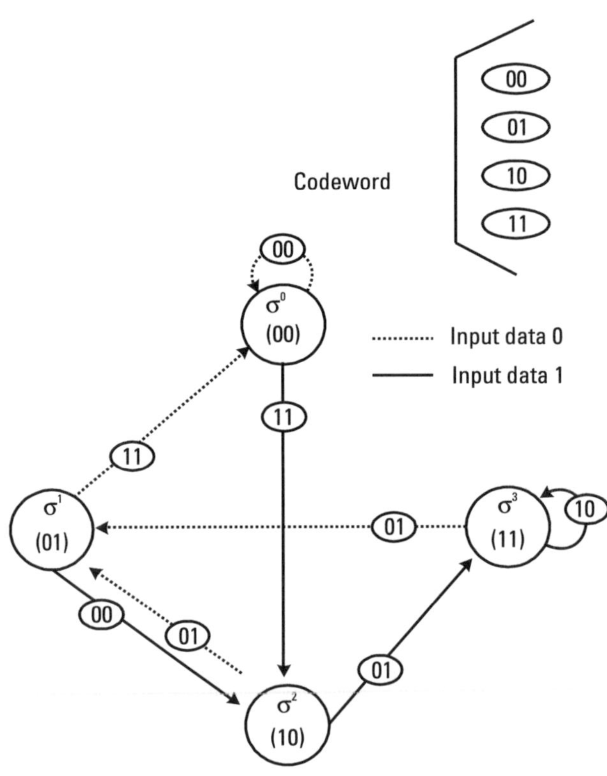

Figure 2.15 State diagram representation for the state transition given by Table 2.2. (After: [3].)

the output codeword. Although this diagram is very suitable for understanding the state transition between two arbitrary states, it is not suitable for expressing a state transition sequence for a certain period.

The trellis diagram is very suitable for expressing the state transition sequence. Figure 2.16 shows the trellis diagram that represents the relationship given by Table 2.2. In this figure, solid lines mean that the input is logical *1* and dashed lines mean logical *0*. Moreover, 2-bit data written near each line indicate the output codeword, and σ_n^j means σ^j for nth timing.

Hence, any information and code sequence pair is uniquely associated with a path through the trellis. Thus, the job of the convolutional decoder is to estimate the path through the trellis that was followed by the encoder.

There are a number of techniques for decoding convolutional codes. The most important of these methods is the Viterbi algorithm, which performs maximum likelihood decoding of convolutional codes. The algorithm was first described in [28] and [30]. Both hard and soft decision decoding can

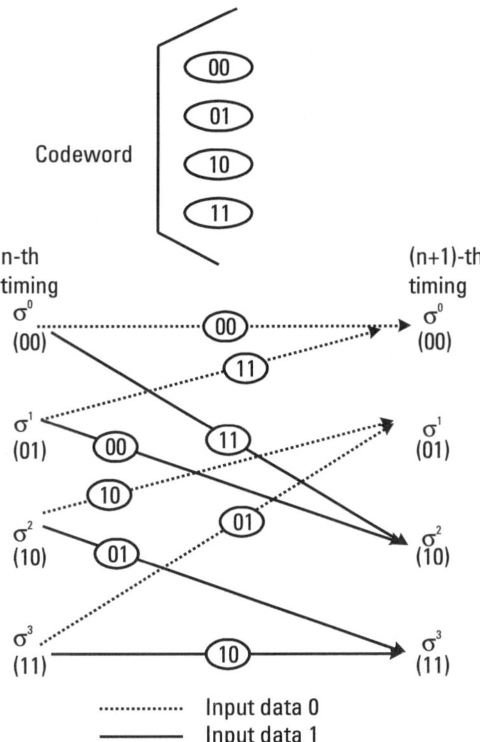

Figure 2.16 Trellis diagram that represents relationships given by Table 2.2. (After: [3].)

be implemented for convolutional codes. Soft decision decoding is superior by about 2 to 3 dB.

Other Decoding Algorithms for Convolutional Codes

Other decoding algorithms for convolutional codes are [25]: The Fano sequential decoding algorithm (Fano), the stack sequential decoding algorithm (Jelinek and Zigangirov), and feedback decoding.

2.3.3.3 Trellis Coded Modulation

This is a technique that combines both coding and modulation to achieve significant coding gains without compromising bandwidth efficiency [29]. Trellis coded modulation (TCM) schemes employ redundant nonbinary modulation in combination with a finite state encoder that decides the selection of modulation signals to generate coded signal sequences. TCM uses signal set expansion to provide redundancy for coding and to design coding and signal mapping functions jointly so as to maximize directly the free distance (minimum euclidean distance) between the coded signals. In the receiver, the signals are decoded by a soft decision maximum likelihood sequence decoder. Coding gains as large as 6 dB can be obtained without any bandwidth expansion or reduction in the effective information rate. In Chapter 4, we shall present the benefits of coding as a distortion mitigation technique in wireless communications. The material in this chapter also confirms the methodology explained in the Preface.

References

[1] Shannon, C., "A Mathematical Theory of Communication," *Bell Systems Technical Journal*, No. 27, 1948, pp. 379–423 and pp. 623–656. Reprinted in Sloane, N.I.A., and A. D. Wyner (eds.), *Claude Elwood Shannon: Collected Papers*, New York: IEEE Press, 1993.

[2] Saunders, S. R., *Antennas and Propagation for Wireless Communication Systems*, New York: John Wiley & Sons, 1999.

[3] Sampei, Seiichi, *Applications of Digital Wireless Technologies to Global Wireless Communications*, Englewood Cliffs, NJ: Prentice-Hall, 1997.

[4] Parsons, J. D., *The Mobile Radio Propagation Channel*, second edition, New York: John Wiley & Sons, 2000.

[5] Okumura, T., E. Ohmori, and K. Fakuda, "Field Strength and Its Variability in VHF/UHF and Mobile Service," *Review of Electrical Communication Laboratory*, Vol. 16, No. 9–10, September-October 1968, pp. 825–873.

[6] Hata, M., "Empirical Formula for Propagation Loss in Land Mobile Radio Services," *IEEE Trans. of Vehicular Tech.*, Vol. VT-29, No. 3, August 1980, pp. 317–325.

[7] Cox, D. C., R. R. Murray, and A. W. Norris, "Measurements of 800-MHz Radio Transmission into Buildings with Metallic Walls," *Bell System Technical Journal (B.S.T.J.)*, Vol. 62, No. 9, November 1983, pp. 695–717.

[8] Toledo, A. R., and A. M. D. Turkmani, "Propagation Into and Within Buildings at 900, 1800 and 2300 MHz," *Proc. 42nd IEEE Veh. Tech. Conf.*, Denver, CO, May 1992, pp. 633–636.

[9] Cox, D. C., "Universal Digital Portable Radio Communications," *Proc. IEEE*, Vol. 75, No. 4, April 1987, pp. 436–477.

[10] Turkmani, A. M. D., and A. F. Toledo, "Radio Transmission at 1800 MHz into and within Multi-story Buildings," *IEEE Proc.-I*, Vol. 138, No. 6, December 1991, pp. 577–584.

[11] Seidel, S. Y., and T. S. Rappaport, "900 MHz Path Loss Measurements and Prediction Techniques for in-building Communication System Design," *IEEE Proc. ICC '91*, Denver, CO, June 1991, pp. 613–618.

[12] Lotse, R., J. E. Berg, and R. Bownds, "Indoor Propagation Measurements at 900 MHz," *IEEE Trans. 42nd Veh. Tech. Conf.*, Denver, CO, May 1992, pp. 629–632.

[13] Rappaport, T. S., *Wireless Communications*, Upper Saddle River, NJ: Prentice-Hall, 1996.

[14] Iwama, T., et al., "Experimental Results of 1.2 GHz Band Premises Data Transmission Using GMSK Modulation," IEEE GLOBECOM '87, Nov. 1987, pp. 1921–1925.

[15] Rappaport, T. S., "Indoor Radio Communications for Factories of the Future," *IEEE Communications Magazine*, Vol. 27, No. 5, May 1989, pp. 15–24.

[16] Lafortune, J. F., and M. Lecours, "Measurement and Modeling of Propagation Losses in a Building at 900 MHz," *IEEE Trans. of Vehicular Tech.*, Vol. 39, No. 2, May 1990, pp. 101–108.

[17] Nakagami, M., "The m-distribution: a General Formula of Intensity Distribution of Rapid Fading," in *Statistical Methods in Radio Wave Propagation*, W. C. Hoffman (ed.), Oxford: Pergamon, 1960.

[18] Saunders, S. R., and F. R. Bonar, "Mobile Radio Propagation in Built-up Areas: A Numerical Model of Slow Fading," *Proc. 41st IEEE VTC*, 1991, pp. 295–300.

[19] Jakes, W. C., *Microwave Mobile Communications*, New York: IEEE Press, 1974.

[20] Simon, M. K., and M. S. Alouini, *Digital Communications over Fading Channels*, New York: John Wiley, 2000.

[21] Ziemer, R. E., and R. L. Peterson, *Digital Communications*, Upper Saddle River, NJ: Prentice-Hall, 1990.

[22] Cavers, J. K., *Mobile Channel Characteristics*, Boston: Kluwer, 2000.

[23] Bracewell, R. N., *The Fourier Transform and its Applications*, second edition, New York: McGraw-Hill, 1986.

[24] Bello, P. A., "Characterization of Randomly Time-invariant Linear Channels," *IEEE Trans, CS-11*, 1963, pp. 360–393.

[25] Rappaport, T. S., *Wireless Communications*, Upper Saddle River, NJ: Prentice-Hall, 1996.

[26]   Hamming, R. W., "Error Detecting and Error Correcting Coding," *Bell Systems Technical Journal*, April 1950.

[27]   Golay, M. J. E., "Notes on Digital Coding," *Proceedings of the IRE*, Vol. 37, June 1949.

[28]   Forney, G.D., "The Viterbi Algorithm," *Proceedings of IEEE*, Vol. 61, No. 3, March 1978, pp. 268–278.

[29]   Ungerboeck, G., "Trellis Coded Modulation with Redundant Signal Sets, Part 1: Introduction," *IEEE Communication Magazine*, Vol. 25, No. 2, Feb. 1987, pp. 5–21.

[30]   Fano, R. M., "A Heuristic Discussion of Probabilistic Coding," *IEEE Trans. Inform Theory*, Vol. IT-9, April 1963, pp. 64–74.

[31]   Bose, R. C., and D. K. Ray-Chaudhuri, "On a Class of Error Correcting Binary Group Codes," *Information and Control*, Vol. 3, Nov. 1960, pp. 68–70.

3

Transmission Systems in an Interference Environment

3.1 Introduction

Following the methodology developed in the preface, we discussed in Chapter 1 the design parameter of wireless systems in use, which can affect or can be affected by interference. We then briefly presented the interference environment in which these systems are called to operate today. In Chapter 2, we discussed the channel characteristics and coding that play a major role in the behavior of these systems and consequently affect performance. In this chapter, we shall present, discuss, and compare the spectra characteristics of the transmitted signals, as well as the access techniques used in multiuser communications. The results of these comparisons allow us to determine the trade-offs that have to be made for the implementation of a specific application that will work satisfactorily within specified criteria of performance.

This chapter is the last necessary link needed in a book on interference reduction before we embark on the discussion of its optimum performance in a realistic interference environment. Both analog and digital transmission systems will be presented. One can say that analog transmission in wireless communications is not relevant anymore, and we agree. However, we must point out that basic techniques of interference analysis and suppression were developed during the analog signal transmission era (satellite communications) [1], and they are still relevant. Actually, most of the modern wireless systems use digital transmission techniques, but their interference behavior is studied using some of the techniques developed during the analog era [2–6]. In this view, the relevance of this chapter in the overall theme of interference is fully justified.

3.2 Analog Transmission

By analog transmission, we mean the systems in which information-bearing waveforms are to be reproduced at the destination without employing digital coding techniques. In other words, the analog form (continuous in time) of the information signal is maintained from transmission to reception. Figure 3.1 depicts an analog transmission system.

Here, $x(t)$ represents the ensemble of probable messages from a given source. Though such messages are not strictly bandlimited, it is safe to assume that there exists some upper frequency, name it W, above which the spectral content is negligible and unnecessary for conveying the information in question. It is called the message bandwidth. For this case $GX(f) = 0$ for $|f| > W$ where $X(f)$ is the spectrum of $x(t)$. For more details see Appendix A.

In Figure 3.1, the transmitter simply becomes an amplifier with power gain G_T, so $S_T = G_T x^2$, and the receiver filter is a nearly ideal low pass filter (LPF) with bandwidth W, so $B_N \approx W$.

We observe that at the output of the receiver, we can measure the signal to noise power ratio, which is used as a quality measure, as we shall see later. A similar ratio, which denotes the signal to interference power ratio, plays a major role in the analysis of the interference behavior of wireless systems.

It can be shown [7] that $(S/N)_D$ can be expressed in terms of some very basic system parameters, namely, the signal power and noise density at the receiver input and the message bandwidth. We define:

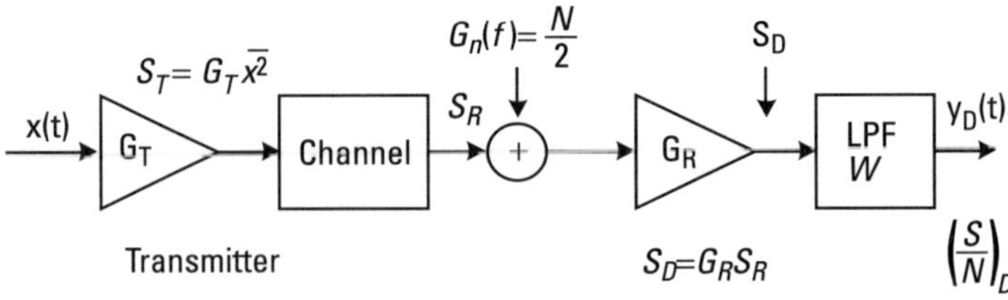

Figure 3.1 Analog transmission system.

$$\gamma \leq \frac{S_R}{\eta \cdot W} \tag{3.1}$$

where γ is equal to $(S/N)_D$ for analog baseband transmission. In (3.1) it is presupposed to have distortionless transmission conditions. With noise and a nearly ideal filter, it is more accurate to notice that:

$$\left(\frac{S}{N}\right)_D \leq \gamma \tag{3.2}$$

It is easy to see that γ presents an upper bound for analog baseband performance that may or may not be achieved in an actual system. Table 3.1 lists representative values of $(S/N)_D$ for selected analog signals along with the frequency range. This ratio, which many times is simply denoted by SNR, plays the role of a fundamental quality measure as far as the performance of wireless systems in an interference environment. This metric is used for digital transmission as well, and it is related to bit error rate (BER), as we shall see in the chapters that follow.

Analog transmission is characterized by the following:

1. *Signal processing.* Processing is performed on the baseband signal before modulation and after demodulation in order to improve the quality of the link.

2. *The number of communication channels supported by the carrier.* In the case of a single communication channel, one refers to single channel per carrier (SCPC) transmission. Several communication channels combined by frequency division multiplexing (FDM) is referred to as FDM transmission.

Table 3.1
Typical Transmission Requirements for Selected Analog Signals

Signal Type	Frequency Range	*S/N* Ratio (dB)
Barely intelligible voice	500 Hz–2 KHz	5–10
Telephone-quality voice	200 Hz–32 KHz	25–35
AM broadcast-quality audio	100 Hz–5 KHz	40–50
High-fidelity audio	20 Hz–20 KHz	55–65
Television video	60 Hz–4.2 MHz	45–55

3. *The type of modulation used.* The most widely used is FM. For this type of modulation, the carrier amplitude is not affected by the modulating signal; thus, it is robust with respect to the nonlinearities of the channel. In Chapter 5 and onward, this conclusion is very important when we will deal with intermodulation effects. On the other hand, for a given quality of link, it offers the useful possibility of a trade-off between the SNR and the bandwidth occupied by the carrier. The station-to-station link is generally identified by the multiplexing/modulation combination, as will be the subject of Chapter 6.

3.3 Analog Modulation Methods

In communication systems, for an information-bearing signal to be easily accommodated and transmitted through a communication channel a modulation process must be utilized. The modulation is commonly the process where the message information is "added" to a radio carrier. The choice of modulation techniques is influenced by the characteristics of the message signal, the characteristics of the channel, the performance desired from the overall communication system, the use to be made of the transmitted data, and economic factors.

The two basic types of analog modulation are:

1. Continues wave;
2. Pulse modulation.

In continuous wave modulation, any combination of amplitude, phase, and frequency of a high-frequency carrier is varied proportionally to the message signal such that a one-to-one correspondence exists between the varying parameter(s) and the message signal. The carrier is usually assumed to be sinusoidal, but this is not a necessary restriction. Many times, however, results obtained using a sinusoidal carrier are used as a basis for extrapolation. For a sinusoidal carrier, a general modulated carrier can be represented mathematically as:

$$x_c(t) = A(t) \cos\left[\omega_c t + \theta(t)\right] \tag{3.3}$$

where

$\omega_c = 2\pi f_c$ and $f_c =$ the carrier frequency;

$A(t) =$ instantaneous amplitude;

$\theta =$ instantaneous phase deviation.

In analog pulse modulation, the message waveform is sampled at discrete time intervals and the amplitude, width, or position of a pulse is varied in one-to-one correspondence with the values of the samples.

The present chapter deals with the transmission of an analog signal by impressing it on either the amplitude, the phase, or the frequency of a sinusoidal carrier or in pulse modulation.

3.3.1 Amplitude Modulation

In amplitude modulation (AM), the message signal is impressed on the amplitude of the carrier signal. The unique feature of AM is that the envelope of the modulated carrier has the same shape as the message waveform. This is achieved by adding the translated message appropriately proportioned to the unmodulated carrier. Hence, the modulated signal can be written as:

$$x_c(t) = A_c \cos \omega_c t + mx(t) A_c \cos \omega_c t \qquad (3.4)$$

$$= A_c [1 + mx(t)] \cos \omega_c t$$

where

$A_c \cos \omega_c t$ is the unmodulated carrier;

$$f_c = \frac{\omega_c}{2\pi}, \text{ carrier frequency;}$$

m is a constant called modulation index. This is a very important parameter for interference analysis, as we shall see later.

Because A_c is the unmodulated carrier amplitude, it can be considered as a linear function of the message. It will be:

$$A_c(t) = A_c [1 + mx(t)] \qquad (3.5)$$

Equation (3.5) underscores the meaning of amplitude modulation. Figure 3.2 shows the spectrum of the AM modulated signal. More details about the special characteristics of AM transmission are given in [7–10].

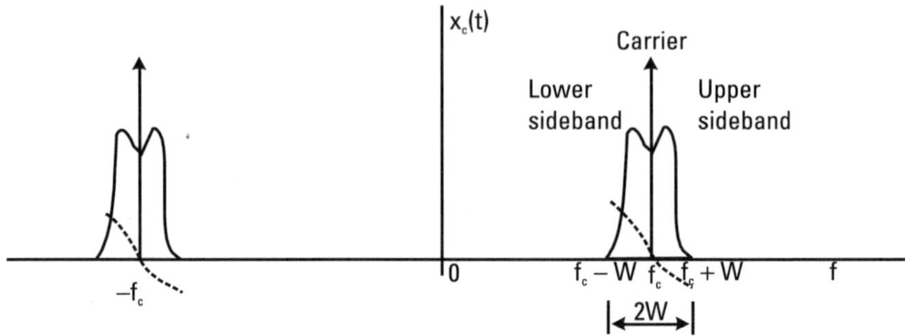

Figure 3.2 An AM spectrum.

3.3.2 Angle Modulation

To generate angle modulation, the amplitude of the modulated carrier is held constant, and either the phase or the time derivative of the phase of the carrier frequency is varied linearly with message signal, $m(t)$. Thus, the general angle modulated signal is given by:

$$x_c(t) = A_c \cos(\omega_c t + \phi(t)) \tag{3.6}$$

The instantaneous phase of $x_c(t)$ is defined as:

$$\theta_i(t) = \omega_c t + \phi_i(t) \tag{3.7}$$

and the instantaneous frequency is defined as:

$$\omega_i(t) = \frac{d\theta_i}{dt} = \omega_c + \frac{d\phi_i}{dt} \tag{3.8}$$

The functions $\phi(t)$ and $d\phi/dt$ are the phase deviation and frequency deviation, respectively, as from (3.7) $\phi(t) = \theta_i(t) - \omega_c(t)$ and from (3.8)

$$\frac{d\phi}{dt} = \omega_i(t) - \omega_c \tag{3.9}$$

The two basic types of angle modulation are phase modulation (PM) and FM. PM implies that the phase deviation of the carrier is proportional to the message signal. Thus, for PM, it is:

$$\phi(t) = k_p m(t) \tag{3.10}$$

where k_p is the deviation constant in radians per unit of $m(t)$. FM implies that the frequency deviation of the carrier is proportional to the modulating signal. This yields:

$$\frac{d\phi}{dt} = k_f m(t) \tag{3.11}$$

The phase deviation of an FM carrier is given by integrating (3.11), which yields:

$$\phi(t) = k_f \int_{t_0}^{t} m(a) \, da + \phi_0 \tag{3.12}$$

where

ϕ_0 is the phase deviation at $t = t_0$;
k_f is the frequency deviation constant in radians per second per unit of $m(t)$.

A deep understanding of the concepts presented thus far is essential for the comprehension of the more advanced material in the later chapters, especially in Chapter 5 and onward.

Because it is often convenient to measure frequency deviation in hertz, we define:

$$k_f = 2\pi f_d \tag{3.13}$$

where

f_d = frequency deviation constant of the modulator in hertz per unit of $m(t)$.

With these definitions, the phase modulator output is:

$$x_c(t) = A_c \cos(\omega_c t + k_p m(t)) \tag{3.14}$$

and the FM output is:

$$x_c(t) = A_c \cos\left(\omega_c t + 2\pi f_d \int^{t} m(a)\, da \right) \qquad (3.15)$$

The lower limit of the integral is typically not specified because to do so would require the inclusion of an initial condition.

Figure 3.3 illustrates the behavior of FM signals. For the case of this figure, $m(t) = A \sin \omega_m t$, β is the modulation index $\left(\beta = \dfrac{f_d A}{f_m} \right)$. Other modulation techniques [7–9] that belong to the same category are pulse modulation, pulse amplitude modulation, pulse width modulation, and pulse position modulation. Details about the basic characteristics of FM modulated signal are given in [7–11].

3.4 Noise and Interference in Analog Transmission

Although a clean, virtually noise-free wave may be transmitted, the signal received at the demodulator is always accompanied by noise, including that

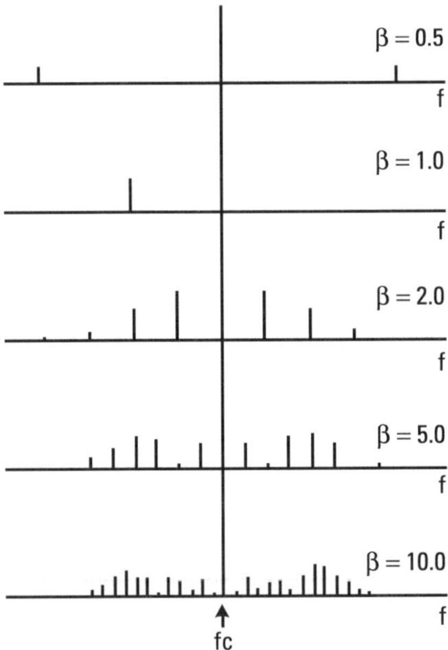

Figure 3.3 Amplitude spectrum of an FM signal as β increases by decreasing f_m. (After: [8].)

generated in preceding stages of the receiver itself. Furthermore, there may be interfering signals in the desired band that are not rejected by a bandpass filter $H_R(f)$. Both noise and interference give rise to undesired components at the detector output. When interference or noise is included, we will write the contaminated signal $u(t)$ in envelope-and-phase or in quadrature-carrier form, given by:

$$u(t) = A(t) \cos\left[\omega_c t + \phi(t)\right] = u_i(t) \cos \omega_c t - u_q(t) \sin \omega_c t \tag{3.16}$$

Equation (3.16) facilitates analysis of the demodulated signal $y(t)$. Specifically, the following idealized mathematical models represent the demodulation operation—idealized in the sense of perfect synchronization and perfect amplitude limiting:

$$y(t) = \begin{cases} u_i(t) & \text{Synchronous detector} & (3.17a) \\ A(t) - \overline{A} & \text{Envelope detector} & (3.17b) \\ \phi(t) & \text{Phase detector} & (3.17c) \\ \dfrac{1}{2\pi} \dfrac{d\phi(t)}{dt} & \text{Frequency detector} & (3.17d) \end{cases}$$

The term $\overline{A} = \langle A(t) \rangle = E[A]$ reflects the DC block (mean value) normally found in an envelope detector.

However, $y(t)$ does not necessarily equate with the final output signal $y_D(t)$. Therefore, assuming the lowpass filter merely removes any out-of-band frequency components, the output signal $y_D(t)$ is given by:

$$y_D(t) = \int_{-W}^{W} Y(f)\, e^{j\omega t}\, df \tag{3.18}$$

3.4.1　Interference

The subject of interference will be analyzed thoroughly in Chapter 5 and onward. It was thought that introducing the concept here along with modulations and demodulations processes will make it easier to comprehend the interference-mitigating techniques that will be discussed later. We begin by considering a very simple case, an unmodulated carrier with an interfering cosine wave (Figure 3.4).

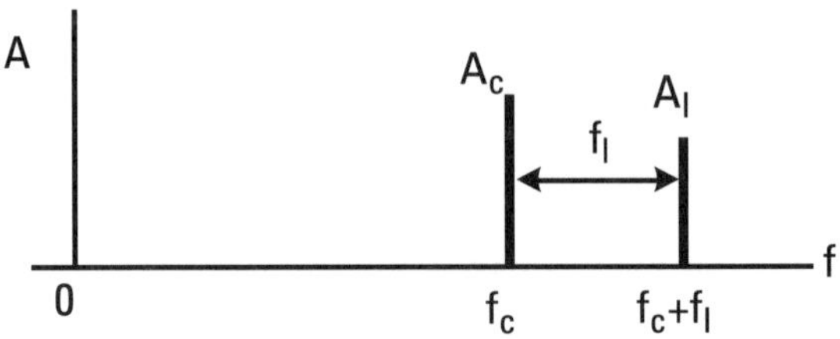

Figure 3.4 Line spectrum for interfering sinusoids.

Let the interference signal have amplitude A_I and frequency $f_c + f_I$. The total signal entering the demodulator is the sum of two sinusoids, given by:

$$u(t) = A_c \cos \omega_c t + A_I \cos (\omega_c + \omega_I) t \qquad (3.19)$$

Following the phasor construction (Figure 3.5), there is:

$$A(t) = \sqrt{(A_c + A_I \cos \omega_I t)^2 + (A_I \sin \omega_I t)^2} \qquad (3.20a)$$

$$\phi(t) = \arctan \frac{A_I \sin \omega_I t}{A_c + A_I \cos \omega_I t} \qquad (3.20b)$$

For arbitrary values of A_c and A_I, these expressions cannot be further simplified. However, if the interference is small compared to the carrier, the phasor diagram shows that the resultant envelope is essentially the sum of the inphase components, while the quadrature component determines the phase angle. That is, if $A_I \ll A$, then:

$$A(t) \approx A_c + A_I \cos \omega_I t \qquad (3.21a)$$

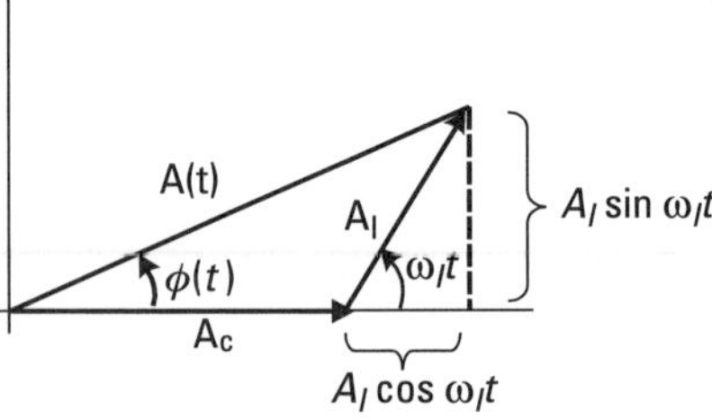

Figure 3.5 Phasor diagram for interfering sinusoids.

$$\phi(t) \approx \frac{A_I}{A_c} \sin \omega_I t \qquad (3.21b)$$

and hence:

$$u(t) = A_c(1 + m_I \cos \omega_I t) \cos (\omega_c t + m_I \sin \omega_I t) \qquad (3.22)$$

where:

$$m_I \equiv \frac{A_I}{A_c} << 1$$

The same result is obtained from first-order expansions of (3.20a) and (3.20b).

At the extreme, if $A_I >> A_c$, the analysis is performed by taking the interference as the reference and decomposing the carrier phasor, which gives:

$$u(t) = A_I(1 + m_I^{-1} \cos \omega_I t) \cos [(\omega_c + \omega_I)t - m_I^{-1} \sin \omega_1 t] \qquad (3.23)$$

From (3.21a) and (3.21b), we can see that the interfering wave performs an AM modulation and phase modulation of a carrier just like a modulating tone of frequency f_I with modulation index m_I. On the other hand, with strong interference, we can consider the carrier to be modulating the interfering wave. In either case, the apparent modulation frequency is the difference frequency f_I.

3.4.1.1 Interference in AM

Suppose there is small amplitude interference in an AM system with envelope detection. Using (3.21), this section, plus (3.17) and (3.18), the output signal becomes:

$$y_D(t) = \begin{cases} A_I \cos \omega_I t & |f_I| < W \\ 0 & |f_I| > W \end{cases} \qquad (3.24)$$

because $\overline{A} = A_c$.

Similarly, for synchronous detection we have:

$$y_D(t) = A_c + A_I \cos \omega_I t \qquad |f_I| < W \qquad (3.25)$$

Because $u(t) = A_c + A_I \cos \omega_I t$, (Figure 3.4), the DC component in (3.25) may or may not be blocked. In either case, any interference in the band $f_c \pm W$ produces a detected signal whose amplitude depends only on A_I, the interference amplitude, providing $A_I \ll A_c$. This is an important observation for the discussion in Chapter 5.

3.4.1.2　Interference in Exponential Modulation

With a phase or frequency detector, the detected interference is found by inserting (3.17b), this section, into (3.17c) and (3.17d). Thus, for $|f_I| < W$, there is:

$$y_D(t) = \frac{A_I}{A_c} \sin \omega_I t \qquad \text{PM} \qquad (3.26)$$

$$y_D(t) = \frac{A_I f_I}{A_c} \cos \omega_I t \qquad \text{FM} \qquad (3.27)$$

where f_I appears as a multiplying factor in (3.21) but not in (3.20), because of the differentiation of $\phi(t)$.

Comparing (3.20) and (3.21) with (3.18) and (3.19), together with $A_I/A_c \ll 1$, one finds that exponential modulation is less vulnerable to small-amplitude interference than linear modulation, all other factors being equal. Moreover, from (3.20) and (3.21), it is shown that FM is less vulnerable than PM when $|f_I|$ is small, as the detected interference is proportional to both the amplitude and frequency of the interfering wave.

In PM systems, like linear modulation, only the amplitude enters the picture. This latter difference can be understood with the aid of simple physical considerations. The strength of a detected signal in FM depends on the maximum frequency deviation. Interfering waves close to the carrier frequency cannot cause significant change in the frequency of the resultant, and therefore produce little effect. The greater the difference between f_c and $f_c + f_I$, the greater the frequency deviation, so we can expect the demodulated output to be proportional to $|f_I|$. But for PM, the maximum phase deviation depends only on relative amplitudes, as shown by the phasor diagram of Figure 3.5.

The performance of FM and PM with respect to interference is best displayed by plotting the amplitude of the unfiltered signal $y(t)$ as a function of $|f_I|$ (Figure 3.6).

It is seen that when the interference is due to cochannel station, then $f_c + f_I \approx f_c$, $|f_I|$ is small, and FM is better than PM, whereas the opposite is true for adjacent channel interference, where $|f_I|$ is relatively large.

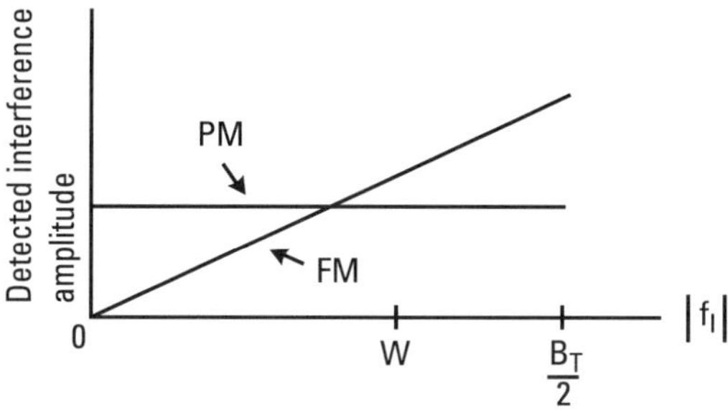

Figure 3.6 Detected interference amplitude as a function of $|f_I|$ for an interfering wave of frequency $f_c + f_I$.

3.4.2 Noise

3.4.2.1 Noise in Amplitude Modulation

To analyze the performance of a linear modulation system in the presence of noise, we need a model for the receiver (Figure 3.7), where the modulated signal plus bandpass noise at the detector input to be [5]:

$$u(t) = K_R x_c(t) + n(t) \tag{3.28}$$

which also holds for exponential modulation with appropriate $x_c(t)$.

Because linear modulation has $B_T = 2W$ or W, depending on whether or not a sideband has been suppressed, the three possible noise spectra $G_n(f)$ are as shown in Figure 3.8, taking $H_R(f)$ to be symmetrical.

The average signal power at the detector input is:

$$K_R^2 \overline{x_c^2} = S_R \tag{3.29}$$

and $\overline{n^2} = N_R = \eta B_T$ is the noise power assuming the noise equivalent bandwidth of the predetection filter equals B_T. The signal and noise are additive in (3.28), so it is meaningful to define predetection SNR, given by:

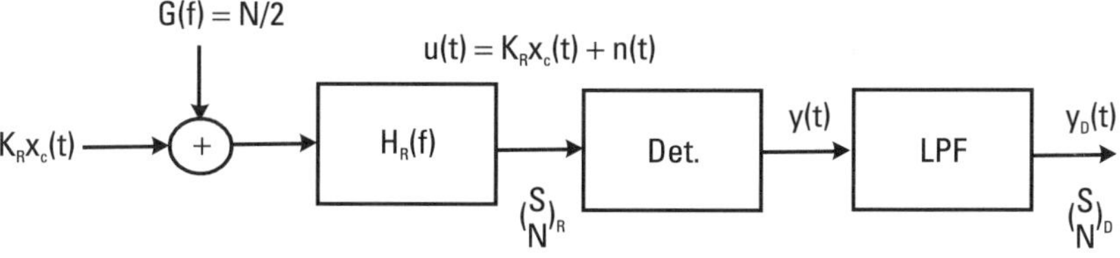

Figure 3.7 Amplitude modulation receiver.

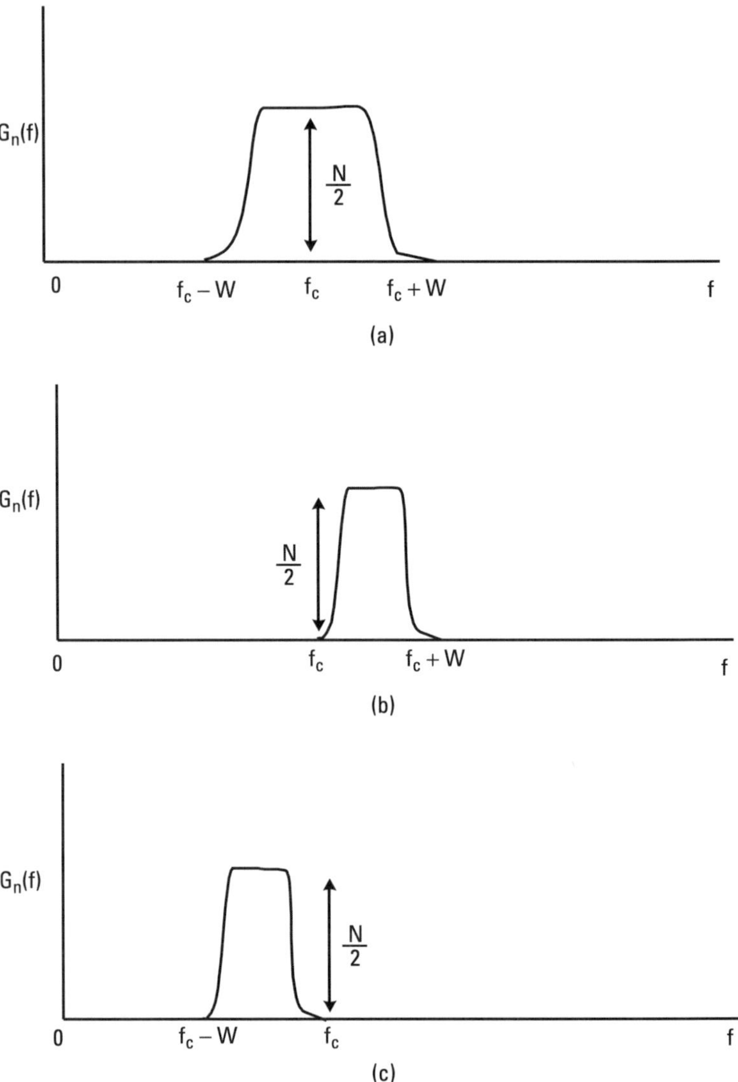

Figure 3.8 Predetection noise spectrum $G_n(f)$ in linear modulation: (a) double sideband, (b) upper sideband, and (c) lower sideband.

$$\left(\frac{S}{N}\right)_R = \frac{S_R}{N_R} = \frac{S_R}{\eta B_T} \tag{3.30}$$

which is suggestive of the parameter given by $\gamma = S_R / \eta W$, where

S_R = signal power at the receiver input;

W = bandwidth;

η = noise density at the receiver input.

Specifically, for equal values of S_R, η, W, we have:

$$\left(\frac{S}{N}\right)_R = \frac{W}{B_T}\gamma \tag{3.31}$$

Hence, $(S/N)_R = \gamma$ for single sideband ($B_T = W$), while $(S/N)_R = \gamma/2$ for double sideband. However, the interpretation to keep in mind is that γ equals to maximum value of the destination signal-to-noise ratio of analog baseband transmission. By the same token, (3.30) and (3.31) actually are upper bounds (e.g., $(S/N)_R < S_R/\eta B_T$ if $B_R > B_T$). $A_r = K_r A_c$ because $x_c(t) = A_c \cos \omega_c t$ and $x_r(t) = K_r A_c \cos \omega_c t$. In other words K_r indicates the power loss through the channel. Finally:

$$S_R = K_R^2 S_T = \left(\frac{A_R}{A_c}\right)^2 S_T \tag{3.32}$$

which relates S_R to S_T.

3.4.2.2 Noise in Exponential Modulation

Turning to the demodulation of FM or PM contaminated by noise, the simulation is the same as Figure 3.7 with:

$$x_c(t) = A_c \cos[\omega_c t + \phi(t)] \tag{3.33}$$

where

$$\phi(t) = \phi_\Delta x(t) \qquad \text{PM} \tag{3.34}$$

$$\frac{1}{2\pi}\frac{d\phi(t)}{dt} = f_\Delta x(t) \qquad \text{FM} \tag{3.35}$$

In either case, the bandpass noise is symmetric about f_c with $N_R = \eta B_T$ while $S_R = A_R^2/2$, so:

$$\left(\frac{S}{N}\right)_R = \frac{A_R^2}{2\eta B_T} \tag{3.36}$$

With $n(t)$ written in envelope-and-phase form, the detector input is:

$$u(t) = A_R \cos\left[\omega_c t + \phi(t)\right] + A_n(t) \cos\left[\omega_c t + \phi_n(t)\right] \tag{3.37}$$

Immediately, there are analytic difficulties in finding the resultant phase $\phi_u(t)$ to insert in the mathematical model of the detector. Let us therefore do as we did with envelope detection—namely, assume the signal component dominates the noise. Figure 3.9 shows the phasor diagram with the phase difference $\psi(t) = \phi_n(t) - \phi(t)$.

Taking the usual small-angle approximation, we have:

$$\phi_u(t) \approx \phi(t) + \frac{A_n(t)}{A_R} \sin \psi(t) \tag{3.38}$$

providing $(S/N)_R \gg 1$.

Not surprisingly, the leading term of (3.38) is the message modulation, but the second term contains both message and noise and is another source of difficulty. Meanwhile, applying (3.17c) and (3.17d) to $\phi_u(t)$ yields the demodulated signal. The analysis so far, even though elementary, will be of great value for the reader later on in Chapters 4 to 6. In order for the material in these chapters to be fully understood, the concepts of Sections 3.1 to 3.5 must be thoroughly comprehended.

3.5 Comparison of Modulation Systems Based on Noise

It is important that these modulation techniques are compared so that the logical choices can be made between the many available systems. This is not

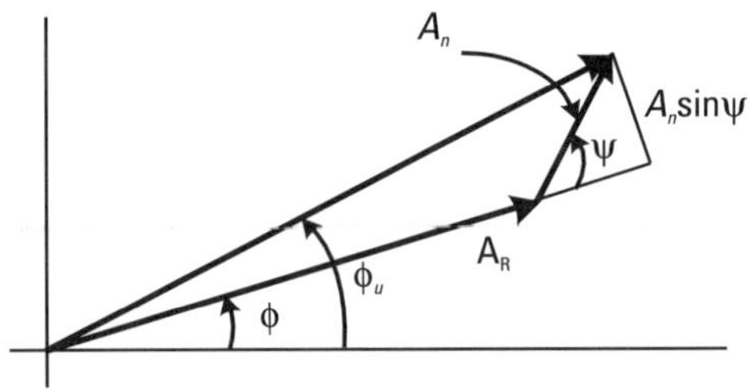

Figure 3.9 Phasor diagram for FM or PM.

possible to be accomplished with any rigor, as the systems have been studied in a highly idealized environment.

In a practical environment, the transmitted signal is subjected to many undesirable distortions by being correlated by noise and interference prior to demodulation. One of the most important distortions is noise, which is inadvertently added to the signal at several points in the system.

The noise performance of modulation systems is often specified by comparing the SNR at the input and output of the demodulator. This is shown in Figure 3.10.

The noise performance of modulation systems is often specified by defining a very powerful quality metric, the SNR at the output of the detector, and using it for comparison. This is accomplished by using as a reference the same ratio as the input after the predetection filters (RF, IF). In Figure 3.10, we have assumed that the signal at the output of the detector is given by $y_0(t)$

$$y_0(t) = m(t) + n(t)$$

where

$m(t)$ = the message at the output of the demodulator corrupted by the noise $n(t)$.

For completeness, the reader is referred to [9], which gives more details as far as the comparison of continuous wave analog systems are concerned.

Comparison of modulated signals based on interference and the ways we mitigate the effects are studied in Chapters 5 and 6. In the next few sections in this chapter, we shall study other modulation techniques that utilize digital signals, as the modern world in wireless communication is becoming entirely digital.

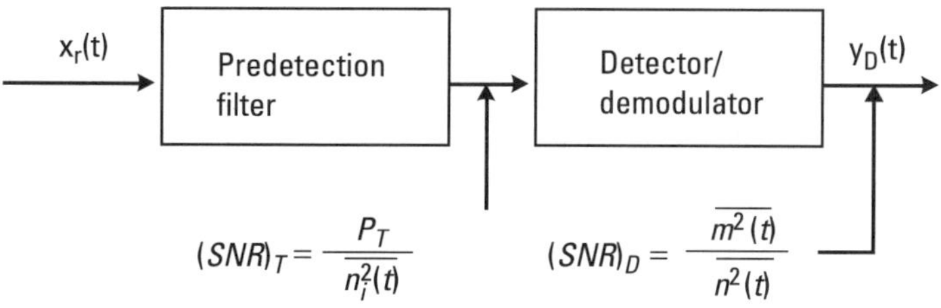

Figure 3.10 Receiver block diagram.

3.6 Digital Transmission

Digital transmission relates to the link for which the user's terminal produces digital signals. In other words, the information signal is in digital form. It is also possible to transmit signals of analog origin (e.g., telephone or sound broadcasting) in digital form by sampling. Figure 3.11 shows the elements of a digital system.

It is often advantageous to transmit messages as digital data. This approach allows greater flexibility in protecting the data from noise and other nonideal effects of the channel through the use of error-control (or channel) coding. This often allows greater efficiency in terms of usage of channel bandwidth through the use of data compression (or source coding). Some types of message sources, such as computer data or text, are inherently digital in nature. However, continuous information sources, such as speech and images, can be converted into digital data for transmission. This process involves analog-to-digital (A/D) conversion—that is, time sampling to convert the continuous-time message signal to a discrete-time message and amplitude quantization to map the continuous amplitudes of the message signal into a finite set of values. To preserve the fidelity of the source, it is necessary that the sampling rate be sufficiently high to prevent loss of information and that the quantization be sufficiently fine to prevent undue distortion. A minimum sampling rate to capture the message in discrete time is twice the bandwidth of the message. A rate known as the Nyquist rate is shown in Appendix A. Once the message is in digital form it can be converted to a sequence of binary words for transmission by encoding its discrete amplitude values. Modulation schemes that transmit analog messages in this way are known as pulse code modulation (PCM) schemes.

A number of forms of modulation can be used to transmit data through a communication channel. The most basic of these transmit a single binary digit (1 or 0) in each of a sequence of symbol intervals of duration T, where T is the reciprocal of the data rate (expressed in bits per second). To use such a scheme, the data sequence of binary words can be converted to a sequence of bits via parallel-to-serial conversion. Most binary transmission schemes can be described in terms of two waveforms, $x^{(0)}(t)$ and $x^{(1)}(t)$, where the transmitted waveform $x(t)$ equals $x^{(0)}(t)$ in a given symbol interval if the corresponding data bit is a 0, and $x(t)$ equals $x^{(1)}(t)$ if the corresponding data bit is a 1.

As with analog modulation, the basic manner in which most binary modulation procedures couple the data sequence to the channel is to impress it onto a sinusoidal carrier. Thus, we can have digital modulation schemes

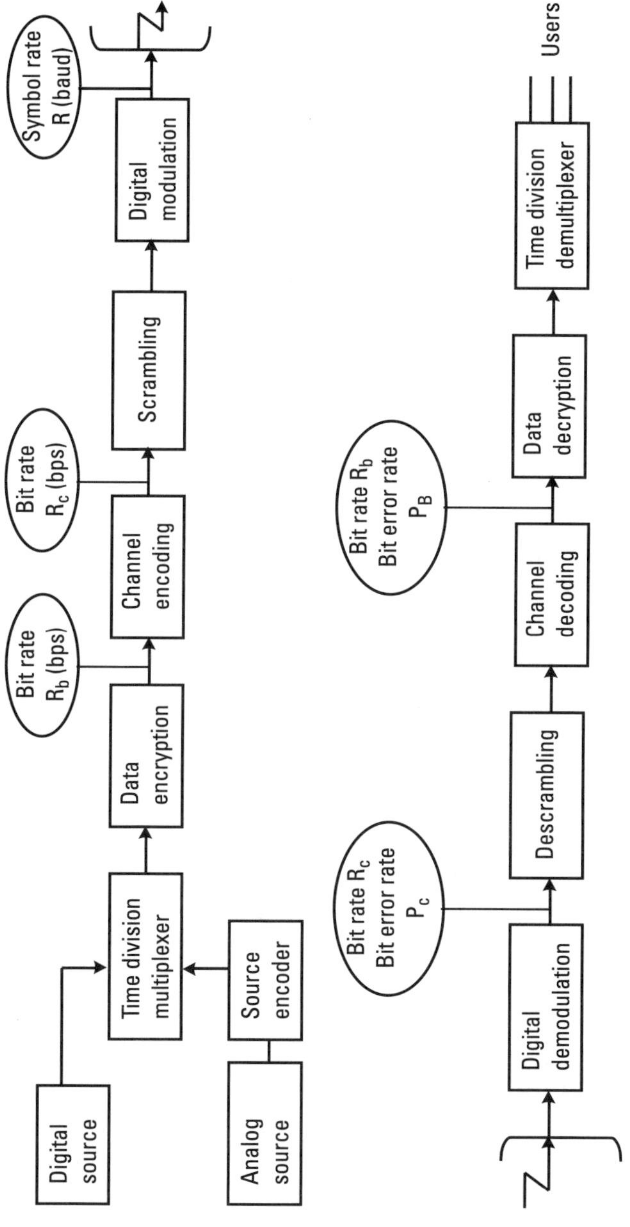

Figure 3.11 The elements of a digital transmission system.

based on amplitude, phase or frequency modulation, and the waveforms $x^{(0)}(t)$ and $x^{(1)}(t)$ are chosen accordingly to modify a basic carrier waveform $A_c \sin(2\pi f_c t + \phi_c)$, where A_c, f_c, and ϕ_c are, respectively, the amplitude, frequency, and phase of the carrier. Several fundamental techniques of this type are described in the following paragraphs. This is the reason why we presented analog modulation techniques in Sections 3.2 to 3.5.

Given the received S/N_0, we can write the received bit-energy to noise-power spectral density E_b/N_0, for any desired data rate R, as follows:

$$\frac{E_b}{N_0} = \frac{ST_b}{N_0} = \frac{S}{N_0}\left(\frac{1}{R}\right) \tag{3.39}$$

Equation (3.39) follows from the basic definitions that received bit energy is equal to received average signal power times the bit duration and that bit rate is the reciprocal of bit duration. Received E_b/N_0 is a key parameter in determining the performance of a digital communication system. Its value indicates the portion of the received waveform energy among the bits that the waveform represents. At first glance, one might think that a system specification should entail the symbol-energy to noise-power spectral density E_b/N_0 associated with the arriving waveforms. We will show, however, that for a given S/N_0, the value of E_b/N_0 is a function of the modulation and coding. The reason for defining systems in terms of E_b/N_0 stems from the fact that E_b/N_0 depends only on S/N_0 and R and is unaffected by any system design choices, such as modulation and coding.

More details about digital modulation or modulation of digital signals are given in [11–13]. In the sections that follow, we shall present the spectra characteristics of the most typical digital modulation techniques that we encounter in wireless communications, because they play a major and direct role in the analysis of the behavior of such systems in an interference environment. If the reader is not familiar with the basic material, he is encouraged to review details in reference [11–13].

3.7 Digital Modulation Techniques

Typical characteristics of an average digital modulation system are the carrier attribute (e.g., amplitude, phase, frequency) that is being modulated, the number of levels assigned to the modulated attribute, and the degree to which the receiver extracts information about the unknown carrier phase in performing the data deflection function (coherent, partially coherent, differentially coherent, noncoherent). In many important applications in wireless transmission, only a single carrier attribute is modulated even though more than one attribute can be modulated for additional degrees of freedom in satisfying the power and bandwidth requirements of the system, as shown in Figure 3.12. The principles of a modulator shown in Figure 3.12 consist of an encoder and a radio frequency signal (carrier) generator [3–6, 14, 15].

The symbol generator generates symbols with M states, where $M = 2^m$, from m consecutive bits of the binary input stream. The encoder establishes a correspondence between the M states of these symbols and M possible states of the transmitted carrier. Two types of coding are used, as explained in the previous chapter.

1. Direct encoding, where one state of the symbol defines one state of the carrier;

2. Encoding of transitions (differential encoding), where one state of the symbol defines a transition between two consecutive states of the carrier.

For a bit rate R_b (bps) at the modulator input, the signaling rate R_S at the modulator output that indicates the number of changes of state of the carrier per second, is given by:

$$R_S = \frac{R_b}{m} = \frac{R_b}{\log_2 M} \tag{3.40}$$

In digital communication systems, discrete modulation techniques are usually used to modulate the source information signal. Discrete modulation includes:

- PCM;

- Differential modulation (DM);

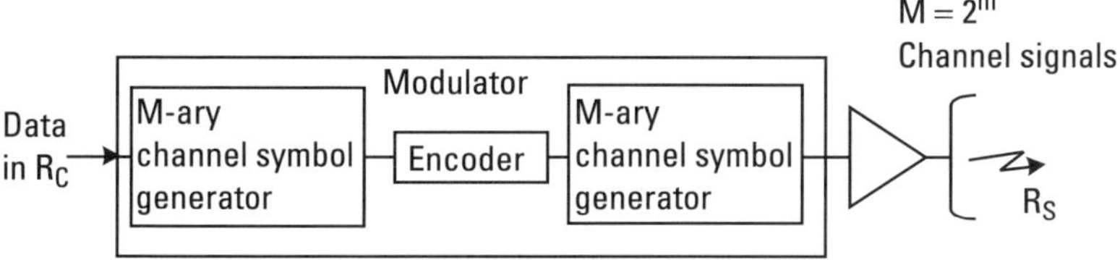

Figure 3.12 The principle of a modulator for digital transmission.

- Differential pulse-code modulation (DPCM);
- FSK;
- PSK;
- DPSK;
- M-ary phase-shift keying (MPSK);
- Quadrature amplitude modulation (QAM).

Several factors influence the choice of a particular digital modulation scheme. A desirable modulation scheme provides low BER at low received signal-to-noise ratios, performs well in multipath conditions, occupies a minimum bandwidth, and is easy and cost effective to implement. None of the existing modulation schemes can simultaneously satisfy all these requirements. Some modulation schemes are better in terms of the BER performance, while others are better in terms of bandwidth efficiency. Depending on the demands of the particular application, trade-offs are made when selecting a digital modulation.

The performance of a modulation scheme is often measured in terms of its power efficiency and bandwidth efficiency.

- Power efficiency describes the ability of a modulation technique to preserve the fidelity of the digital message at low power levels. In a digital communication system, in order to increase the noise immunity, it is necessary to increase the signal power. However, the amount by which the signal power should be increased to obtain a certain level of fidelity depends on the particular type of modulation employed. The power efficiency, sometimes called energy efficiency of a digital modulation scheme, is a measure of how favorably the trade-off between fidelity and signal power is made. It is often expressed as the ratio of the signal energy per bit to noise power spectral density (E_b/N_0) required at the receiver input for a certain probability of error. The power efficiency typical for some cases is (10^{-5}).

- Bandwidth efficiency describes the ability of a modulation scheme to accommodate data within a limited bandwidth. In general, increasing the data rate implies decreasing the pulse width of a digital symbol, which increases the RF bandwidth of the signal. Thus, there is an unavoidable trade-off between data rate and RF bandwidth occupancy. Some modulation schemes perform better than the others in making this trade-off. Bandwidth efficiency is expressed as:

$$\eta_B = \frac{R_b}{B} \text{ bps/Hz} \tag{3.41}$$

The system capacity of a digital mobile communication system is directly related to the bandwidth efficiency of the modulation scheme used. This is because a modulation with a greater value of η_B will transmit more data in a given spectral allocation. The more bandwidth efficient the modulation scheme used, the greater will be the capacity of the system [3]. The same criteria were used to compare various detection mechanisms in the previous chapter.

3.7.1 Linear Modulation Techniques

With linear modulation techniques, the amplitude of the transmitted signal $x(t)$, varies linearly with the modulating digital signal $m(t)$. Linear modulation techniques are bandwidth efficient and hence are very attractive for use in wireless communication systems, where there is an increasing demand to accommodate more and more users within a limited spectrum. The bandpass complex transmitted modulated signal can take the form:

$$s(t) = S(t) e^{j(2\pi f_c t + \theta_i)} \tag{3.42}$$

where $S(t)$ is the baseband equivalent signal and takes the following form:

$$S(t) = A_c \alpha(t)$$
$$S(t) = A_c e^{j\theta(t)} \tag{3.43}$$
$$S(t) = A_c e^{jf(t)t}$$

For amplitude, phase, and frequency modulation, when more than one attribute of the carrier is modulated (e.g., amplitude and phase), the transmitted signal would have the form:

$$s(t) = A_c a(t) e^{j(2\pi f_c t + \theta_c + \theta(t))}$$

When perfect knowledge of phase and frequency of the phase and frequency of the carrier is possible, the receiver generates a signal used for demodulation given by:

$$c_t(t) = e^{j(2\pi f_c t + \theta_c)} \tag{3.44}$$

A generic form of such a detector is shown in Figure 3.13. The total received signal as shown here is given by:

$$r(t) = \alpha_{ch}\, s(t) + n(t) \tag{3.45}$$

where

α_{ch} is a random variable dependent on the particular channel used (fading characteristics);
$n(t)$ is the bandpass form of the Gaussian noise process.

The output of the demodulation process $x(t)$ is given by

$$x(t) = r(t)\, c_r^*(t) = S(t) + n(t)\, c_r^*(t) \tag{3.46}$$

where $*$ is the complex conjugate operation and

$$c_r(t) = e^{j(2\pi f_c t + \theta_c)} \tag{3.47}$$

The optimum receiver then performs matched filtering operation on $x(t)$ during each successive transmitted interval and proceeds to make a decision based on the largest of the resulting M outputs. Depending on the particular form of modulation corresponding to the three simple cases presented here, we obtain:

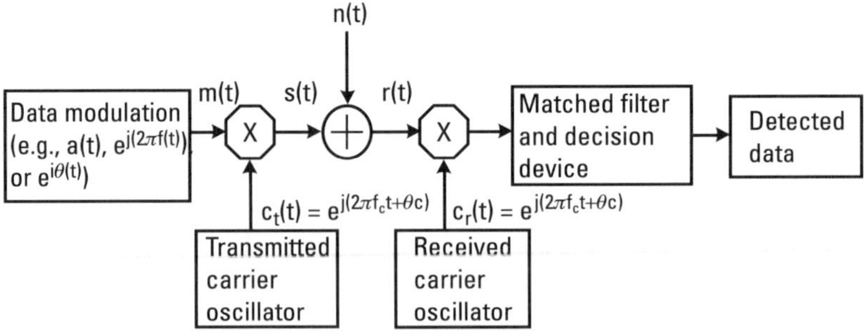

Figure 3.13 Ideal coherent detector over additive white Gaussian noise. (After: [16]. © 2000 John Wiley & Sons, Inc.)

$$x(t) = A_c a(t) + n(t) c_r^*$$

$$x(t) = A_c e^{j\theta(t)} + n(t) c_r^* \tag{3.48}$$

$$x(t) = A_c e^{j(2\pi f(t)t)} + n(t) c_r^*$$

3.7.1.1 On-Off Keying

The on-off keying (OOK) is the simplest type of binary modulation. It transmits the signal

$$x_{OOK}^{(1)}(t) = A_c \sin(2\pi f_c t + \phi_c) \tag{3.49}$$

in a given symbol interval if the corresponding data bit is a 1, and it transmits nothing—that is,

$$x_{OOK}^{(0)}(t) = 0 \tag{3.50}$$

if the corresponding data bit is a 0.

An OOK waveform is illustrated in Figure 3.14. OOK is a form of *amplitude-shift keying* (ASK) because it "keys" (i.e., modulates) the carrier by shifting its amplitude by an amount depending on the polarity of the data bit. ASK waveforms other than the on-off version described here can also be used.

OOK can be demodulated either coherently (i.e., with knowledge of the carrier phase), as shown in Figure 3.15(a), or noncoherently (i.e., without knowledge of the carrier phase), as shown in Figure 3.15(b). In either case, the output of the detector (coherent or noncoherent) is sampled at the end of each symbol interval and compared with a threshold. If this output exceeds the threshold for a given sampling time, then the corresponding data symbol is detected as a 1; otherwise, this symbol is detected as a 0. (In a coherent detector, the integrator is *quenched*—that is, reset to 0, as it is sampled.)

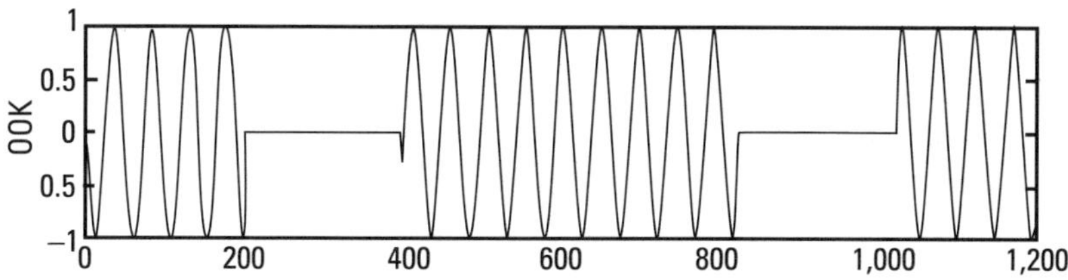

Figure 3.14 Digital OOK modulation waveform for transmitting the bit sequence 101101 (T = 200).

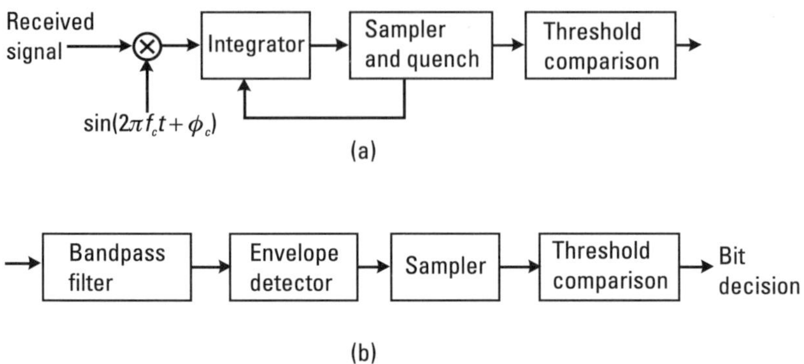

Figure 3.15 Demodulation of OOK: (a) coherent demodulator (BPSK) and (b) noncoherent demodulator.

Errors occur in these systems because the noise in the channel can move the output of the detector to the incorrect side of the threshold. The proper mechanism for choosing the threshold to minimize this effect, and the corresponding rate of bit errors, are shown graphically next [6, 14, 15, 17–20].

The power spectral density (PSD) of this complex envelope is proportional to that for the unipolar signal. We find that this PSD is given by:

$$P_{OOK}(f) = \frac{A_c^2}{2}\left[\delta(f) + T_b\left(\frac{\sin\,\pi f T_b}{\pi f T_b}\right)^2\right] \qquad (3.51)$$

where $\delta(f)$ is the Fourier transform of a delta function. For positive frequencies, it is seen that the null-to-null bandwidth is $2R$. That is, the transmission bandwidth of the OOK signal is $B_T = 2B$ where B is the baseband bandwidth because OOK is AM-type signaling.

3.7.1.2 Multiple Amplitude Shift Keying

In the multiple amplitude shift keying (M-ASK) case, the amplitude of the modulated signal takes the form [16]:

$$s(t) = A_c\,\alpha_n\,e^{j(2\pi f_c t + \theta_c)} \qquad (3.52a)$$

where α_n is the information (data) amplitude in the nth symbol interval $nT_s \leq t \leq (n+1)T_s$ ranging over the set of M possible values $\alpha_i = 2i - 1 - M$ where $i = 1, 2, \ldots, M$.

Because we usually use each symbol to modulate a rectangular pulse shape during the nth symbol interval, the transmitted signal is given by

$$s(t) = A_c \, \alpha_n \, e^{j(2\pi f_c t + \theta_c)} \qquad (3.52b)$$

and the baseband signal is given by:

$$S(t) = A_c \, \alpha_n \qquad (3.53)$$

At the receiver the signal is given by:

$$s(t) = A_c \, \alpha_n \, e^{j(2\pi f_c t + \theta_c)} + n(t) \qquad (3.54)$$

Multiplying (3.54) by $c_r^*(t) = e^{-j(2\pi f_c t + \theta_c)}$ we obtain

$$x(t) = A_c \, \alpha_n + N(t)$$

where

$$N(t) = n(t) \, c_r^*(t)$$

Passing $x(t)$ through M matched filters (integrate and dump) results in the M outputs:

$$y_{nk} = \alpha_k \, \alpha_n \, A_c \, T_s + \alpha_k N_n \qquad (3.55)$$

where: $k = 1, 2, \ldots, M$ and

$$N_n = \int_{nT_s}^{(n+1)T_s} N(t) \, dt$$

Decision about which of the data α_n were received is made by examining and finding the maximum of $Re\{y_{nk}\}$, as shown in Figure 3.16(a, b) [16].

A similar procedure leads to coherent detectors for QAM, where the transmitted waveform during the nth symbol interval is given by:

$$s(t) = A_c (\alpha_{In} + j\alpha_{Qn}) \, e^{j(2\pi f_c t + \theta_c)}$$

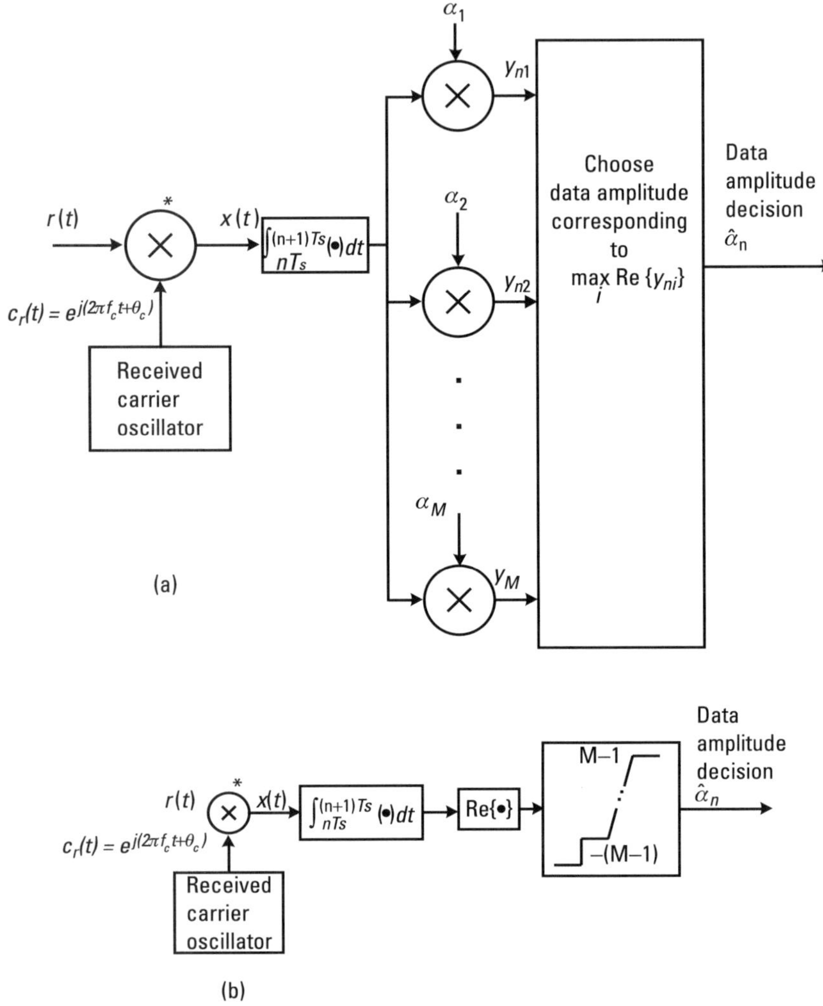

Figure 3.16 (a) Maximum likelihood of coherent detector of multiple amplitude modulation. (b) Threshold coherent detector of multiple amplitude modulation. (Source: [16]. © 2000 John Wiley & Sons, Inc.)

and the decision about the data α_n can be made either by a maximum likelihood operation or a threshold decision operation [16] as shown in Figure 3.17(a, b), respectively.

3.7.1.3 PSK

As its name suggests, PSK uses the phase of the carrier to encode the binary data to be transmitted. The basic forms of PSK are described in the following sections [3, 17–19, 21, 22].

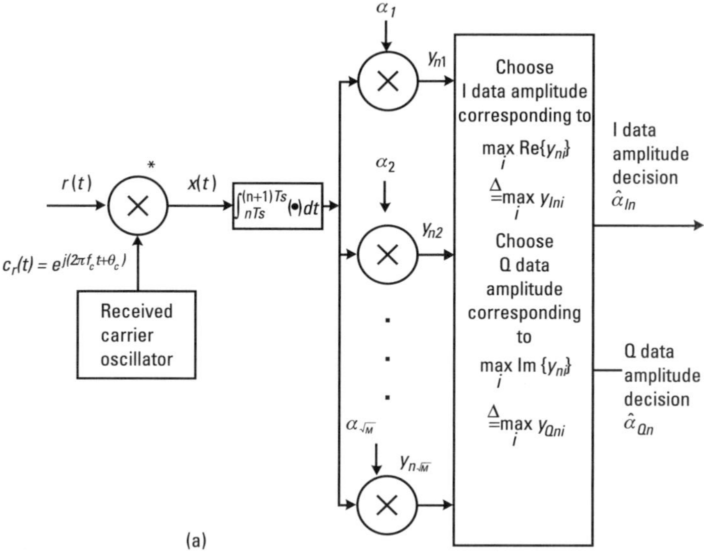

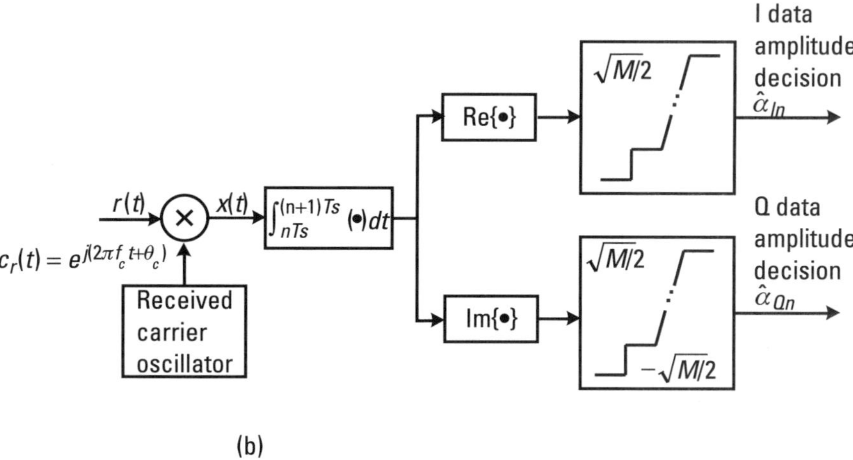

Figure 3.17 (a) Maximum likelihood coherent detector of QAM. (b) Threshold detector of QAM (Source: [16]. © 2000 John Wiley & Sons, Inc.)

Binary PSK

In binary phase shift keying (BPSK), the phase of a constant amplitude carrier signal is switched between two values according to the two possible signals, corresponding to binary 1 and 0, respectively. Usually, the two phases are separated by 180°, and if the sinusoidal carrier has an amplitude A_c and energy per bit $E_b = \frac{1}{2}A_c^2 T_b$, then the transmitted BPSK signal can be one of the following waveforms:

$$x_{BPSK}^{(1)}(t) = A_c \sin(2\pi f_c t + \phi_c) \qquad 0 \leq t \leq T_b \tag{3.56}$$

$$x_{BPSK}^{(0)}(t) = A_c \sin(2\pi f_c t + \phi_c + \pi) = -A_c \sin(2\pi f_c t + \phi_c) \tag{3.57}$$

$$\text{for } 0 \leq t \leq T_b$$

A BSPK waveform is illustrated in Figure 3.18.

This type of modulation uses antipodal signalling (i.e., $x^{(0)}(t) = -x^{(1)}(t)$). Note that BPSK is also a form of ASK, in which the two amplitudes are ± 1. Because the information in BPSK is contained in the carrier phase, it is necessary to use coherent detection in order to have an accurately demodulated BPSK. The block diagram of a BPSK receiver along with the carrier recovery circuits is shown in Figure 3.19.

The PSD of the complex envelope of the signal can be shown to be:

$$P(f) = 2E_b \left(\frac{\sin \pi f T_b}{\pi f T_b} \right)^2 \tag{3.58}$$

Hence the PSD of a BPSK signal is given by:

$$P_{BPSK} = \frac{E_b}{2} \left[\left(\frac{\sin \pi(f - f_c)T_b}{\pi(f - f_c)T_b} \right)^2 + \left(\frac{\sin \pi(-f - f_c)T_b}{\pi(-f - f_c)T_b} \right)^2 \right] \tag{3.59}$$

For more details the reader is referred to [23].

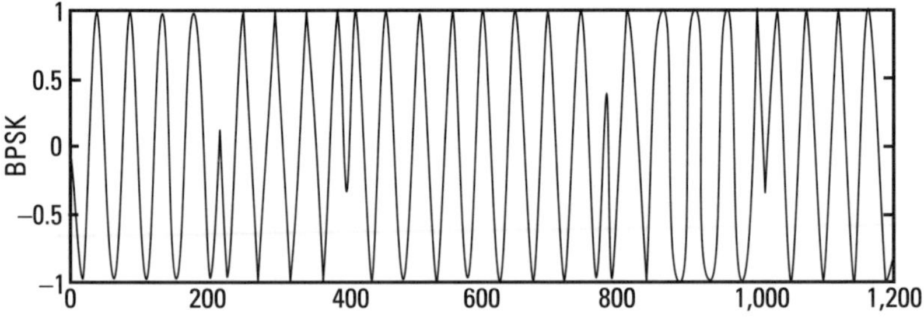

Figure 3.18 Digital BPSK modulation waveform for transmitting the bit sequence 101101 (T = 200).

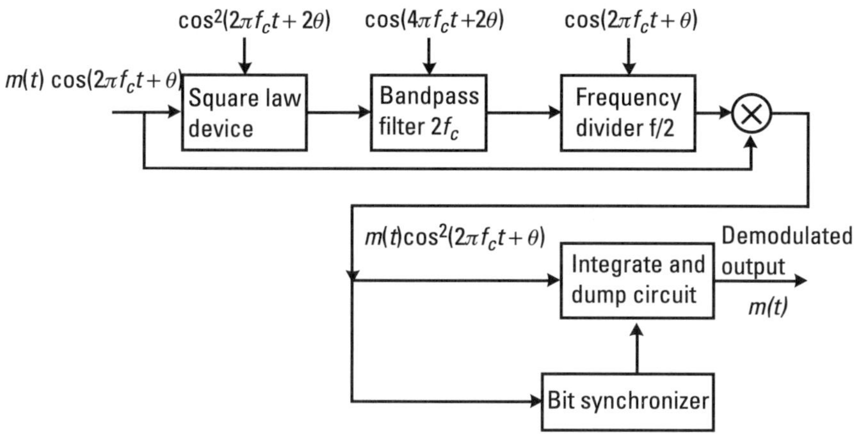

Figure 3.19 Block diagram of a BPSK receiver with carrier recovery circuits. (After: [23].)

DPSK

The necessity of knowing the carrier for demodulation of BPSK is a disadvantage that can be overcome by the use of DPSK. In a given bit interval (say the kth one), DPSK uses the following waveforms:

$$x_{DPSK}^{(1)}(t) = A_c \sin\left(2\pi f_c t + \phi_{k-1}\right) \tag{3.60}$$

$$x_{DPSK}^{(0)}(t) = A_c \sin\left(2\pi f_c t + \phi_{k-1} + \pi\right) \tag{3.61}$$

where ϕ_{k-1} denotes the phase transmitted in the *preceding* bit interval (i.e., the $(k-1)$th bit interval). Thus, the information is encoded in the difference between the phases in succeeding bit intervals rather than in the absolute phase, as illustrated in Figure 3.20 (DPSK requires an initial reference bit, which is taken to be 1 in the illustration).

This step allows for noncoherent demodulation of DPSK, as shown in Figure 3.21.

Note that the demodulator in Figure 3.21(a) does not require knowledge of the carrier phase or frequency, whereas that in Figure 3.21(b) requires knowledge of the carrier frequency but not its phase. The block marked "decision logic" in Figure 3.21(b) makes each bit decision based on two successive pairs of outputs of the two channels that provide its inputs. In particular, the kth bit is demodulated as a 1 if $p_k p_{k-1} + q_k q_{k-1} > 0$ and as a 0 otherwise, where p_k and p_{k-1} are the outputs of the upper channel (known as *in-phase* channel) at the end of the kth and $(k-1)$th bit intervals,

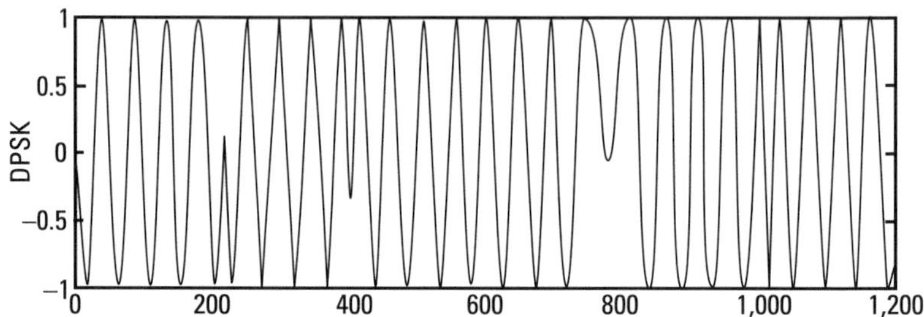

Figure 3.20 Digital DPSK modulation waveform for transmitting the bit sequence 101101 (T = 200).

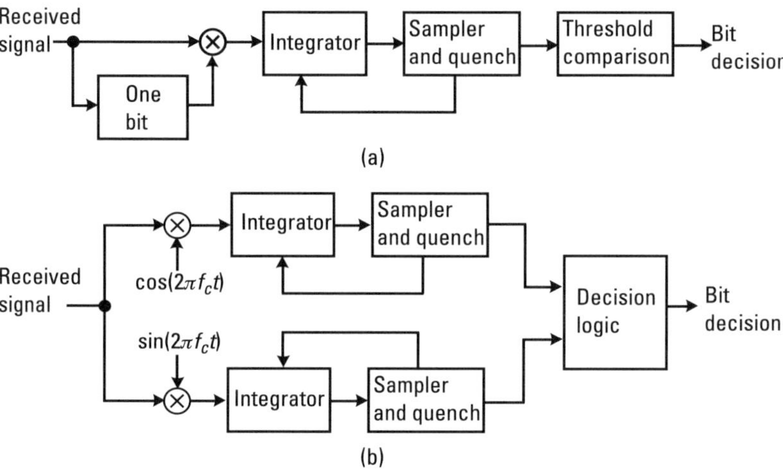

Figure 3.21 Demodulation of DPSK: (a) suboptimum demodulator for DPSK, and (b) coherent optimum demodulator for DPSK. (After: [23].)

respectively, and where q_k and q_{k-1} are the corresponding outputs of the lower channel (the *quadrature* channel). The second of these demodulators is actually the optimum for demodulating DPSK and, as such, exhibits performance advantages over the first. This performance comes in exchange for the obvious disadvantage of requiring carrier-frequency reference signals at the receiver.

Quadrature PSK

The bandwidth efficiency of BPSK can be improved by taking advantage of the fact that there is another pair of antipodal signals, namely [14, 22–24]

$$x^{(0)}(t) = A_c \cos\left(2\pi f_c t + \phi_c\right) \tag{3.62}$$

$$x^{(1)}(t) = A_c \cos\left(2\pi f_c t + \phi_c + \pi\right) \tag{3.63}$$

Those have the same frequency as the two signals used in BPSK (i.e., $x_{BPSK}^{(1)}(t)$ and $x_{BPSK}^{(0)}(t)$ of (3.56) and (3.57)), while being completely orthogonal to those signals. By using all four of these signals, two bits can be sent in each symbol interval, thereby doubling the transmitted bit rate. Such a signaling scheme is known as quadrature PSK (QPSK) because it involves the simultaneous transmission of two BPSKs in quadrature (i.e., 90 degrees out of phase). Although the performance in terms of bit error of rate of QPSK is the same as that for BPSK, QPSK has the advantage of requiring half the bandwidth needed by BPSK to transmit at the same bit rate. This situation is directly analogous to that involving DSB and SSB analog modulation, the latter of which uses two quadrature signals to transmit the same information as the former does, while using only half the bandwidth [24].

The block diagram of a typical QPSK transmitter is shown in Figure 3.22:

The input unipolar binary stream at a bit rate of R_b is first converted into a bipolar nonreturn-to-zero (NRZ) sequence using a unipolar-to-bipolar converter. The bit stream $m(t)$ is then split into two bit streams $m_I(t)$ and $m_Q(t)$ (in-phase and quadrature streams), each having a bit rate of $R_S = R_b/2$, the symbol rate, and consisting of odd and even bits, respectively, by means of a serial-to-parallel converter. The two binary sequences are separately modulated by two carriers $\phi_1(t)$ and $\phi_2(t)$, which are in quadrature. The filter at the output of the modulator confines the power spectrum of the QPSK signal within the allocated band. This prevents spillover of signal energy into adjacent channels and also removes out-of-band spurious signals generated during the modulation process.

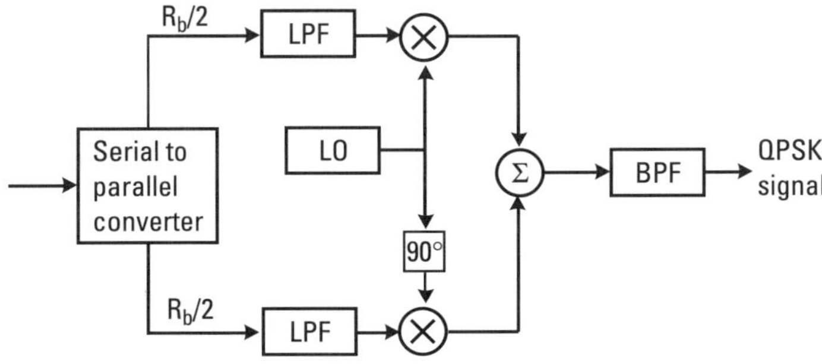

Figure 3.22 Block diagram of a QPSK transmitter. (After: [23].)

The block diagram of a coherent QPSK receiver is shown in Figure 3.23 [23]:

The front-end bandpass filter removes the out-of-band noise and adjacent channel interference. The filtered output is split into two parts, and each part is coherently demodulated using the in-phase and quadrature carriers. The coherent carriers used for demodulation are recovered from the received signal using carrier recovery circuits of the type described in Figure 3.19. The outputs of the demodulators are passed through decision circuits, which generate the in-phase and quadrature binary streams. The two components are then multiplexed to reproduce the original binary sequence with a minimum of error.

The PSD of a QPSK signal can be obtained in a manner similar to that used for BPSK, with the bit periods T_b replaced by symbol periods T_S. Hence, the power spectral density of a QPSK signal using rectangular pulses can be expressed as [24]:

$$P_{QPSK} = E_b \left[\left(\frac{\sin 2\pi (f - f_c T_S)}{2\pi (f - f_c T_S)} \right)^2 + \left(\frac{\sin 2\pi (-f - f_c T_S)}{2\pi (-f - f_c T_S)} \right)^2 \right]$$

$$(3.64)$$

More details are given in [23] regarding the power spectral density of a QPSK signal from (3.64).

Offset QPSK

The amplitude of a QPSK signal is ideally constant. However, when QPSK signals are pulse shaped, they lose the constant envelope property. The occasional phase shift of π radians can cause the signal envelope to go to zero for just an instant. Any kind of nonlinear amplification can cause sidelobe

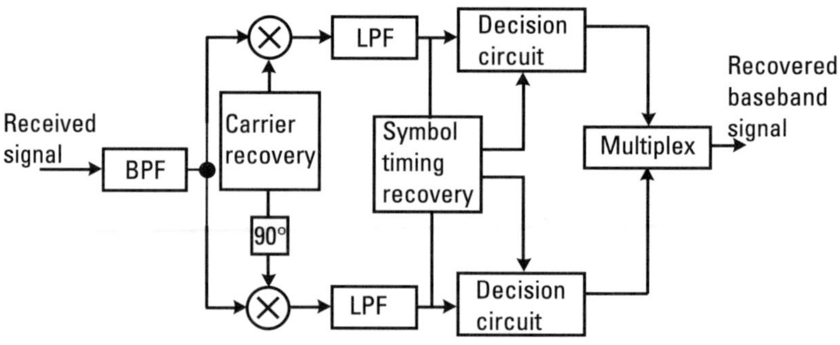

Figure 3.23 Block diagram of a QPSK receiver.

regeneration. A modified form of QPSK, called offset QPSK (OQPSK) or staggered, is less susceptible to these effects [24]. This is achieved by staggering the relative alignments of even and odd bit streams by one bit period. This way at any given time only one of the two bit streams can change values. This implies that the maximum phase shift of the transmitted signal at any given time is limited to $\pm 90°$. Thus by switching phases more frequently, OQPSK eliminates the $180°$ phase transitions.

$\pi/4$ QPSK

A compromise between QPSK and OQPSK is the $\pi/4$ QPSK with maximum phase change limited to $135°$. A bit advantage of this modulation is that it can be demodulated noncoherently and performs better in multipath spread and fading.

3.7.2 Nonlinear Modulation Techniques

Many mobile communication systems use nonlinear modulation methods as opposed to the linear modulation techniques. In this case, amplitude of the carrier is constant, regardless of the variation in the modulating signal. The constant envelope families of modulations have the advantage of satisfying a number of conditions:

1. Power amplifications can be used without introducing degradation in the spectrum performance of the transmitted signal.

2. Low out-of-band radiation on the order of -60 to -70 dB can be achieved.

3. Limiter-discriminator detection can be adopted, which simplifies receiver design and provides high immunity against random FM noise and level fluctuations due to Rayleigh fading [24].

3.7.2.1 FSK

FSK transmits binary data by sending one of two distinct frequencies $f_c + f_\Delta$ and $f_c - f_\Delta$ in each bit interval, depending on the polarity of the bit to be transmitted. This scheme can be described in terms of the two signaling waveforms:

$$x_{FSK}^{(1)} = A_c \sin\left(2\pi(f_c + f_\Delta)t + \phi_c\right) \tag{3.65}$$

$$x_{FSK}^{(0)} = A_c \sin\left(2\pi(f_c - f_\Delta)t + \phi_c\right) \tag{3.66}$$

where f_Δ is a constant. An FSK waveform is illustrated in Figure 3.24.

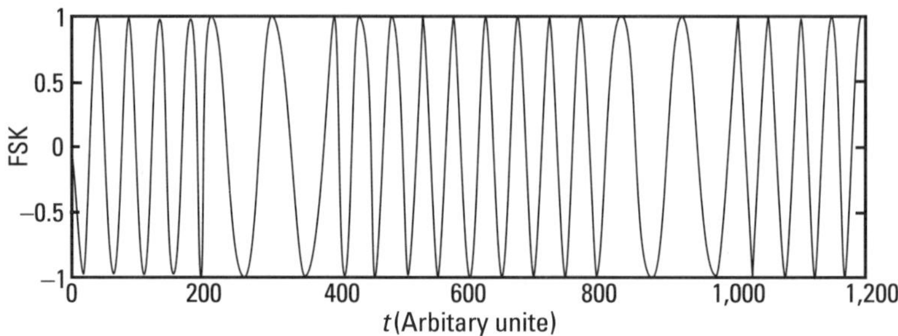

Figure 3.24 Digital FSK modulation waveform for transmitting the bit sequence 101101 (T = 200).

FSK can be demodulated either coherently or noncoherently, as shown in Figure 3.25. In these demodulators, the block marked "comparison" chooses the bit decision as a 1 if the upper-channel output is larger than the lower-channel output, and as a 0 otherwise. Note that when noncoherent demodulation is to be used (as is very commonly the case), it is not necessary for the carrier phase to be maintained from bit interval to bit interval. This

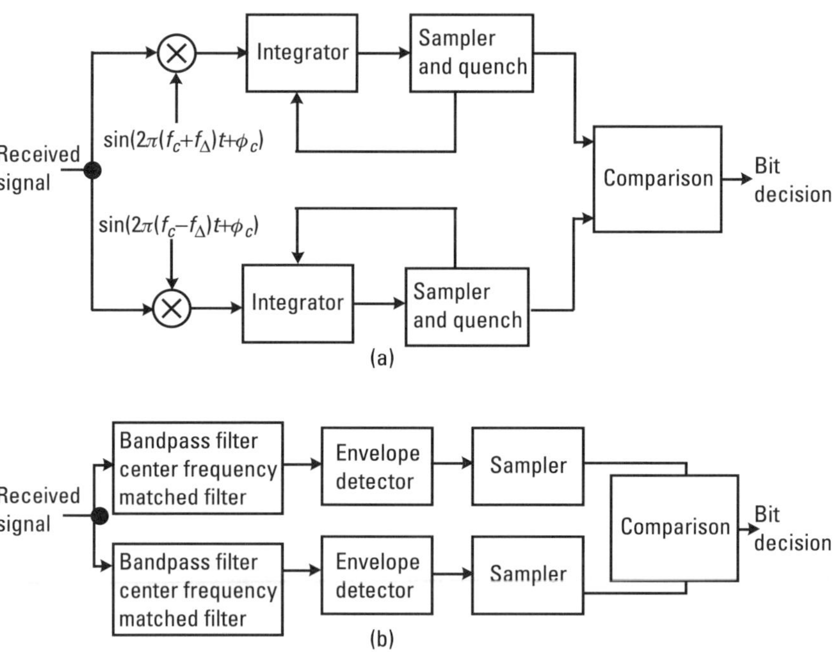

Figure 3.25 Demodualtion of FSK (a) coherent detection, and (b) noncoherent detection. (After: [23].)

simplifies the design of the modulator and makes noncoherent FSK one of the simplest types of digital modulation. It should be noted, however, that FSK generally requires greater bandwidth than do the other forms of digital modulation.

The exact PSD for continuous-phase FSK signals is difficult to evaluate for the case of random data modulation. However, it can be done with the use of some elegant statistical techniques. The resulting power spectral density for the complex envelope of the FSK signal is given by the following expression:

$$P(f) = \frac{A_c^2 T_b}{2} \times \tag{3.67}$$

$$\left\{ A_1^2(f)[1 + B_{11}(f)] + A_2^2(f)[1 + B_{22}(f)] + 2B_{12}(f)A_1(f)A_2(f) \right\}$$

where

$$A_n(f) = \frac{\sin\left[\pi T_b(f - \Delta F(2n - 3))\right]}{\pi T_b(f - \Delta F(2n - 3))} \tag{3.68}$$

$$B_{nm}(f) = \frac{\cos\left[2\pi f T_b - 2\pi \Delta F T_b(n + m - 3)\right] - \cos\left(2\pi \Delta F T_b\right)\cos\left[2\pi \Delta F T_b(n + m - 3)\right]}{1 + \cos^2\left(2\pi \Delta F T_b\right) - 2\cos\left(2\pi \Delta F T_b\right)\cos\left(2\pi f T_b\right)} \tag{3.69}$$

where ΔF is the peak frequency deviation, $R = 1/T_b$ is the BER, the modulation index is $h = 2\Delta F/R$, and $n = 1, 2$ and $m = 1, 2$.

3.7.2.2 MSK

MSK is a special type of continuous phase frequency shift keying (CPFSK) wherein the peak frequency deviation is equal to half the bit rate. In other words, MSK is continuous phase FSK with a modulation index of 0.5. The modulation index of an FSK signal is similar to the FM modulation index and is defined as:

$$k_{PSK} = \frac{2\Delta F}{R_b} \tag{3.70}$$

where

$2\Delta F$ = the peak-to-peak frequency shift;

R_b = bit rate.

A modulation index of 0.5 corresponds to the minimum frequency spacing that allows two FSK signals to be coherently orthogonal. The name minimum shift keying implies the minimum frequency separation that allows orthogonal detection. The block diagram of an MSK modulator is shown in Figure 3.26.

Multiplying a carrier signal with $\cos(\pi t/2T)$ produces two phase coherent signals at $f_c + 1/4T$ and $f_c - 1/4T$. These two signals are separated using two narrow bandpass filters and appropriately combined to form the in-phase and quadrature carrier components $x(t)$ and $y(t)$, respectively. These carriers are multiplied with the odd and even bit streams $m_I(t)$ and $m_Q(t)$ to produce the MSK modulated signal.

The received signal $s_{MSK}(t)$ in the absence of noise and interference is multiplied by the respective in-phase and quadrature carriers $x(t)$ and $y(t)$. The output of the multipliers are integrated over two bit periods and dumped to a decision circuit at the end of each two bit periods. Based on the level of the signal at the output of the intergrator, the threshold detector decides whether the signal is a 0 or a 1. The output data streams to $m_I(t)$ and $m_Q(t)$, which can be offset combined to obtain the demodulated signal. The block diagram of an MSK receiver is shown in Figure 3.27.

The normalized PSD for MSK is given by [23]:

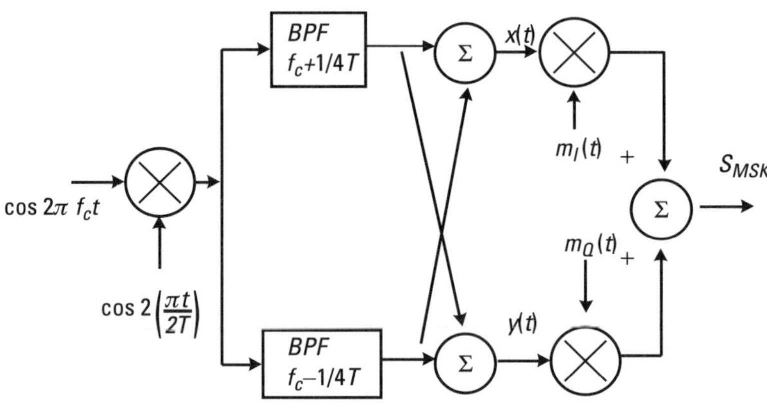

Figure 3.26 Equivalent real forms of precoded MSK transmitters. (After: [23].)

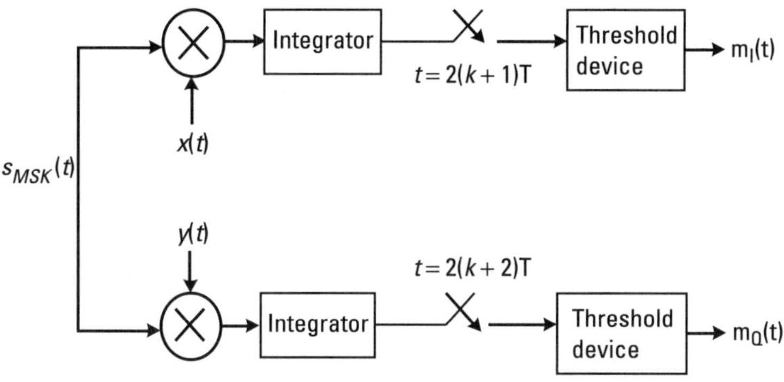

Figure 3.27 Block diagram of an MSK receiver. (After: [23].)

$$P_{MSK} = \frac{16}{\pi^2} \left(\cos \frac{2\pi(f+f_c)T}{1.16f^2T^2} \right)^2 + \frac{16}{\pi^2} \left(\cos \frac{2\pi(f-f_c)T}{1.16f^2T^2} \right)^2$$

$$(3.71)$$

The PSD of an MSK signal is shown.

3.7.3 Spread Spectrum Systems

Spread spectrum systems transmit the information signal after spectrum spreading to a bandwidth N times larger, where N is called processing gain. It is given by

$$N = \frac{B_s}{B}$$

where B_s is the bandwidth of the spread spectrum signal and B is the bandwidth of the original information signal. As we shall see in Section 3.9.3, in conjunction with CDMA, this unique technique of spreading the information spectrum is the key to improving its detection in an interference environment. It also allows narrowband signals exhibiting a significantly higher spectral density to share the same frequency band. There are basically two main types of spread spectrum systems:

1. Direct sequence (DS);

2. Frequency hopping (FH).

3.7.3.1   FH Spread Spectrum

In FH spread spectrum, the narrowband signal is transmitted using different carrier frequencies at different times. A conceptual FH spread spectrum transmitter and receiver as well as the signal spectrum are depicted in Figures 3.28 and 3.29.

Frequency hopping is accomplished by using a digital frequency synthesizer that is driven by a pseudonoise (PN) sequence generator. Each information symbol is transmitted on one or more hops. The most commonly used modulation with frequency hopping is M-ary frequency shift keying (MFSK). With MFSK, the complex envelope is given by:

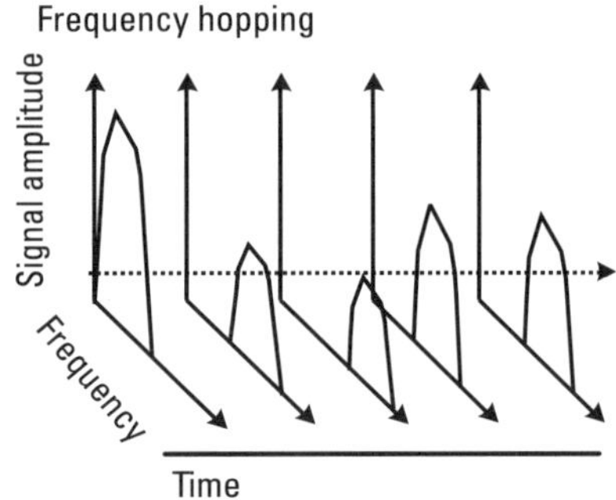

Figure 3.28 Signal spectrum using frequency hopping.

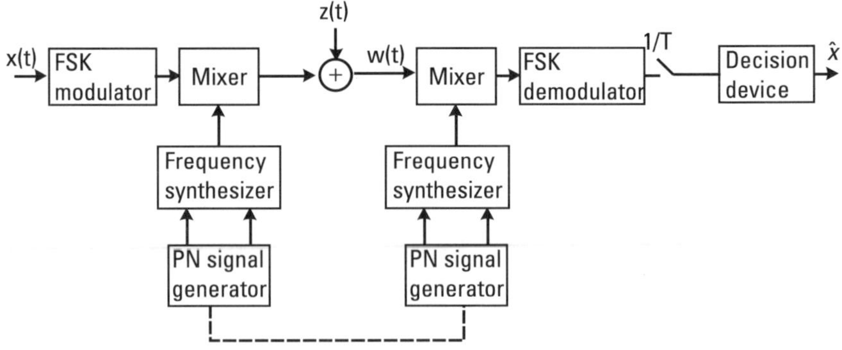

Figure 3.29 Simplified FH system operating on an AWGN channel. (After: [24].)

$$u(t) = A \sum_n e^{x_n 2\pi f_\Delta t} u_T(t - nT) \tag{3.72}$$

where

$$x_n \in \{\pm 1, \pm 3, \ldots, \pm M - 1\}$$

Usually, the frequency separation $f_\Delta = 1/2T$ is chosen so that the waveforms:

$$u_i(t) = A e^{x_n 2\pi f_\Delta t}, \qquad 0 \leq t \leq T \tag{3.73}$$

are orthogonal.

Using a PN sequence to select a set of carrier frequency shifts generates a FH/MFSK signal. There are two basic types of FH spread spectrum:

1. Fast frequency hop (FFH);
2. Slow frequency hop (SFH).

With SFH one or more (in general L) source symbols are transmitted per hop. The complex envelope in this case can be written as:

$$u(t) = A \sum_n \sum_i e^{x_{n,i} 2\pi f_\Delta t + 2\pi f_n t} u_T(t - nT) \tag{3.74}$$

where

$$f_n = \text{the } n\text{th hop frequency;}$$

$$x_{n,i} = \text{the } i\text{th source symbol that is transmitted on the } n\text{th hop.}$$

FFH systems, on the other hand, transmit the same source symbol on multiple hops. In this case, the complex envelope is:

$$u(t) = A \sum_n \sum_i e^{x_n 2\pi f_\Delta t + 2\pi f_{n,i} t} u_T(t - nT) \tag{3.75}$$

where $f_{n,i}$ is the ith hop frequency for the nth source symbol.

Detection of FH/MFSK is usually performed noncoherently using a square-law detector. With SFH, the error probability on an AWGN channel is given by:

$$P_b = \frac{1}{2}\, e^{-\gamma/2} \tag{3.76}$$

3.7.3.2 DS Spread Spectrum

A simplified quadrature DS/QPSK spread spectrum system is shown in Figure 3.30(a).

The PN sequence generator produces the spreading waveform, given by:

$$\alpha(t) = \sum_k \alpha_k h_a(t - kT_c) \tag{3.77}$$

where

$\alpha = \{\alpha_k : \alpha_k \in (\pm 1, \pm j)\}$ is the complex spreading sequence;

T_c = the PN symbol or chip duration;

h_α = a real chip amplitude shaping function.

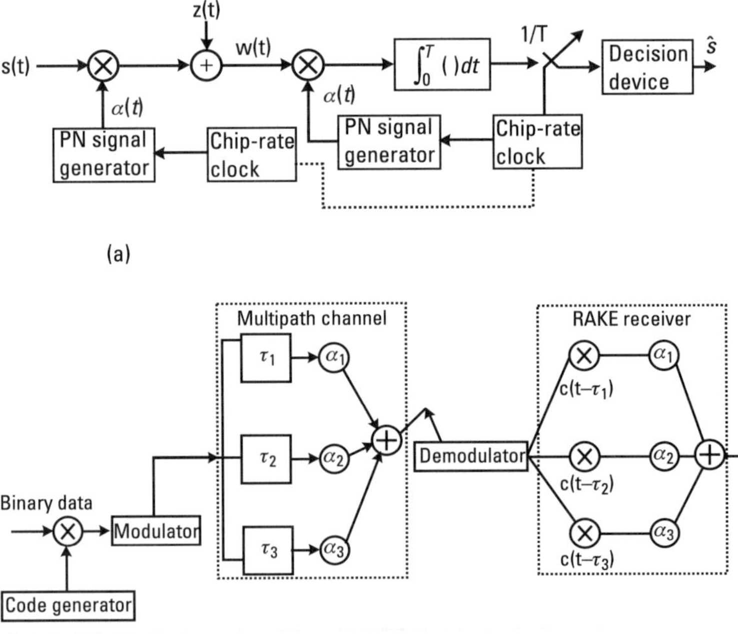

Figure 3.30 (a) Simplified quadrature DS system operating on an AWGN channel. (After: [24].) (b) RAKE receiver. (After: [25].)

The energy per chip is:

$$E_c = \frac{1}{2} \int_0^{T_c} h_\alpha^2(t) \, dt \tag{3.78}$$

The data sequence can be represented by the waveform:

$$x(t) = A \sum_n x_n u_T(t - nT) \tag{3.79}$$

where

$x = \{x_n : x_n \in (\pm 1, \pm j)\}$ is the complex spreading sequence;

$A =$ the amplitude;

$T =$ symbol duration.

It is necessary that T be an integer multiple of T_c, and the ratio $G = T/T_c$ is called the processing gain and is defined as the ratio of spread-to-unspread bandwidth. The complex envelope is obtained by multiplying $\alpha(t)$ and $x(t)$. Then we have that:

$$u(t) = A \sum_n \sum_{k=1}^{G} x_n \alpha_{nG+k} h_\alpha[t - (nG + k) T_c] \tag{3.80}$$

This waveform is applied to a quadrature modulator to produce the bandpass waveform:

$$s(t) = \sum_n \sum_{k=1}^{G} \{ x_n^R \alpha_{nG+k}^R h_\alpha[t - (nG + kT_c) \cos(2\pi f_c t)] \tag{3.81}$$

$$- x_n^I \alpha_{nG+k}^I h_\alpha[t - (nG + kT_c) \sin(2\pi f_c t)] \}$$

where

$$\alpha_k = a_k^R + j\alpha_k^I \tag{3.82}$$

$$x_n = x_n^R + jx_n^I \tag{3.83}$$

The complex envelope $u(t)$ appears like that for ordinary QPSK, except that the signaling rate is G times faster. If the sequences α and x above are completely random, then the power spectral density of $u(t)$ can be obtained directly as:

$$PSD_u(f) = \frac{A^2}{T_c} |H_\alpha(f)|^2 \tag{3.84}$$

In general, the DS spread spectrum receiver must perform three functions: synchronize with the incoming spreading sequence, dispread the signal, and detect the data. Multiplying the received complex envelope $m(t) = u(t) + z(t)$ by $\alpha(t)$, integrating over the nth data symbol interval, and sampling, yields the decision variable:

$$\mu = x_n \int_0^T \sum_{k=1}^G h_a^2 [t - (nG + k)T_c]\, dt + \int_0^T z(t) \sum_{k=1}^G h_a^2 [t - (nG + k)T_c]\, dt$$

$$= 2GE_c x_c + z_n \tag{3.85}$$

$$= 2Ex_n + z_n$$

where

$$E = GE_c$$

$z(t)$ = zero-mean Gaussian random variable with variance $\frac{1}{2} E\left[|z_n|^2\right] = 2N_0 E$.

Because $x_n \in \{\pm 1, \pm j\}$, it follows that the probability of decision error is exactly the same as QPSK on an AWGN channel, which is given by:

$$P_b = Q\left(\sqrt{2\gamma}\right) \tag{3.86}$$

where $\gamma = E_b/N_0$ is the received bit energy-to-noise ratio. The use of spread spectrum signaling does not improve the bit error performance on an AWGN channel. However, spread spectrum signaling will be shown to offer significant performance gains against interference, multipath fading, and other types of channel impairments. Actually a DS spread spectrum is an ideal interference- and multipath-mitigating device, asexplained in Section 3.9.3.1.

In a multipath environment, the receiver receives several copies of the original signal with different delays, and thus each signal can be considered as an interferer to all others. This effect can be eliminated by the processing gain of the system itself in a multiuser/multiple access system, in which we achieve the equivalent of a multipath diversity. This is achieved by a receiver that looks like a RAKE (it has a finger for each multipath component). It is called a RAKE receiver. It is shown in Figure 3.30(b) for the case of three multipath components.

We observe that maximum ratio combining is used for detection by multiplying each signal with the conjugated path gain. It is necessary, however, for delays and attenuations to be reevaluated because the multipath environment changes. The RAKE finger must then be readjusted.

3.7.3.3 Performance of DS and FH Spread Spectrum

Both DS and FH spread spectrum have been proposed for cellular radio application, and one has a number of advantages and disadvantages with respect to the other. For frequency selective fading channels, DS spread spectrum can obtain diversity by exploiting the correlation properties of the spreading sequences to resolve and combine the signal replicas that are received over multiple independently faded paths. Sometimes this is called multipath diversity or spread spectrum diversity. In practice, multipath diversity is obtained by using a RAKE receiver. Also, during the dispreading operation, unwanted narrowband interference is spread throughout the spread spectrum bandwidth, which will reduce its effect on the desired signal.

For frequency-selective fading channels, FFH can obtain frequency diversity provided that the channel coherence bandwidth is much greater than the instantaneous bandwidth of the FH signal. Under this condition, FFH transmits the same data bit on multiple, independently faded hops. FFH can also reduce the effect of multiple access interference because multiple hops have to be hit to destroy a data bit. The actions of hopping from one carrier frequency to the next places a limit on the amount of interference that a narrowband signal can inflict on the spread spectrum signal. That is, frequency hopping rejects narrowband interference by avoidance.

The advantages of the DS spread spectrum can be disadvantages for the FH spread spectrum and vice versa. Regarding radio-location, detection, processing gain, and electromagnetic compatibility, the DS spread spectrum systems have an advantage on performance and respectively in power control, in multiple access interference, and in coding gain and flexibility the FH spread spectrum systems surpass DS in performance.

3.8 BERs and Bandwidth Efficiency

The various binary modulation/demodulation types described here can be compared by analyzing their BERs or bit error probabilities—the probabilities with which errors occur in detecting the bits. As in the analysis of analog modulation/demodulation, this comparison is commonly done by assuming that the channel is corrupted by AWGN. In this case and under the further assumption that the symbols 0 and 1 are equally likely to occur in the message, expressions for the bit error probabilities of the various schemes described here are shown in Table 3.2. These results are given as functions of the SNR parameter, E_b/N_0, where E_b is the signal energy received per bit and N_0 is the spectral density of the AWGN. In some cases, the expressions involve the function Q, which denotes the tail probability of a standard normal probability distribution:

$$Q(x) \equiv \frac{1}{\sqrt{2\pi}} \int_x^\infty e^{-y^2/2} \, dy \qquad (3.87)$$

In Chapters 5 and 6, we will point out how these expressions are derived. The expressions for OOK assume that the decision threshold in the demodulators have been optimized. This optimization requires knowledge

Table 3.2
Bit Error and Bandwidth Efficiencies for Digital Modulation/Demodulation Techniques

Modulator/Demodulator	Bit Error Probabilities	Bandwidth Efficiency (bps/hertz)
BPSK	$Q\left(\sqrt{2E_b/N_0}\right)$	1/2
QPSK	$Q\left(\sqrt{2E_b/N_0}\right)$	1
Optimum DPSK	$\frac{1}{2}e^{-E_b/N_0}$	1/2
Coherent OOK	$Q\left(\sqrt{E_b/N_0}\right)$	1/2
Coherent FSK	$Q\left(\sqrt{E_b/N_0}\right)$	1/3
Noncoherent OOK	$\frac{1}{2}e^{-E_b/2N_0}$	1/2
Noncoherent BPSK	$\frac{1}{2}e^{-E_b/2N_0}$	1/3

of the received SNR, which makes OOK the only one of these techniques that requires this information for demodulation. The expression for noncoherent OOK is an approximation that is valid for large SNRs. The result for DPSK corresponds to the optimum demodulator depicted in Figure 3.21(b). The suboptimum DPSK demodulator of Figure 3.21(a) requires approximately 2 dB higher values of E_b/N_0 in order to achieve the same performance as the optimum demodulator. The quantities of Table 3.2 are plotted in Figure 3.31.

From this figure, we see that BPSK is the best performing of these schemes, followed, in order, by DPSK, coherent OOK and FSK, and noncoherent OOK and FSK. The superiority of BPSK is due to its use of antipodal signals, which can be shown to be an optimum choice in this respect for signaling through an AWGN channel. DPSK exhibits a small loss relative to BPSK, which is compensated for by its simpler demodulation. OOK and FSK are both examples of *orthogonal* signaling schemes (i.e., schemes in which $\int x^{(0)}(t) \cdot x^{(1)}(t)\, dt = 0$, where the integration is performed over a single bit interval). This explains why they exhibit the same performance. Orthogonal signaling is less efficient than antipodal signaling, which is evident from Figure 3.31. Finally, note that there is a small loss in performance

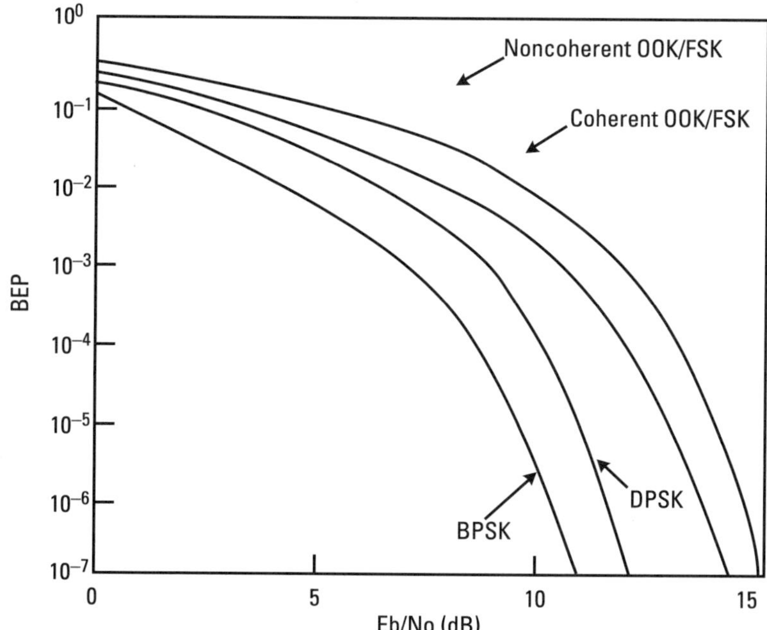

Figure 3.31 Bit error probabilities for digital communication systems. (After: [13].)

for these two orthogonal signaling schemes when they are demodulated noncoherently. Generally speaking, DPSK and noncoherent FSK are seen from this comparison to be quite effective means of acceptable performance. At the same time, they have simple demodulation. They also compare closely to the best antipodal and orthogonal signaling, respectively.

In addition to BER, digital communication systems can also be compared in terms of bandwidth efficiency, which is often quantified in terms of the number of bits per second that can be transmitted per hertz of bandwidth. Bandwidth efficiencies for the various signaling schemes are also shown in Table 3.2.

3.9 Access Techniques

The radio channel is fundamentally a broadcast communication medium. Therefore, signals transmitted by one user can potentially be received by all other users within range of the transmitter. Although this high connectivity is very useful in some applications, like broadcast radio or television, it requires stringent access control in wireless communication systems to avoid, or at least to limit, interference between transmissions. Throughout this book, the term *wireless communication systems* is taken to mean communication systems that facilitate two-way communication between a fixed or portable radio communication terminal and the fixed network infrastructure. Such systems range from satellite, mobile cellular systems through personal communication systems (PCS), to cordless telephones, as we saw in Chapter 1.

Design criteria for such systems include capacity, cost of implementation, and quality of service [26–28]. All of these measures are influenced by the method used for providing multiple-access capabilities. However, the opposite is also true: the access method should be chosen carefully in light of the relative importance of design criteria as well as the system characteristics. Multiple access in wireless radio systems is based on insulating signals used in different connections from each other. The support of parallel transmissions on the uplink and downlink, respectively, is called multiple access, whereas the exchange of information in both directions of a connection is referred to as duplexing. Hence, multiple access and duplexing are methods that facilitate the sharing of the broadcast communication medium. The necessary insulation is achieved by assigning to each transmission different components of the domains (space, frequency, time, code) that contain the signals.

1. *Spatial domain.* Using directional antennas mainly in mobile systems, we are allowed to reuse signals and maintain the required isolation between them.

2. *Frequency domain.* Signals, which occupy nonoverlapping frequency bands, can be easily separated using appropriate bandpass filters. Hence, signals can be transmitted simultaneously without interfering with each other. This method of providing multiple access capabilities is called FDMA.

3. *Time domain.* Signals can be transmitted in nonoverlapping time slots in a round-robin fashion. Thus, signals occupy the same frequency band but are easily separated based on their time of arrival. This multiple access method is called TDMA.

4. *Code domain.* In CDMA, different users employ signals that are coded by codes of little correlation. The same technique is used to extract individual signals from a mixture of signals, even though they are transmitted simultaneously and in the same frequency band. The term *code division multiple access* is used to denote this form of channel sharing. Two forms of CDMA introduced earlier, FH and DS, are most widely employed and will be further described in detail subsequently.

System designers have to decide in favor of one, or a combination, of the latter three domains to facilitate multiple access. The three access methods are illustrated in Figure 3.32. The principal idea in all three of these access methods is to employ signals that are orthogonal or nearly orthogonal to provide the necessary separation—having always the interference effects in mind.

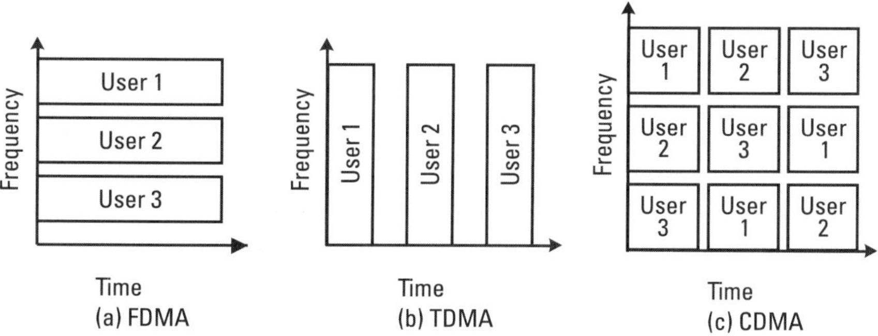

Figure 3.32 Multiple-access methods for wireless communication systems.

3.9.1 FDMA

As mentioned, in FDMA (Figure 3.32), nonoverlapping frequency bands are allocated to different users on a continuous-time basis. Hence, signals assigned to different users are clearly orthogonal, at least ideally. In practice, out-of-band spectral components cannot be completely suppressed, leaving signals not quite orthogonal. Another parameter that is important to the system designer is the type of modulation to be used. This is the reason we placed some emphasis on the spectrum characteristics of the various modulations techniques in Sections 3.6 to 3.8. This concept will come up again in Chapters 5 and 6. This necessitates the introduction of guard bands between frequency bands to reduce adjacent channel interference.

It is advantageous to combine FDMA with TDD to avoid simultaneous reception and transmission that would require insulation between receives and transmits antennas. In this scenario, the base station and portable take turns using the same frequency band for transmission. Nevertheless, combining FDMA and FDD is possible in principle, as is evident from the analog FM-based systems deployed throughout the world since the early 1980s. We must point out that the methods of interference suppression that have been developed have their origin in this type of classical technique.

3.9.2 TDMA

In TDMA systems, the receiver filters are simply time windows instead of the bandpass filters required in FDMA (see Figure 3.32(b)). As a consequence, the guard time between transmissions can be made as small as the synchronization of the network permits. Guard times of 30–50 μs between time slots are commonly used in TDMA-based systems. As a consequence, all users must be synchronized with the base station to within a fraction of the guard time. This is achievable by distributing a master clock signal on one of the base station's broadcast channels.

TDMA can be combined with TDD or FDD. The former duplexing scheme is used, for example, in the DECT standard and is well suited for systems in which base-to-base and mobile-to-base propagation paths are similar (i.e., systems without extremely high base station antennas). In the cellular application, the high base station antennas make FDD the more appropriate choice. In these systems, separate frequency bands are provided for uplink and downlink communication. Note that it is still possible and advisable to stagger the uplink and downlink transmission intervals such that they do not overlap, to avoid the situation in which the portable must transmit and receive at the same time. With FDD, the uplink and downlink channel are not identical; hence, signal-processing functions cannot be implemented in the base-station as it is done in the FDMA/TDD case for downlink-uplink separation. Other techniques such as antenna diversity and equalization have to be realized in the portable, as we shall see in the chapters to follow.

3.9.3 CDMA

A third technique for dividing the radio spectrum into channels is code division. Used as a multiple access technique almost always related to spread spectrum systems, the physical channels are created by encoding different users with different user signature sequences or simply different codes.

CDMA systems employ wideband signals with good cross-correlation properties. A large body of work exists on spreading sequences that lead to signal sets with small cross correlations [21, 26–28]. Because of their noise-like appearance, such sequences are often referred to as PN sequences, and because of their wideband nature, CDMA systems are often called spread-spectrum systems. Spectrum spreading can be achieved in two main ways: through frequency hopping, which is accomplished by using a digital fre-quency synthesizer that is driven by a PN sequence generator, or through direct sequence spreading. In direct-sequence spread spectrum, a high-rate, antipodal pseudorandom spreading sequence modulates the transmitted sig-nal such that the bandwidth of the resulting signal is roughly equal to the rate of the spreading sequence. The cross correlation of the signals is then largely determined by the cross-correlation properties of the spreading signals. Clearly, CDMA signals overlap in both time and frequency domains, but they are separable based on their spreading waveforms.

Capacity considerations do not say much about the spreading codes, except that they should have low cross correlations. Essentially they should look like Gaussian noise to all but the intended receiver. They should also have low, ideally zero, autocorrelation between nonadjacent bits of the sequence. Other system considerations, however, dictate many additional properties of the codes. For the case of mobile communications, they have many advantages, such as:

1. Timing in the subscriber stations (mobiles) is to be established, at least in part, by synchronizing with the code transmitted by the base stations. The goal is to eliminate any need for accurate time-keeping in the mobiles when they are idle.

2. The mobiles identify base stations, at least in part, by correlating with a priori known base station spreading codes.

3. The process of synchronization in the mobiles should be rapid enough that the placement of a call from a "cold start" takes no more than a few seconds.

4. Access to base stations by mobiles should not require any prearrangement. That is, it should not be necessary for the base station to have a database of authorized users in order to establish radio communications. The base station, once physical layer access has been achieved, may choose to deny service for administrative reasons, such as nonpayment of the bill, but communication through the air interface should always be possible to cover emergency access.

5. In the CDMA forward link, the fact that each base station is transmitting multiple channels Walsh coding can be used beneficially to decrease mutual interference. We shall see more of this in later chapters.

6. The acquisition search rate for reverse CDMA channel signals in the base stations can be speeded up if the mobiles can precorrect their timing so that their signal arrives at the base station as close to system time as possible.

An immediate consequence of this observation is that CDMA systems do not require tight synchronization between users, as do TDMA systems. By the same token, frequency planning and management are not required, as frequencies are reused throughout the coverage area. While it appears that we have many parameters available and free to change, we must not forget that in this book our main objective is the behavior of the system to be designed in the context of the channel characteristics studied in Chapter 2 —that any wireless system can be suitably optimized to yield a competitive spectral efficiency regardless of the multiple access technique being used. CDMA offers a number of advantages along with some disadvantages.

The advantages of CDMA for cellular applications include:

- Universal one-cell frequency reuse;
- Narrowband interference rejection;
- Inherent multipath diversity in DS CDMA;
- Ability to exploit silent periods in speech voice activity;
- Soft handover capability;

- Soft capacity limit;
- Inherent message privacy.

The disadvantages of CDMA include:

- Stringent power control requirements with DS CDMA;
- Handoffs in dual-mode systems;
- Difficulties in determining the base station power levels for deployments that have cells of differing sizes;
- Pilot timing.

3.9.3.1 Principles of CDMA

The goal of spread spectrum is a substantial increase in bandwidth of an information-bearing signal, far beyond that needed for basic communication. The bandwidth increase, while not necessary for communication, can mitigate the harmful effects of interference, either deliberate, like a military jammer, or inadvertent, like cochannel users. The interference mitigation is a well-known property of all spread spectrum systems. However, the cooperative use of these techniques to optimize spectral efficiency in a commercial, nonmilitary environment was a major conceptual advance. Figures 3.33 and 3.34 present the interference-mitigating properties of CDMA systems graphically.

The noise and interference, being uncorrelated with the PN sequence, become noise-like and increase in bandwidth when they reach the detector. Narrowband filtering that rejects most of the interference power can enhance the SNR. We define by processing gain W/R, the value by which the SNR

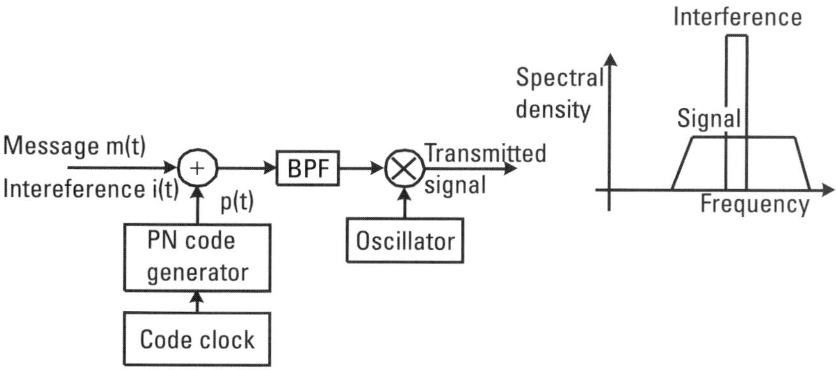

Figure 3.33 Spread spectrum of modulator.

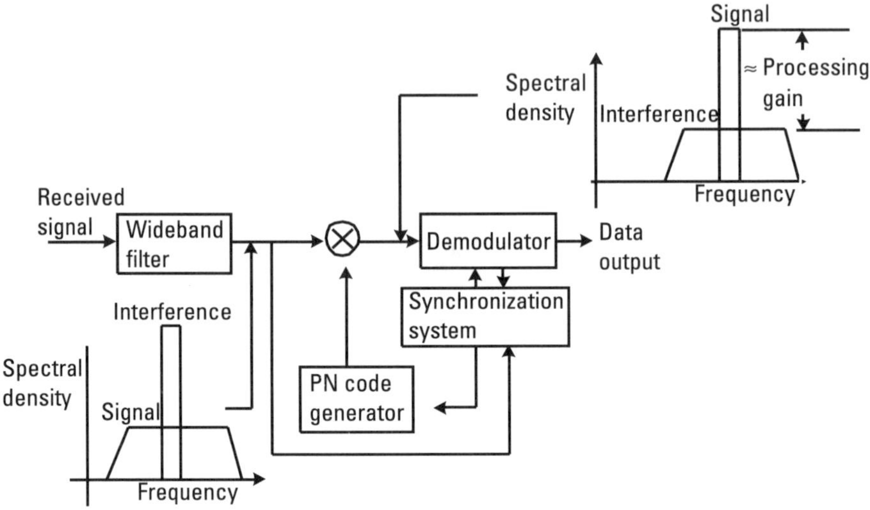

Figure 3.34 Spread spectrum of demodulator. (After: [23].)

is enhanced through this process, where W is the spread bandwidth and R is the data rate. A careful analysis is needed, however, to accurately determine the performance. In IS-95A CDMA, $W/R = 10 \log(1.2288 \text{ MHz}/9,600 \text{ Hz}) = 21$ dB for the 9,600 bps rate.

3.9.3.2 Forward CDMA Spread Channel

If all base stations transmit a common, universal code, then the mobiles need no prior knowledge of where they are in order to know what to search for—they always search for the same code. Second, search time is roughly proportional to the number of timing hypotheses that must be tested [28].

Does a common, universal code work? The answer is yes. Don't the stations interfere with each other so that they cannot be distinguished from one another? The answer is no, and for the same reason that communication works in this environment.

The advantages of linear feedback shift registers (LFSRs) are:

1. LFSR sequences are easily generated by very simple binary logic circuits.

2. Very-high-speed generators are possible because of the simple logic.

3. Maximal-length sequence generators are easily designed using finite (Galois) field mathematics.

4. The full period autocorrelation functions of maximal-length LFSR sequences are binary valued, facilitating synchronization searching.

It is easily shown that any linear feedback binary state machine that generates a maximum-length output sequence must be equivalent to some maximal length LFSR [21, 23].

If the tap weights are identical and configured as shown in the figures, then the two implementations will produce exactly the same sequence (this can be verified by simple arguments). Initial conditions required to produce the same phase of the sequence are obviously not identical, however. There are actually two sequences produced by each of these generators. One is the trivial one, of length one, that occurs in both cases when the initial state of the generator is all zeros. The other, the useful one, has length $2^m - 1$. Together these two sequences account for all 2^m states of the m-bit state register.

In the case where each base station radiates a family of 64 orthogonal cover code channels, thus each base station must serve in the neighborhood of 40 mobiles, there must be some way of creating independent communication channels. Moreover, because these channels all come from the same site, they can share precise timing and *must* somehow share the common short code spreading.

This is easily accomplished because the number of spreading chips per code symbol is fairly large. Suppose, for example, that the FEC code rate is r. Code rates from perhaps $r = 1/3$ to $r = 3/4$ are good design choices in most terrestrial communication systems. Toll-quality vocoders now exist that can operate at data rates from $R = 8$ to $R = 16$ Kbps. Then the symbol rate from the FEC encoder, R/r, assuming a binary alphabet, ranges from about 10 Kbps to 50 Kbps. With the 1.2288-MHz chip rate, there are about 25 to 125 chips per code symbol. This suggests an orthogonal cover technique that can be applied to each symbol. The orthogonal cover technique is based on the so-called Hadamard-Walsh sequences. These are binary sequences, powers-of-two long, that have the property that the *dot product* of any two of them is zero. The Walsh sequences of order 8, for example, are:

$$H_8 = \begin{bmatrix} + & + & + & + & + & + & + & + \\ + & - & + & - & + & - & + & - \\ + & + & - & - & + & + & - & - \\ + & - & - & + & + & - & - & + \\ + & + & + & + & - & - & - & - \\ + & - & + & - & - & + & - & + \\ + & + & - & - & - & - & + & + \\ + & - & - & + & - & + & + & - \end{bmatrix}$$

If we represent each + as a positive amplitude, and each − by a negative amplitude, then take the dot product of any two rows as the sum of the products of the amplitudes in corresponding columns. That dot product is zero for any two distinct rows. Walsh functions of order 64 are used in the forward CDMA channel to create 64 orthogonal channels. There is exactly one period of the Walsh sequence per code symbol: 64 * 19.2 Kbps = 1.2288 Mbps. These channels are readily generated by the binary logic shown in Figure 3.35. The "impulse modulators" generate a discrete ± 1 outputs in response to binary (0, 1) inputs.

Summing the code symbols, the Walsh cover, and the two short code sequences as shown here, and changing to the bipolar ± 1 representation, result in a quadrature (I, Q) sequence of elements from the set $(\pm 1, \pm j)$. These elements drive a modulator that generates the appropriately bandlimited analog output.

One of the Walsh codes, numbered zero by tradition, has all 64 symbols the same. It is the universal pilot sequence that all mobile use as their search target. Those searches are done for several purposes:

1. Initiation of handoff;

2. Initial acquisition of an appropriate serving station;

3. RAKE finger assignment.

The common, universal pilot code facilitates the implementation of all these processes.

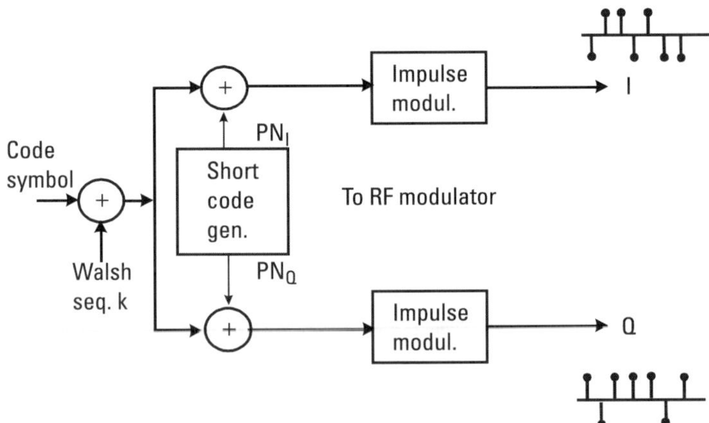

Figure 3.35 Forward spreading logic. (After: [28].)

3.9.3.3   Reverse CDMA Spread Channel

Two different criteria apply to the reverse link spreading, as shown in Figure 3.36. When a mobile is engaged in user traffic (i.e., in a conversation), it is desirable that that mobile use a unique code that is distinct from all others. A mobile-unique code, rather than a base station–associated code, facilitates handoff. With a mobile-unique code, nothing needs to change about the mobile's modulation or coding when handoff occurs [21, 28].

The second situation occurs when a mobile is attempting to gain the attention of a base station. Initially the base station has no knowledge that any particular mobile is in its service area. It is wildly impractical for each base station to search simultaneously for millions of potential subscriber codes. For these initial accesses, or any other nontraffic uses of the air interface, it is desirable to have some reverse spreading codes that *are* base station associated. If there are only a few associated codes for each base station, then it *is* practical for the base station to search for them continuously and simultaneously, awaiting the arrival of any user who wants service.

The mobile applies its unique logical connection manager (LCM) to the long-code generator, and modulo-2 adds the output (i.e., the unique-phase long code) to the universal short code. As in the forward CDMA channel, the spreading modulation is quadrature, so as to homogenize the phase of the interference. Again, both short code sequences are used.

3.9.3.4   Comparison of FDMA, TDMA, CDMA, FDD, and TDD

We have seen that access techniques play a major role in both capacity and performance of wireless systems. A comparison of these methods will be briefly discussed on the basis of their behavior in an interference environment [26].

From the viewpoint of system configuration, FDMA is the simplest access scheme of the three. However, it is not suitable for achieving high-

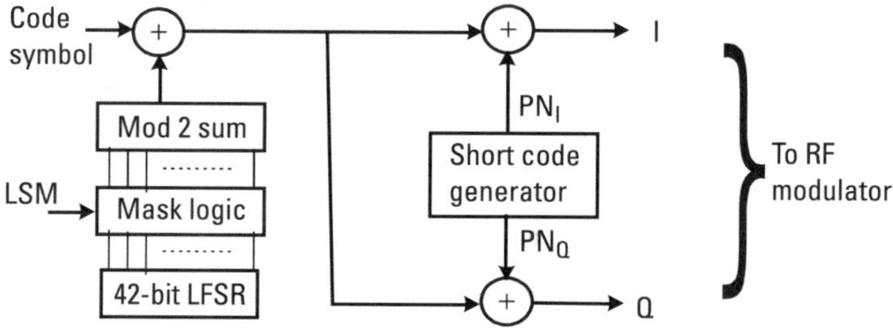

Figure 3.36 Reverse CDMA channel spreading logic. (After: [28].)

capacity voice transmission systems using low-bit-rate codec and spectral-efficient modulation schemes because it requires very high stability of the oscillator. Moreover, variable transmission rate control is very difficult in the case of FDMA because it requires K-set of modems to achieve variable transmission rate control from R_b bps to K R_b bps. As a result, no second-generation cellular system applies the FDMA scheme at present. Furthermore, it is very difficult for FDMA systems to monitor the received signal level of the adjacent cells for channel reassignment or handover processes.

When we apply TDMA, although we can mitigate the requirement for carrier frequency stability and achieve variable transmission rate control using a modem, we need a highly accurate slot, frame, or superframe synchronization. Moreover, we have to develop antifrequency-selective fading techniques if the number of slots in each frame (N_{ch}) is large. Furthermore, the transmitter amplifier should be operated at K times higher peak power than the average power. Fortunately, we can solve these problems at present thanks to extensive developments in timing-control techniques, adaptive equalizing techniques, and high-power-efficient power amplifier techniques. Another important advantage of TDMA systems is that we can measure the received signal level of adjacent cells during idle time slots. Such received signal level measurement is very effective for the handover process as well as for developing dynamic channel assignments using the carrier to interference (C/I) ratio.

FDMA Versus TDMA

In comparison to an FDMA system supporting the same user data rate, the transmitted data rate in a TDMA system is larger by a factor equal to the number of users sharing the frequency band. This factor is eight in the pan-European global system for mobile communicatons (GSM) and three in the advanced mobile phone service (D-AMPS) system. Thus, the symbol duration is reduced by the same factor and severe intersymbol interference results, at least in the cellular environment.

To illustrate, consider the earlier example, where each user transmits 25K symbols per second. Assume eight users per frequency band leads to a symbol duration of 5 μs. Even in the cordless application with delay spreads of up to 1 μs, an equalizer may be useful to combat the resulting interference between adjacent symbols. In cellular systems, however, the delay spread of up to 20 μs introduces severe intersymbol interference spanning up to five symbol periods. As the delay spread often exceeds the symbol duration, the channel can be classified as frequency selective, emphasizing the observation that the channel affects different spectral components differently.

The intersymbol interference in cellular TDMA systems can be so severe that linear equalizers are insufficient to overcome its negative effects. Instead, more powerful, nonlinear decision feedback or maximum-likelihood sequence estimation equalizers must be employed. Furthermore, all of these equalizers require some information about the channel impulse response that must be estimated from the received signal by means of an embedded training sequence. Clearly, the training sequence carries no user data and, thus, wastes valuable bandwidth.

In general, receivers for cellular TDMA systems will be fairly complex. On the positive side of the argument, however, the frequency selective nature of the channel provides some built-in diversity that makes transmission more robust to channel fading. The diversity stems from the fact that because the multipath components of the received signal can be resolved at a resolution roughly equal to the symbol duration and the different multipath, the equalizer can combine components during the demodulation of the signal. To further improve robustness to channel fading, coding, and interleaving, slow frequency hopping and antenna diversity can be employed, as discussed in connection with FDMA.

As far as channel assignment in both FDMA and TDMA systems, channels should not be assigned to a mobile on a permanent basis. A fixed assignment strategy would either be extremely wasteful of precious bandwidth or highly susceptible to cochannel interference. Instead, channels must be assigned on demand. Clearly, this implies the existence of a separate uplink channel on which mobiles can notify the base station of their need for a traffic channel. This uplink channel is referred to as the random-access channel because of the type of strategy used to regulate access to it.

The successful procedure for establishing a call that originates from the mobile station is outlined in Figure 3.37. The mobile initiates the procedure by transmitting a request on the random-access channel. Because this channel is shared by all users in range of the base station, a random access protocol, like the ALOHA protocol, has to be employed to resolve possible collisions. Once the base station has received the mobile's request, it responds with an immediate assignment message that directs the mobile to tune to a dedicated control channel for the ensuing call setup. Upon completion of the call setup negotiation, a traffic channel (i.e., a frequency in FDMA systems or a time slot in TDMA systems) is assigned by the base station, and all future communication takes place on that channel. In the case of a mobile-terminating call request, the sequence of events is preceded by a paging message alerting the base station of the call request.

In cellular systems, such as GSM or the North-American D-AMPS, TDMA is combined with FDMA. Different frequencies are used in neigh-

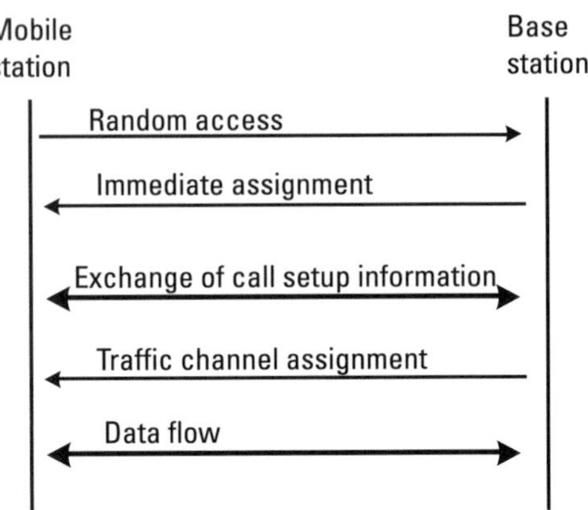

Figure 3.37 Mobile-originating call establishment. (After: [28].)

boring cells to provide orthogonal signaling without the need for tight synchronization of base stations. Furthermore, channel assignment can then be performed in each cell individually. Within a cell, users in the time domain share one or more frequencies.

From an implementation standpoint, TDMA systems have the advantage that common radio and users communicating on the same frequency can share signal-processing equipment at the base station. A somewhat more subtle advantage of TDMA systems arises from the possibility of monitoring surrounding base stations and frequencies for signal quality to support mobile-assisted handovers.

CDMA Versus FDMA and TDMA

In case of CDMA, the most serious problem is the near-far problem, as we have discussed before. The near-far problem is now solved by fast power control techniques. In addition to mitigating the near-far problem, the fast power control technique is also effective for improving receiver sensitivity because it makes the received signal level constant. Moreover, the following CDMA-specific techniques can further improve receiver sensitivity.

- Low-coding-rate FEC is applicable.
- Peak power is the same as the average power.
- Soft and softer handover is applicable.

Therefore, CDMA has the potential to achieve lower power consumption than TDMA or FDMA, provided that very accurate power control is applicable.

Another advantage for CDMA is that we can easily compensate for frequency-selective fading by using the path-diversity technique. Furthermore, we can easily monitor the received signal of adjacent cells just by changing a reference code at the correlator for the channel delay profile monitor. This is because all the base stations use the same carrier frequency and chip rate. This feature is actively applied to the soft handover process.

On the other hand, in the case of CDMA, smaller zone radius is preferable for power control because larger zones require a wider dynamic range of the power control. Even in the case of TDMA, a larger zone radius requires longer guard time or accurate time alignment if smaller guard time is necessary. On the other hand, zone radius is limited only by the requirement of the transmitter power, in the case of FDMA. Table 3.3 summarizes the results of the comparison of FDMA, TDMA, and CDMA systems.

3.9.4 FDD

FDD is the most popular duplex scheme for two-way radio communication systems because it can easily discriminate between uplink and downlink signals by filters. Actually, most of the land mobile communicaton systems other than the DECT and PHS employ FDD.

Figure 3.38 shows an example of spectrum allocation and the modem configuration of FDD systems. In the FDD systems, a different frequency band with its bandwidth of W_{sys} is employed for uplink and downlink. Moreover, transmission and reception are carried out through the same antenna. Therefore, a duplexer that discriminates the spectrum for uplink and downlink is inserted in both the base station and the terminal. In this case, the carrier frequency spacing should be sufficiently large from the hardware implementation point of view because shorter carrier spacing requires higher Q-value for the duplex filter. In PDC systems, 130 MHz is used for an 800- to 900-MHz band and 48 MHz is used for a 1.5-GHz band.

3.9.5 TDD

TDD is another duplex scheme for two-way radio systems. In this scheme, both the base station and terminal transmit a signal over the same radio frequency channel but at different segments in time. Figure 3.39 shows an example of spectrum allocation and the modem configuration of TDD systems. In TDD systems, the uplink and downlink alternatively use the same spectrum. Because each signal has to transmit data during half a period

Table 3.3
Comparison of the Features of FDMA, TDMA, and CDMA Systems

	FDMA	**TDMA**	**CDMA**
Timing control Carrier frequency stability	Not required High stability is required	Required Low stability is acceptable if large number of channels are multiplexed	Required Low stability is acceptable if chip rate is sufficiently high
Near-far problem	Not affected	Not affected	Fast power control is required
Peak/average power ratio	1	K	1
Variable transmission rate	Difficult	Easy	Easy
Antimultipath fading technique	Diversity, high coding rate FEC	Diversity, high coding rate FEC— adaptive equalizer (if N_{ch} is large)	RAKE diversity, low coding rate FEC, fast power control
Received signal level monitoring	Difficult	Easy	Easy
Suitable zone radius	Any size is OK	Any size is OK (time alignment required)	Large size is not suitable

(From: [26].)

for FDD systems, the occupied bandwidth for each link is twice as wide as that for FDD systems, although the total bandwidth for FDD and TDD are the same bandwidth.

One of the most important features of the TDD systems is that it does not require a duplexer that occupies a relatively large mass in the FDD modem because uplink and downlink signals are discriminated in the time domain. However, the TDD system requires guard space or time alignment, as in the case of TDMA.

3.9.6 Comparison of FDD and TDD

Table 3.4 shows a comparison of the features for FDD and TDD systems [25]. The most important feature of the FDD system is that it does not

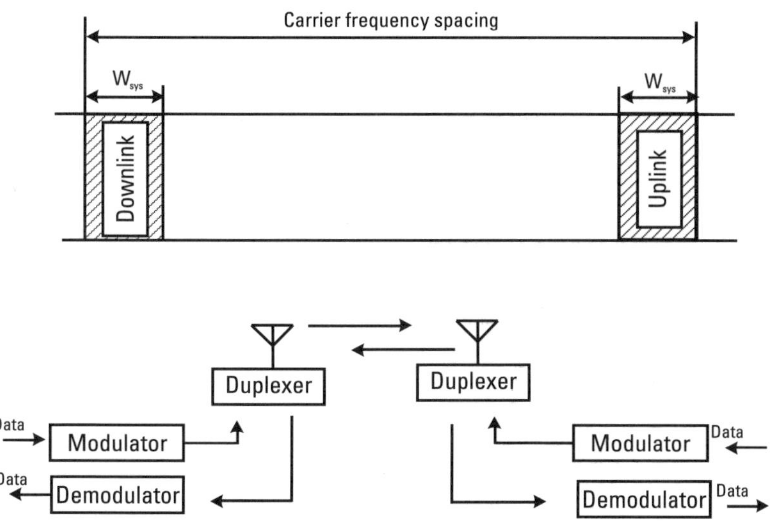

Figure 3.38 An example of spectrum allocation and the modem configuration of FDD systems. (After: [26].)

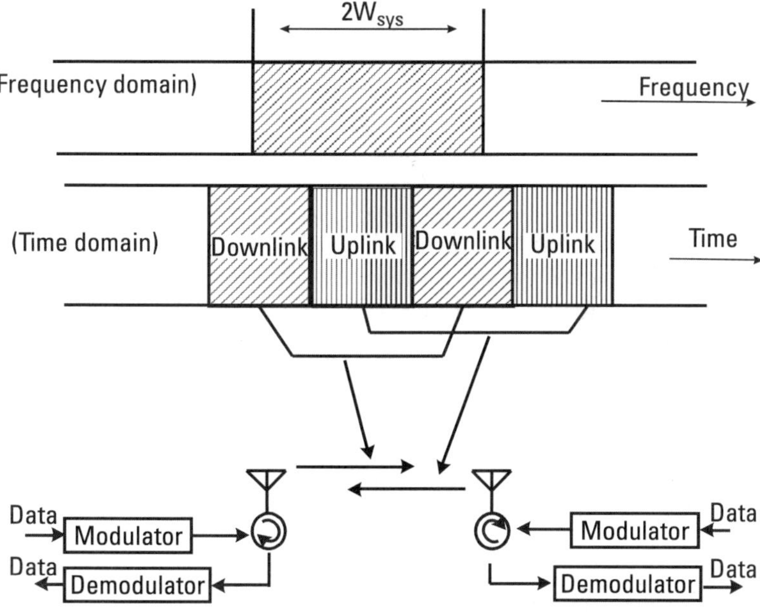

Figure 3.39 An example of spectrum allocation and the modem configuration of TDD systems. (After: [26].)

Table 3.4
Comparison Between FDD and TDD Systems

Items	FDD System	TDD System
Required total bandwidth	Same as TDD	Same as FDD
Symbol rate	R_s	$2R_s$
Duplexer	Necessary	Not necessary
Flexibility of radio resource management	A pair of spectrums required	Flexible
Immunity to multipath fading	More robust	Less robust
Requirement to synchronization	No synchronization required	Uplink and downlink timing synchronization required
Requirement to zone radius	Applicable to either small cell or large cell systems	Preferable to smaller cell systems
Reciprocity between uplink and downlink channels	Not satisfied	Satisfied for the desired signal
Transmission diversity	Impossible	Possible
Direct communication between terminals	Possible	Possible (easy)

(From: [26].)

require any timing synchronization. This advantage is more important if the coverage area for each base station becomes large because a larger zone radius requires a larger dynamic range of the time alignment or a longer guard space. Moreover, FDD is more robust than delay spread because TDD requires twice as much symbol rate as FDD.

On the other hand, TDD does not require an RF duplexer, which occupies a large amount of volume of the modem. Moreover, spectrum management will be more flexible if we employ TDD because we do not have to prepare a pair of spectra, as in the case of FDD. Especially, it is a very important advantage for systems using discontinuous radio spectrum. In Table 3.4, we summarize the comparative features of FDD and TDD systems.

3.9.7 Orthogonal Frequency Division Multiplex

Orthogonal frequency division multiplex (OFDM) is a special case of multi-carrier transmission, where a single data stream is transmitted over a number of lower rate subcarriers. It is worth mentioning here that OFDM can be seen as either a modulation technique or a multiplexing technique. One of the main reasons to use OFDM is to increase the robustness against frequency selective fading or narrowband interference. In a single carrier system, a single fade or interferer can cause the entire link to fail, but in a multicarrier system, only a small percentage of the subcarriers will be affected. Error correction coding can then be used to correct for the few erroneous subcarriers. The concept of using parallel data transmission and frequency division multiplexing is published in [26–28].

3.9.7.1 Generation of OFDM Signals

An OFDM signal consists of a sum of subcarriers that are modulated by usually using PSK or QAM, as shown in Figure 3.40. If d_i is the complex QAM symbol, N_s is the number of subcarriers, T_s is the symbol duration and f_c the carrier frequency, then one OFDM symbol starting at $t = t_0$ can be written as

$$s(t) = \sum_{i=(N_s/2)}^{(N_s/2)-1} d_{i+N_s/2} \exp\left(j2\pi \frac{i}{T_s}(t - t_0)\right), \qquad t_0 \leq t \leq t_0 + T_s \quad (3.88)$$

$$s(t) = 0, \, t \leq t_0 \qquad \text{and} \qquad t > t_0 + T_s$$

As an example, Figure 3.41 shows four subcarriers from one OFDM signal. In this example, all subcarriers have the same phase and amplitude, but in practice the amplitudes and phases may be modulated differently for each subcarrier. Note that each subcarrier has exactly an integer number of cycles in the interval T, and the number of cycles between adjacent subcarriers

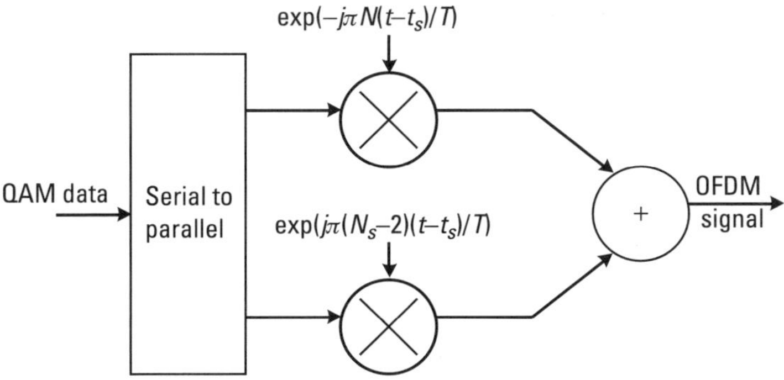

Figure 3.40 OFDM modulator. (After: [25].)

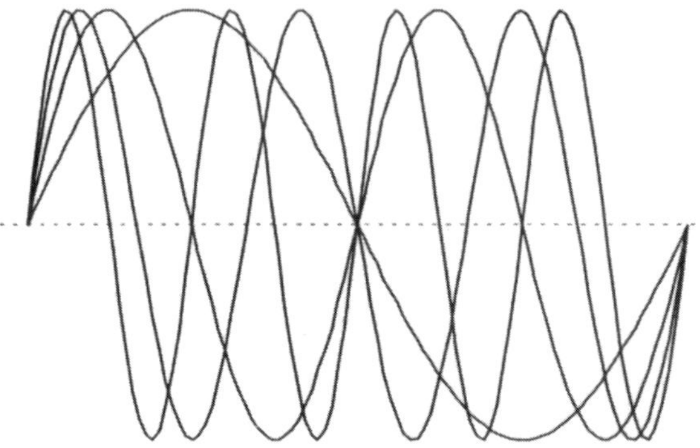

Figure 3.41 Example of four subcarriers within one OFDM symbol. (After: [25].)

differs by exactly one. This property accounts for the orthogonality between the subcarriers. For instance, if the jth subcarrier from (3.88) is demodulated by downconverting the signal with a frequency of j/T and then integrating the signal over T seconds, the result is as written in (3.89). For the demodulated subcarrier j, this integration over T seconds gives the desired output $d_j + N_s/2$ (multiplied by a constant factor T), which is the QAM value for that particular subcarrier. For all other subcarriers, the integration is zero because the frequency difference $(i - j)/T$ produces an integer number of cycles within the integration interval T, such that the integration result is always zero, having thus proved the orthogonality of sucbarriers of OFDM as the name indicates

$$\int_{t_0}^{t_0+T} \exp\left(-j2\pi \frac{j}{T}(t - t_0)\right) \sum_{i=-(N_s/2)}^{(N_s/2)-1} d_{i+N_s/2} \exp\left(j2\pi \frac{i}{T}(t - t_0)\right) dt$$

$$= \sum_{i=-(N_s/2)}^{(N_s/2)-1} d_{i+N_s/2} \int_{t_0}^{t_0+T} \exp\left(j2\pi \frac{i-j}{T}(t - t_0)\right) dt = d_{j+N_s/2} T$$

$$(3.89)$$

The complex baseband OFDM signal as defined by (3.88) is in fact nothing more that the inverse Fourier transform of N_s QAM input symbols. The time-discrete equivalent is the inverse discrete Fourier transform (IDFT), which is given by (3.90), where the time t is replaced by a sample number n. In practice, this transform can be implemented very efficiently by the inverse fast Fourier transform (IFFT). Thus from (3.88)

$$s(n) = \sum_{i=0}^{N_s-1} d_i \exp\left(j2\pi\frac{in}{N}\right) \tag{3.90}$$

One might think that by dividing the input datastream in N_s subcarriers. the symbol is made N_s times smaller, and this reduces the relative multipath delay spread, relative to the symbol time. This is true. A guard time for each OFDM symbol chosen larger than the expected delay spread might eliminate intersymbol interference. This guard time, however, could consist of no signal and the problem of intercarrier interference could arise. Also, when the multipath delay becomes larger than the guard time, the orthogonality is lost, and the summation of the time waves of the first path with the phase-modulated waves of the delayed path no longer gives a set of orthogonal pure time waves. Also, certain level of interference is caused. In general, OFDM has the ability to deal with large delay spread with a reasonable implementation complexity. A fading channel might cause, however, deep fades to the weakest subcarriers, which contributes and dominates BER. In such cases, proper coding might elevate the problem.

In most cases, data bits are modulated on the subcarriers by some form of phase shift keying or QAM. To estimate the bits at the receiver, knowledge is required about the reference phase and amplitude of constellation on each subcarrier.

To cope with these unknown phase and amplitude variations, two different approaches exist. The first one is coherent detection, which uses estimates of the reference amplitudes and phases to determine the best possible decision boundaries for the constellation of each subcarrier. The main issue with coherent detection is how to find the reference values without introducing too much training overhead. The second approach is differential detection, which does not use absolute reference values, but only looks at the phase and/or amplitude differences between two QAM values, as we saw in Chapter 2 with interleaving and in Sections 3.5 to 3.8 of this chapter. Differential detection can be done in the time domain or in the frequency domain. In the first case, each subcarrier is compared with the subcarrier of the previous OFDM symbol. In the case of differential detection in the frequency domain, each subcarrier is compared with the adjacent subcarrier within the same OFDM symbol. A generic system for coherent detection is shown in Figure 3.42.

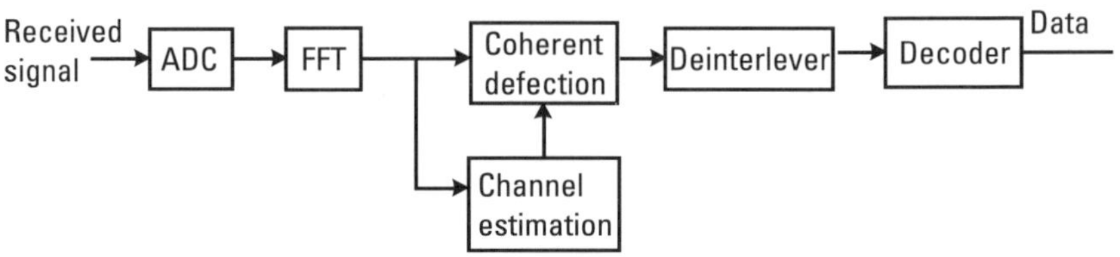

Figure 3.42 OFDM receiver. (After: [25].)

3.9.7.2 Peak-to-Average Power

A large peak-to-average power (PAP) ratio brings disadvantages, such as an increased complexity of the analog-to-digital and digital-to-analog converters and a reduced efficiency of the RF power amplifier. To reduce the PAP ratio, several techniques have been proposed, which can be divided into three categories. First, there are digital distortion techniques, which reduce the peak amplitudes simply by nonlinearly distorting the OFDM signal at or around the peaks. Examples of distortion techniques are clipping, weak widowing, and peak cancellation. The second category is coding techniques that use a special forward-error correcting code set, which includes OFDM symbols with a large PAP ratio. The third technique is based on scrambling each OFDM symbol with different scrambling sequences and selecting the sequence that gives the smallest PAP ratio.

3.9.7.3 Combination of CDMA and OFDM

In an OFDM scheme alone, the transmission performance becomes more sensitive to time-selective fading as the number of subcarriers N_s increases, because a longer symbol duration means an increase in the amplitude and phase variation during a symbol. This causes an increased level of intercarrier interference (ICI). As N_s decreases, the modulation becomes more robust to fading in time, but it becomes more vulnerable to delay spread as the ratio of delay spread and symbol time increases. The latter is not necessarily true if the guard time is kept at a fixed value, but as the symbol duration decreases, a fixed guard interval (Δ) means an increased loss of power.

The OFDM scheme is robust to frequency-selective fading. But it has some disadvantages, such as difficulty in subcarrier synchronization and sensitivity to frequency offset and nonlinear amplification. This is because it is composed of many subcarriers, with their overlapping power spectra, and it exhibits a nonconstant nature in its envelope. In contrast to this, DS-CDMA is quite robust to frequency offsets and nonlinear distortion. The combination of OFDM signaling and CDMA scheme has one major advantage, however, in that it can lower the symbol rate in each subcarrier so that a longer symbol duration makes it easier to synchronize the transmission.

Thus, combining OFDM transmission with CDMA allows us to exploit the wideband channel's inherent frequency diversity by spreading each symbol across multiple subcarriers. In [26], various methods of combining with two techniques are compared, identifying three different structures: multicarrier CDMA (MC-CDMA), multicarrier direct sequence CDMA (MC-DS-CDMA), and multitone CDMA (MT-CDMA). Like nonspread OFDM transmission, OFDM/CDMA methods suffer from high peak-to-mean power ratios, which are dependent on frequency domain spreading scheme [25, 29–31].

References

[1]   Stavroulakis, P., *Interference Analysis of Communication Systems,* New York: IEEE Press, 1980.

[2]   Wozencraft, M. J., and M. I. Jacobs, *Principles of Communication Engineering,* New York: John Wiley, 1965.

[3]   Smith, D. R., *Digital Transmission Systems,* second edition, New York: Van Nostrand Reinhold, 1993.

[4]   Sklar, B., *Digital Communications, Fundamental and Applications,* second edition, Upper Saddle River, NJ: Prentice-Hall, 2001.

[5]   Couch, L. W., *Digital Analog Communication Systems,* fourth edition, Hampshire, UK: MacMillan, 1993.

[6]   Fugin, Xiong, *Digital Modulation Techniques,* Norwood, MA: Artech House, 2000.

[7]   Haykin, Simon, *Communications Systems,* New York: John Wiley, 1978.

[8]   Taub, H., and L. D. Schilling, *Principles of Communications,* New York: McGraw-Hill, 1971.

[9]   Ziemer, E. R., and H. W. Tranter, *Principles of Communications,* Boston: Houghton-Mifflin Company, 1976.

[10] Carlson, Bruce A., *Communication Systems,* New York: McGraw-Hill, 1968.

[11] Roden, S. M., *Analog and Digital Communication Systems,* Upper Saddle River, NJ: Prentice-Hall, 1979.

[12] Lee, A. E., and G. D. Messerschmitt, *Digital Communicatons,* Boston: Kluwer Academic Publishers, 1994.

[13] Proakis, G. J., *Digital Communications,* New York: McGraw-Hill, 1995.

[14] Benedetto, S., and E. Biglieri, *Principles of Digital Transmission with Wireless Applications,* Boston: Kluwer Academic, 1999.

[15] Anderson, R. R., and J. Salz, "Spectra of Digital FM," *Bell System Technical Journal,* Vol. 44, July–Aug. 1965.

[16] Simon, M. K., and M. S. Alouini, *Digital Communication over Fading Channels,* New York: John Wiley, 2000.

[17] Ziemer, R. E., and R. L. Peterson, *Introduction to Digital Communications,* Hampshire, UK: MacMillan, 1992.

[18] McGillem, C., and G. Cooper, *Continuous and Discrete Signal and System Analysis,* third edition, London: Saunders College Publishing, 1991.

[19] Simon, K. M., S. M. Hinedi, and M. C. Lindsy, *Digital Communication Techniques, Signal Design and Detection,* Upper Saddle River, NJ: Prentice-Hall, 1995.

[20] Meyr, H., M. Moeneclaey, and S. A. Fechtel, *Digital Communication Receivers,* New York: John Wiley, 1998.

[21] Hanzo, L., W. Webb, and T. Keller, *Single and Multi-Carrier Quadrature Amplitude Modulation,* New York: John Wiley, 2000.

[22] Gerakoulis, D., and E. Geraniotis, *CDMA Access and Switching,* New York: John Wiley, 2001.

[23] Rappaport, T. S., *Wireless Communications,* Upper Saddle River, NJ: Prentice Hall, 1996.

[24] Stüber, G. L., *Principles of Mobile Communication,* Boston: Kluwer, 1996.

[25] Van Nee, R., and R. Prasad, *OFDM for Wireless Multimedia Communication,* Norwood, MA: Artech House, 2000.

[26] Sampei, S., *Application of Digital Wireless Technologies to Global Wireless Communications,* Upper Saddle River, NJ: Prentice Hall, 1997.

[27] Prasad, R., and S. Hara, "Overview of Multi-Carrier CDMA," *IEEE Communications Magazine,* Dec. 1997, pp. 126–133.

[28] Groe, J. B., and L. E. Larson, *CDMA Mobile Radio Design,* Norwood, MA: Artech House, 2000.

[29] Chang, R. W., "Synthesis of Band Limited Orthogonal Signal for Multichannel Data Transmission," *Bell System Technical Journal,* Vol. 45, Dec. 1996, pp. 1775–1796.

[30] Choi, B. J., E. L. Kuan, and L. Hanzo, "Crest Factor Study of MC-CDMA and OFDM," *Proc. VTC 99,* Sept. 1999, Amsterdam, pp. 233–237.

[31] Salzberg, B. R., "Performance of an Efficient Parallel Data Transmission System," *IEEE Trans. Comm.,* Vol. COM-15, Dec. 1967, pp. 805–813.

4

Optimal Detection in Fading Channels

4.1 Introduction

In Chapter 3, we examined the most typical transmission modulation/demodulation techniques encountered in wireless communications. We have emphasized that there are three different ways to modulate digital data on a carrier to be transmitted. This can be amplitude, phase, or frequency modulation. A generic form of a transmitter and receiver for ideal coherent detection of an AWGN channel was shown in Figure 3.14. We also analyzed, discussed, and compared in Chapter 3 the access techniques that play a major role in the BER performance.

Depending on the modulation technique used, the particular service implemented, and the specific channel over which the data are transmitted [1–4], we design the appropriate decision scheme. Almost all of the time, it is a variation of a form of a matched filter, in conjunction with a maximum likelihood operation or a threshold decision operation, as was shown in Figure 3.15.

For optimal reception, we compute the set of a positeriori probabilities $P(s_k(t)/r_l(t))$ and choose the message whose signal $s_k(t)$ corresponds to the largest of these probabilities. Because these messages are equiprobable, this maximization is equivalent to the maximum likelihood decision rule. In other words, we choose as optimal this signal $s_k(t)$, which corresponds to the largest of the conditional probabilities $p(r_l(t)/s_k(t))$. In this chapter, we shall analyze and consider fading as an interfering agent and present the most popular techniques for mitigating its effects and greatly improving the performance of wireless systems. In other words, we will present the modern tools that are used to suppress the effects of multipath interferers.

4.2 Received Signal Conditional Probability Density Function

If the symbol period is T_s seconds, then the transmitter sends a real bandpass signal of the form [1]:

$$s_k(t) = \text{Re}\{\tilde{s}_k(t)\} = \text{Re}\{\tilde{S}_k(t)\,e^{j2\pi f_c t}\} \tag{4.1}$$

where $\tilde{s}_k(t)$ is the kth complex bandpass signal and $\tilde{S}_k(t)$ are the complex baseband signals chosen from a set of M equiprobable messages where $k = 1, 2, \ldots, M$ and $M = 2^m$. The message transmitted in a generalized fading channel will be affected in amplitude in a multiplicative manner, in phase. Plus, it will be time delayed and corrupted by AWGN. Thus, the received signal will be given by

$$r_l(t) = \text{Re}\{a_l \tilde{s}_k(t - \tau_l)\,e^{j\theta_l} + \tilde{n}_l(t)\} \tag{4.2}$$

$$= \text{Re}\{a_l \tilde{S}_k(t - \tau_l)\,e^{j(2\pi f_c t + \theta_l)} + \tilde{N}_l(t)\,e^{j\pi f_c t}\} \tag{4.3}$$

$$= \text{Re}\{\tilde{R}_l(t)\,e^{j2\pi f_c t}\}$$

For the case when the amplitudes, phases, and delays are known, then $p(r_l(t)/s_k(t))$, because of the independence assumptions on the additive noise components, it can be written as [1]:

$$p(r_l(t)/s_k(t)) = \prod_{l=1}^{L_p} K_l \exp\left[-\frac{1}{2N_l}\int_{T_l}^{T_s+T_l} |\tilde{r}_l(t) - a_l s_k(t - \tilde{\tau}_l)\,e^{j\theta_l}|^2 dt\right] \tag{4.4}$$

$$p(r_l(t)/s_k(t)) = \prod_{l=1}^{L_p} K_l \exp\left[-\frac{1}{2N_l}\int_{\tau_l}^{T_s+T_l} |\tilde{R}_l(t) - a_l \tilde{S}_k(t - \tau_l)\,e^{j\theta_l}|^2 dt\right] \tag{4.5}$$

where K_l are integration (normalization) constants. The square of the absolute value of a complex number in the equation above can be written as

$$|\tilde{R}_l - a_l \tilde{S}_k(t - \tau_l) e^{j\theta_l}|^2 = (\tilde{R}_l - a_l \tilde{S}_k(t - \tau_l) e^{j\theta_l})(\tilde{R}_l^* - a_l \tilde{S}_k^*(t - \tau_l) e^{-j\theta_l}) \tag{4.6}$$

Hence, the conditional probability density function $p(r_l(t)/s_k(t))$ becomes

$$P(r_l(t)/s_k(t)) = \prod_{l-1}^{L_p} \tilde{K}_l \exp\left[-\frac{1}{2N_l} \int_{\tau_l}^{T_s+\tau_l} \hat{R}_l(t)\,\tilde{R}_l^*(t)\;dt\right]$$

$$\cdot \exp\left[\mathrm{Re}\left\{\frac{a_l}{N_l} e^{-j\theta_l} \rho_{kl}(\tau_l)\right\} - \frac{a_l^2 E_k}{N_l}\right] \tag{4.7}$$

$$= K \cdot \prod_{l-1}^{L_p} \exp\left[\left\{\frac{a_l}{N_l} e^{-j\theta_l} \rho_{kl}(\tau_l)\right\} - \frac{a_l^2 E_k}{N_l}\right]$$

$$= K \exp\left[\sum_{l-1}^{L_p} \mathrm{Re}\left\{\frac{a_l}{N_l} e^{-j\theta_l} \rho_{kl}(\tau_l)\right\} - \sum_{l=1}^{L_p} \frac{a_l^2 E_k}{N_l}\right]$$

where

$$\rho_{kl}(\tau_l) = \int_{\tau_l}^{T_s+\tau_l} \tilde{R}_l(t)\, S_k^*(t - \tau_l)\;dt$$

and

$$E_k = \frac{1}{2} \int_0^{T_s} |\tilde{S}_k(t)|^2\;dt \tag{4.8}$$

and K is a constant that can absorb all $K_l \cdot \exp\left[\sum_{l=1}^{L_p} -\frac{1}{2N_l} \int |R_l(t)|^2\,dt\right]$ that are independent of k. Thus, it does not contribute to the maximization of conditional probability density function $p(r_l(t)/s_k(t))$. If we take the natural logarithm of $p(r_l(t)/s_k(t))$ given by (4.7), we obtain

$$\Lambda_\kappa = \ln p(r_l(t)/s_k(t)) \tag{4.9}$$

$$= \sum_{l=1}^{L_p} \left[\mathrm{Re}\left\{ \frac{a_l}{N_l}\, e^{-j\theta_l}\, \rho_{kl}(\tau_l) \right\} - \frac{a_l^2 E_k}{N_l} \right]$$

We observe that in (4.9), we ignored lnK because it is independent of k. Maximization of Λ_k, as the natural logarithm is a monotonic function, is equivalent to maximizing $p(r_l(t)/s_k(t))$. Maximization of Λ_k implies, for this reason, optimization of

$$\Lambda_k = \sum_{l=1}^{L_p} \left[\mathrm{Re}\left\{ \frac{a_l}{N_l}\, e^{-j\theta_l}\, \rho_{kl}(\tau_l) \right\} - \frac{a_l^2 E_k}{N_l} \right] \tag{4.10}$$

Putting the process followed so far in a schematic form, we obtain the structure of a receiver which was shown in the Figure 3.30(b). It was referred to as a RAKE receiver because of its structural similarity with the teeth of a garden RAKE. This receiver is also, by implementation, considered to act as a maximum-ratio combiner, as it is known in the diversity systems. It will be explained later. For the cases when the amplitudes or the phases or amplitudes and phases or phases and delays are unknown, with random variables, we must average the conditional probability over the probability density of these unknown random variable(s) jointly. For example, for the case when both the amplitudes and the phases are unknown, we proceed as follows. Because the amplitudes and phases are assumed to be independent, we can average over each one separately and start with the phase. Following this procedure, we obtain

$$p(r_l(t)/s_k(t)) = K \prod_{l=1}^{L_p} \int_0^{2\pi} \exp\left[\frac{a_l}{N_l}\, \mathrm{Re}\, e^{-j\theta_l}\, \rho_{lkl}(\tau_l) - \frac{a_l^2 E_k}{N_l} \right] \cdot p_{\theta_l}(\theta_l)\, d\theta_l$$

$$\tag{4.11}$$

For uniformly distributed phases where

$$p_{\theta_l}(\theta_l) = \frac{1}{2\pi}$$

the previous expression becomes

$$p(r_l(t)/s_k(t))$$

$$= K \prod_{l=1}^{L_p} \exp\left(-\frac{a_l^2 E_k}{N_l}\right) \int_0^{2\pi} \frac{1}{2\pi} \exp\left[\frac{a_l}{N_l} \operatorname{Re}\{e^{-j\theta_l} \rho_{kl}(\tau_l)\}\right] d\theta_l \qquad (4.12)$$

$$= K \prod_{l=1}^{L_p} \exp\left(-\frac{a_l^2 E_k}{N_l}\right) \cdot \frac{1}{2\pi} \int_0^{2\pi} \exp\left[\frac{a_l}{N_l} |\rho_{kl}(\tau_l)| \cos(\theta_l - \angle\rho_{kl}(\tau_l))\right] d\theta_l$$

Hence

$$P(r_l(t)/s_k(t)) = K \prod_{l=1}^{L_p} \exp\left(-\frac{a_l^2 E_k}{N_l}\right) I_0\left(\frac{a_l}{N_l} |\rho_{kl}(\tau_l)|\right) \qquad (4.13)$$

because

$$I_0\left(\frac{a_l}{N_l} |\rho_{kl}(\tau_l)|\right) \equiv \frac{1}{2\pi} \int_0^{2\pi} \exp \frac{a_l}{N_l} |\rho_{kl}(\tau_l)| \cos(\theta_l - \angle\rho_{kl}(\tau_l)) \, d\theta_l$$

where $\int I_0(\cdot)$ is the Bessel function of the first kind and zeroth order, and $<\rho_{kl}(\tau_l)$ is the phase of $\rho_{kl}(\tau_l)$.

Now we take the average of the probability density of the amplitudes. Using the previous expression for $p(r_l(t)/s_k(t))$ given by (4.13), we obtain

$$p(r_l(t)/s_k(t)) = K \prod_{l=1}^{L_p} \exp\left(-\frac{a_l^2 E_k}{N_l}\right) I_0\left(\frac{a_l}{N_l} |\rho_{kl}(\tau_l)|\right) p_{a_l}(a_l) \, da_l$$

$$(4.14)$$

To proceed from this point on, we must take into consideration the mathematical form of $p_{a_l}(a_l)$ from the previous equation. To do that, we must consider the characteristics of the particular case at hand. In other words, if we take the case when the observations interval of the received signal is one symbol in duration, we refer to noncoherent receivers, whereas when the observations interval is over two symbols, we refer to the case of differentially coherent receivers and over N_s symbols, we refer to multiple symbol differentially coherent detection.

For the first case and for a Rayleigh fading, (4.14) becomes

$$p(r_l(t)/s_k(t)) = K \prod_{l=1}^{L_p} \exp\left(-\frac{a_l^2 E_k}{N_l}\right) I_0\left(\frac{a_l}{N_l}\rho_{kl}(\tau_l)\right) \cdot \frac{2a_l}{A_l} e^{-(a_l^2/A_l)} \, da_l \tag{4.15}$$

where $A_l = E\{a_l^2\}$.

It can be shown [1] that the Λ_k for this case is given by

$$\Lambda_k = -\sum_{l=1}^{L_p} \ln(1 + \overline{\gamma}_{kl}) + \sum_{l=1}^{L_p} \frac{E_k}{4N_l}\left(\frac{\overline{\gamma}_{kl}}{1 + \overline{\gamma}_{kl}}\right) \cdot \left[\frac{1}{E_\kappa}\rho_{kl}(\tau_l)\right]^2 \tag{4.16}$$

where $\overline{\gamma}_{kl} = \dfrac{A_l E_k}{N_l}$ is the average SNR of the kth signal over the lth path.

The realization of this receiver and the schematic of its structure is similar to that given by Figure 3.30(b). Similar procedures can be applied to the other aforementioned cases [1]. In other words, following exactly the same procedure, we can study all possible cases and combinations among the various possibilities that could arise as the amplitude, phase, and delays enter the picture [1].

4.3 Average BER Under Fading

It was shown and discussed in Chapter 3 that in every AWGN channel, a direct relationship exists between the ultimate metric of quality of the transmission system, which is the average bit error probability (BEP) or the symbol error probability (SEP), and the *SNR*.

This ratio, as we saw in Chapter 3 but also just before, is a stochastic process presented by the parameter γ. If we take the typical case of multiple amplitude modulation (M-AM) system, the SEP is given by [1].

$$P_s(E) = 2\frac{M-1}{M} Q\left(\sqrt{\frac{6E_s}{N_0(M^2-1)}}\right) \tag{4.17}$$

where E_s is the average symbol energy related to carrier amplitude A_c by

$$E_s = A_c^2 T_s \frac{M^2 - 1}{3}$$

and for the binary AM where $M = 2$, the BEP is given by

$$P_b(E) = Q\left(\sqrt{\frac{2E_b}{N_0}}\right) \tag{4.18}$$

where

$$Q(\cdot) \equiv \int_{(\cdot)}^{\infty} \frac{1}{2\pi} \exp\left(-\frac{y^2}{2}\right) dy \tag{4.19}$$

$$= \frac{1}{\pi} \int_0^{\pi/2} \exp\left(-\frac{(\cdot)^2}{2 \sin^2 \theta}\right) d\theta$$

and

$$E_b = A_c^2 T_s$$

It is most common to consider the case of large SNR, for which the only significant symbol errors are those that occur in adjacent signal levels. For such cases, the average BEP is directly related to SEP by the same equation.

$$P_b(E) \cong \frac{P_s(E)}{\log_2 M} \tag{4.20}$$

When fading is present, the received carrier amplitude is attenuated by the fading amplitude α, which is a random variable with mean square value $\overline{a^2} = A$, and in that case the average BEP is given by

$$P_b(E) = \int_0^{\infty} P_b(E/\gamma) p_\gamma(\gamma) \, d\gamma \tag{4.21}$$

where $P_b(E/\gamma)$ is the conditional average BEP based on a specific value of γ. If we replace $\dfrac{E_s}{N_0} = \gamma \log_2 M$ in (4.21) for $P_b(E)$, we obtain [1]:

$$P_b(E/\gamma) = 2\,\frac{M-1}{M}\,Q\!\left(\sqrt{\frac{6\gamma \log_2 M}{M^2 - 1}}\right) \tag{4.22}$$

and if we deal with a Rayleigh channel, we substitute $p_\gamma(\gamma) = \dfrac{1}{\overline{\gamma}}\,e^{-(\gamma/\overline{\gamma})}$.

It can be shown [1] that this integral given (4.21) yields

$$P_b(E) = \frac{M-1}{M}\left(1 - \sqrt{\frac{3\overline{\gamma}_s}{M^2 - 1 + 3\overline{\gamma}_s}}\right) \tag{4.23}$$

$$\overline{\gamma}_s \triangleq \overline{\gamma} \log M \tag{4.24}$$

For the case when $M = 2$

$$P_b(E) = \frac{1}{2}\left(1 - \sqrt{\frac{\overline{\gamma}}{1 + \overline{\gamma}}}\right) \tag{4.25}$$

To obtain the SEP, we need to go back to $P_b(E)$ and evaluate the integral $\int_0^\infty Q\!\left(a\sqrt{\gamma}\right) p_\gamma(\gamma)\,d\gamma$ for the fading channel of interest. We then multiply it by $\dfrac{2(M-1)}{M}$ and substitute $\dfrac{6 \log_2 M}{M^2 - 1}$ for a^2 in the integral (4.21) and then use (4.20). Similarly, if the channel we are dealing with is a Nakagami-m fading channel

$$P_b(E) = \int_0^\infty 2\,\frac{M-1}{M}\,Q\!\left(\frac{6\gamma \log_2 M}{M^2 - 1}\right)\left[\frac{m^m \gamma^{m-1}}{\overline{\gamma}^m\,\Gamma(m)}\,e^{-(m\gamma/\overline{\gamma})}\right] d\gamma \tag{4.26}$$

where

m is the Nagakami-m fading parameter ranging from $\dfrac{1}{2}$ to ∞;

$\Gamma(\cdot)$ is the gamma function.

It is shown in [1] that this integral yields

$$P_b(E) = \left(\frac{M-1}{M}\right)\left[1 - \mu \sum_{k=0}^{m-1} \binom{2k}{k}\left(\frac{1-\mu^2}{4}\right)^k\right] \qquad (4.27)$$

where

$$\mu = \sqrt{\frac{3\overline{\gamma}_s}{m(M^2-1)+3\overline{\gamma}_s}}, \quad m \text{ integer} \qquad (4.28)$$

and thus

$$P_b(E) = \frac{M-1}{M}\left[1 - \sqrt{\frac{3\overline{\gamma}_s}{m(M^2-1)+3\overline{\gamma}_s}}\sum_{k=0}^{m-1}\binom{2k}{k}\left(\frac{1-\mu^2}{4}\right)^k\right] \qquad (4.29)$$

We observe that when $m = 1$ and $M = 2$, the previous equation reduces to (4.23) and thus

$$P_b(E) = \frac{1}{2}\left(1 - \sqrt{\frac{\overline{\gamma}_s}{1+\overline{\gamma}_s}}\right) \qquad (4.30)$$

as expected.

The analysis so far assumes that the detector used is an ideal coherent detector. In other words, the attributes of the local carrier and especially the phase used to demodulate the received signal were perfectly matched to those of the transmitted carrier. In practical systems this is rarely the case, and in such situations we try to evaluate the average BEP by alsoaveraging over the various sets of values of the difference between the carrier phase and the phase of the locally produced carrier. For the very simple case of BPSK systems, $P_b(E)$ is given by

$$P_b(E/\phi_c) = \int_{-(\pi/2)}^{(\pi/2)} P_b(E/\phi_c)\,p(\phi_c)\,d\phi_c \qquad (4.31)$$

where as shown in [1] is given by

$$P_b(E/\phi_c) = Q\left(\sqrt{\frac{2E_b}{N_0}}\, \cos\, \phi_c\right) \tag{4.32}$$

where as $p(\phi_c)$ is the PDF of carrier phase difference ϕ_c given by

$$p(\phi_c) = \frac{\exp\,(\gamma_{eq}\, \cos\, \phi_c)}{2\pi I_0(\gamma_{eq})} \tag{4.33}$$

where γ_{eq} is the equivalent loop signal to noise ratio SNR which is related to the parameters of the phase tracking loop under consideration. For more details the reader is referred to [1–6]. Using (4.33), we can calculate the $P_b(E)$. For the noncoherent case, there is no way to partially track the transmitted carrier phase, and in those situations we use a particular modulation for which noncoherent detection is possible. The most appropriate modulation of this case is M-frequency shift keying. For matched filter outputs and the assumption of orthogonal signals corresponding to a minimum frequency spacing $\Delta f_{\min} = 1/T_s$, then it is shown in [1] that

$$P_s(E/\gamma) = \sum_{n=1}^{M-1} (-1)^{n+1} \binom{M-1}{n} \frac{1}{n+1} \exp\left(\frac{-n}{n+1} \cdot \frac{E_s}{N_0}\right) \tag{4.34}$$

and

$$P_b(E) = \frac{1}{2}\left(\frac{M}{M-1}\right) P_s(E)$$

For noncoherent detection of binary FSK this equation reduces to

$$P_b(E) = \frac{1}{2}\, \exp\left(-\frac{E_b}{2N_0}\right) \tag{4.35}$$

For the case that we have Rayleigh fading, (4.34) for $P_s(E/\gamma)$ yields

$$P_s(E) = \sum_{n=1}^{M-1} (-1)^{n+1} \binom{M-1}{n} \frac{1}{1+n(1+\overline{\gamma}_s)} \tag{4.36}$$

This equation for binary FSK simplifies to give

$$P_s(E) = \frac{1}{2 + \overline{\gamma}} \qquad (4.37)$$

A summary of results for $P_b(E)$ for various cases of modulations and fading is given in Table 4.1 [1]. This table somehow shows how we can proceed to evaluate BEPs, which is a metric of quality for any wireless channel, including fading channels. The method is summarized as follows. We must first determine the signal of BER based on the knowledge of a specific SNR as random variables and then calculate the mean BEP by integrating over the probability density function of SNR in a particular fading channel. In the simplest cases, the result can be given in a closed form. In some practical cases, however, (as explained in [1]), such as M-ary FSK, only bounds can be obtained. In the next sections, we shall show how we can improve the performance of wireless systems over fading channels by implementing some kind of compensation. The same concept will be discussed again in Chapter 6 in the context of interference and signal distortion reduction.

4.4 Flat Fading Compensation Techniques

So far in this book, we have examined the characteristics of wireless systems/channels and the way they perform in a fading environment. In the sections to follow, we shall study how this performance can further be improved by employing antifading techniques. If this book were written before the mid-1980s, this chapter and Chapter 2 would not have been necessary because most of the transmission systems used were FM, which can operate without the need for fading compensation, as we saw in Chapter 3. The information signals were mainly analog and the FM discriminators used for demodulation were sufficient. With the advent of coherent detection and the digitization of the information signals, we are able to optimize signal detection under the condition of fading by employing fading compensation for the coherent demodulators.

Coherent detection, which utilizes fading compensation, can roughly be divided into two categories: those that don't employ pilot signal-aided techniques and those that do employ pilot signals. The mathematical structures of coherent demodulators were presented in Chapter 2. Here we will show how this structure can be improved, as far as detection is concerned, in a fading channel. Figure 4.1 shows a typical configuration for a coherent demodulator.

Table 4.1
Coherent Detection in Fading Channels Calculation of BEP

1. Modulation QAM

$$P_b(E/\gamma) \approx 4 \frac{\sqrt{M}-1}{\sqrt{M}} \left(\frac{1}{\log_2 M}\right) \sum_{i=1}^{\sqrt{M}/2} Q\left((2i-1)\sqrt{\frac{3E_b \log_2 M}{N_0(M-1)}}\right)$$

then

$$P_s(E) = \frac{4}{\pi} \frac{\sqrt{M}-1}{\sqrt{M}} \int_0^{\pi/2} \exp\left[-\frac{E_s}{N_0}\frac{3}{2(M-1)\sin^2\theta}\right] d\theta$$

$$-\frac{4}{\pi}\left(\frac{\sqrt{M}-1}{\sqrt{M}}\right)^2 \int_0^{\pi/4} \exp\left(-\frac{E_s}{N_0}\frac{3}{2(M-1)\sin^2\theta}\right) d\theta$$

and for Rayleigh fading

$$P_b(E) \approx 2 \frac{\sqrt{M}-1}{\sqrt{M}} \frac{1}{\log_2 M}$$

$$\sum_{i=1}^{\sqrt{M}/2} \left(1 - \sqrt{\frac{1 \cdot 5(2i-1)^2 \bar{\gamma} \log_2 M}{M-1+1 \cdot 5(2i-1)^2 \bar{\gamma} \log_2 M}}\right)$$

2. Modulation M-PSK

$$P_s(E) = \frac{1}{\pi} \int_0^{\left(\frac{\pi(M-1)}{M}\right)} \exp\left(-\frac{E_s}{N_0}\frac{\sin^2\frac{\pi}{M}}{\sin^2\theta}\right) d\theta$$

and for Rayleigh fading

$$P_b(E) = \frac{1}{\max(\log_2 M, 2)}$$

$$\sum_{i=1}^{\max\left(\frac{M}{4},1\right)} \left(1 - \sqrt{\frac{0.5\bar{\gamma} \log_2 M \sin^2\frac{(2i-1)\pi}{M}}{1 + 0.5\bar{\gamma} \log_2 M \sin^2\frac{(2i-1)\pi}{M}}}\right)$$

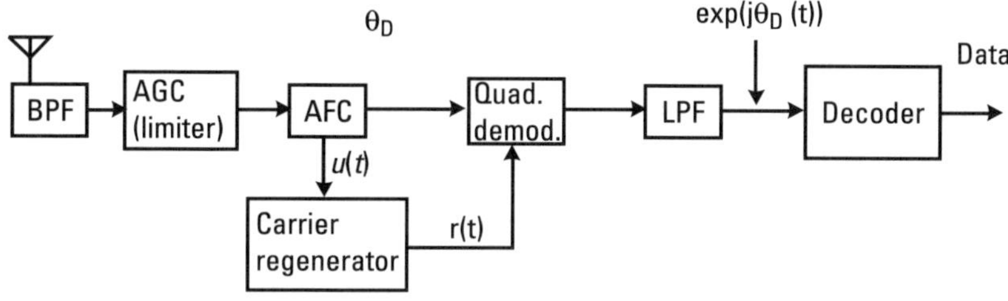

AGC = Automatic gain controller
AFC = Automatic frequency controller (synthesizer)
θ_D = Phase variation by phase modulation
r(t) = Reference carrier

Figure 4.1 M-ary PSK coherent demodulator. (After: [2].)

4.4.1 Nonpilot Signal–Aided Techniques

4.4.1.1 Phase Lock Loop–Based Carrier Generation

The received signal is first filtered by a BPF to pick up the spectrum around the desired signal. For the M-ary PSK case, because the information resides only in the phase component of the transmitted signal, we usually remove amplitude variation using an automatic gain controller (AGC) or a hard limiter. Furthermore, the frequency of the received signal is controlled at a proper frequency by the automatic frequency controller (AFC). In Figure 4.2, a carrier regeneration circuit is shown using a phase lock loop (PLL) with an M × M multiplier for M-ary PSK signals. Other carrier regeneration circuits of the same category are using a Costas loop [2, 5, 6].

PLLs sometimes lose synchronization when phase errors in the PLL are very large. The modified PLL has an instantaneous phase error monitoring function to detect a large phase error that could cause out of lock conditions. When it detects a very large instantaneous phase error, the voltage control oscillator (VCO) output phase is compulsorily shifted to reduce the phase error and thus operates as an adaptive carrier tracking (ACT) circuit, which

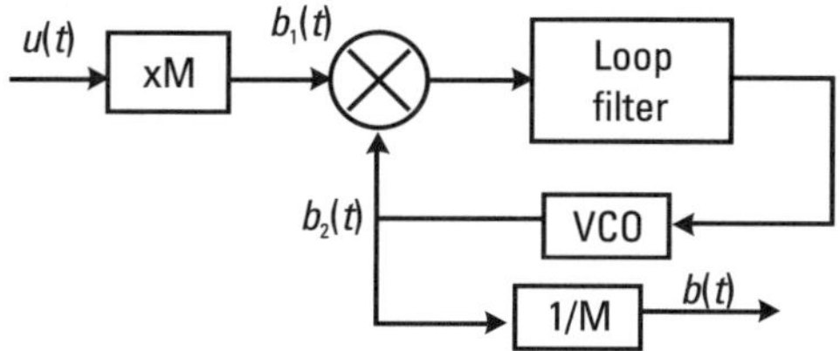

Figure 4.2 Carrier regeneration circuit for M-ary PSK using PLL. (After: [2].)

performs better under a flat Rayleigh fading condition at high E_s/N_0. At lower E_s/N_0, the PLL with ACT performs worse than a conventional Costas loop. To overcome this problem, a dual mode carrier recovery (DCR) controller that selects the appropriate PLL mode is used.

4.4.1.2 Least Mean Square–Based Carrier Regeneration

Figure 4.3 shows a receiver configuration based on a least mean square (LMS) estimation fading compensator.

The received signal is picked up by the BPF, its envelope variation is suppressed by the AGC, its frequency drift is compensated for by using an AFC, and then the received signal is downconverted to the baseband using a local oscillator. Using these techniques (i.e., modified PLL and LMS estimation), it is shown in [2] that substantial improvement is achieved over nonfading compensated channels.

4.4.2 Pilot Signal–Aided Techniques

When we want to apply phase-encoding schemes, we have to estimate carrier frequency as well as its phase variation due to fading with no ambiguity. Moreover, we also have to estimate amplitude variation if we want to employ amplitude modulation as a modulation scheme.

To accurately estimate fading variation, pilot signal–aided calibration techniques are widely used in wireless communication systems. There are three types of the pilot signal–aided techniques:

1. Pilot tone–aided techniques in which one or more tone [continuous wave (CW)] signal(s) and the information signal are multiplexed

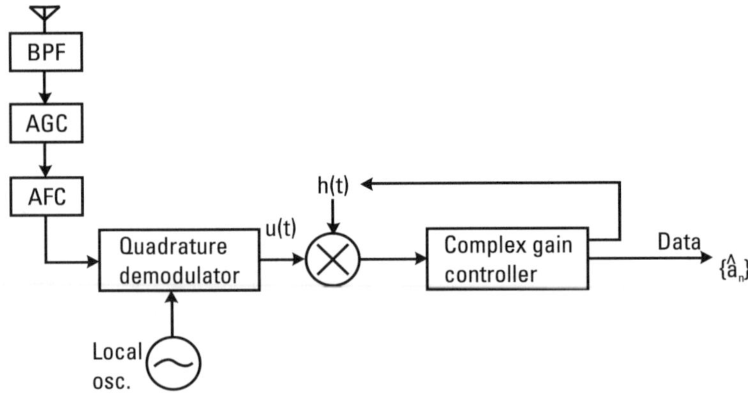

Figure 4.3 LMS-based fading compensator. (After: [2].)

in the frequency domain (frequency division multiplexing type, or FDM);

2. Pilot symbol–aided techniques in which a known pilot symbol sequence and the information symbol sequence are multiplexed in the time domain (time division multiplexing type, or TDM);

3. Pilot code–aided techniques in which a spread-spectrum signal using a spreading code orthogonal to that for the information (traffic) channel(s) and the traffic channel are multiplexed (code division multiplexing type, or CDM).

Figures 4.4(a–c) show classification of the pilot signal–aided calibration techniques. We will discuss each technique in detail in the following sections. We saw that fading compensation techniques exist but require differential decoding due to phase ambiguity on the part of PLL.

4.4.2.1 FDM Pilot Signal

For this type of pilot-assisted fading compensation, we transmit a carrier component simultaneously with the modulated signal and thus regenerate a reference signal of the received signal with no phase ambiguity. It is, however, necessary to make them orthogonal with each other because we have to discriminate these two components at the receiver. An example is shown in Figure 4.5, which achieves the required orthogonality between a modulated signal and its pilot tone.

Depending on the modulation scheme used for transmission, various specialized techniques have been developed [2], such as tone calibration technique more applicable to QPSK and the transparent tone in band (TTIB) scheme more applicable to QAM.

4.4.2.2 TDM Pilot Signal

As shown in Figure 4.4(b), we insert a pilot symbol every $(N-1)$ information symbols. If the symbol rate of the information sequence is R_s and the pilot symbol sequence with symbol rate is R_p, their relationship is given by

$$R_s = (N - 1)R_p \qquad (4.38)$$

The information rate is reduced after multiplexing and becomes $\frac{N}{N-1}R_s$, because for every $N-1$ information symbols, one pilot symbol was inserted. When in-phase and quadrature-phase fading components are changing, as shown is Figure 4.6(a) for the particular QAM case, we can

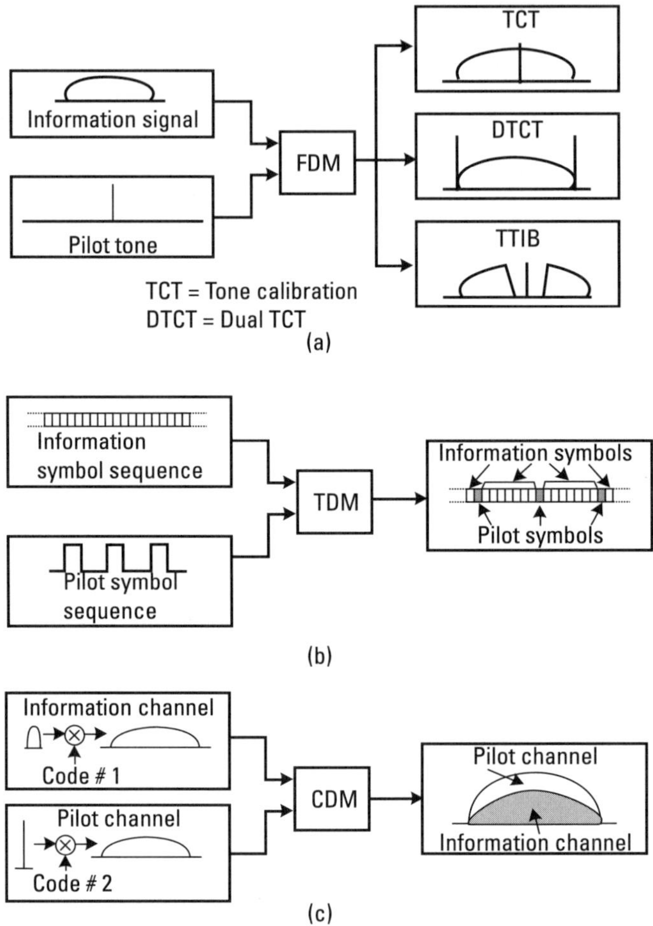

Figure 4.4 (a) FDM-type pilot signal, (b) TDM-type pilot signal, and (c) CDM-type pilot signal. (After: [2].)

sample these components and obtain a sample data sequence of Figure 4.6(b). Because the fading variation is subject to the band-limited Gaussian random process and thus very smooth, it can be estimated by interpolation techniques. It is shown in [2] that two interpolation techniques, such as the Nyquist and Gaussian, give satisfactory results for a variety of E_b/N_0, as well as information signal and pilot symbol rates and fading spectra for QAM-modulation schemes.

4.4.2.3 CDM Pilot Signal

Pilot code–aided techniques applicable to DS-CDMA use the orthogonal codes, like Walsh codes, to multiplex the channels that carry information

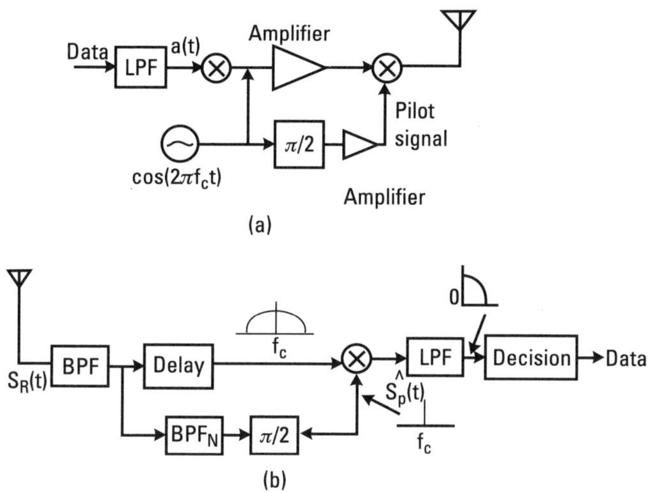

Figure 4.5 FDM pilot signal–aided transmitter and receiver. (After: [2].)

with the pilot channel. The transmission from the base station to the receiver is shown in the Figure 4.7 [2].

The baseband signal is given by

$$s(t) = \sum_{-\infty}^{\infty} a_k\, \delta(t - \eta\, T_k) \tag{4.39}$$

where

$a_k = a_{Ik} + j a_{Qk}$;

a_{Ik}, a_{Qk} are the in phase and quadrature components of symbol k;

T_s is the symbol duration;

$\delta(t)$ is the Delta function.

In this case, we use Walsh spreading codes for spreading in the form given here (i.e., for the information channel and the pilot channel, respectively).

$$w(t) = \sum_{\eta=-\infty}^{\infty} w_{(n\,\mathrm{mod}\,N)}\, \delta(t - nT_c) \tag{4.40}$$

$$w^p(t) = \sum_{\eta=-\infty}^{\infty} w^p_{(n\,\mathrm{mod}\,N)}\, \delta(t - nT_c)$$

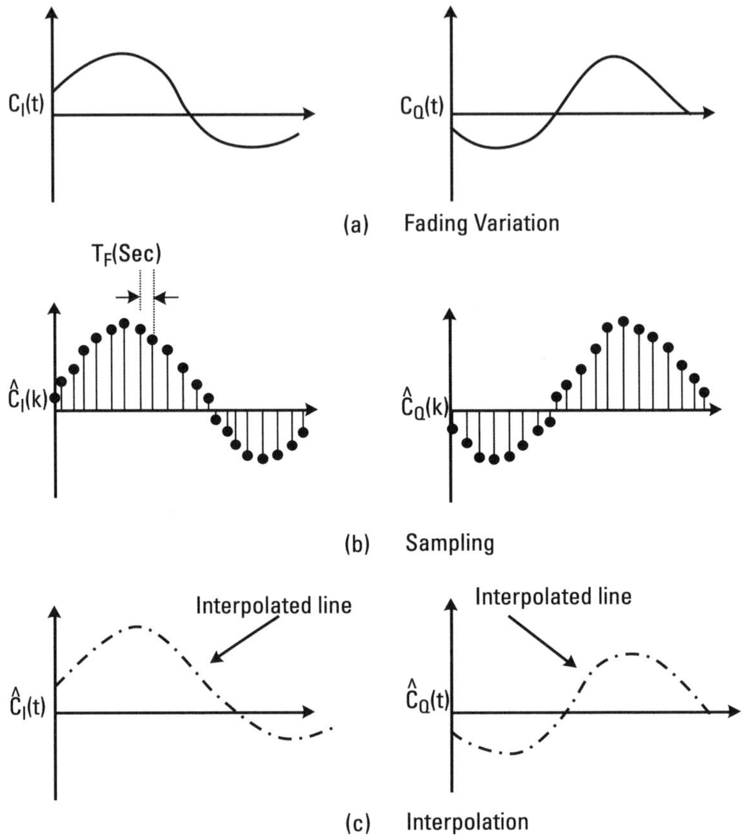

Figure 4.6 Fading estimation/compensation using pilot symbol–aided techniques: (a) fading variation, (b) sampling, and (c) interpolation. (After: [2].)

where

N is the number of symbols in the Walsh code;

T_c is a symbol duration of the Walsh code, which is assumed to be the same as the chip duration of the Walsh code multiplied baseband signals.

The spreading PN sequence used to spread the Walsh-coded information data is given in the form

$$n(t) = \sum_{l=-\infty}^{\infty} n_{(l \bmod N_p)}\, \delta(t - lT_c) \tag{4.41}$$

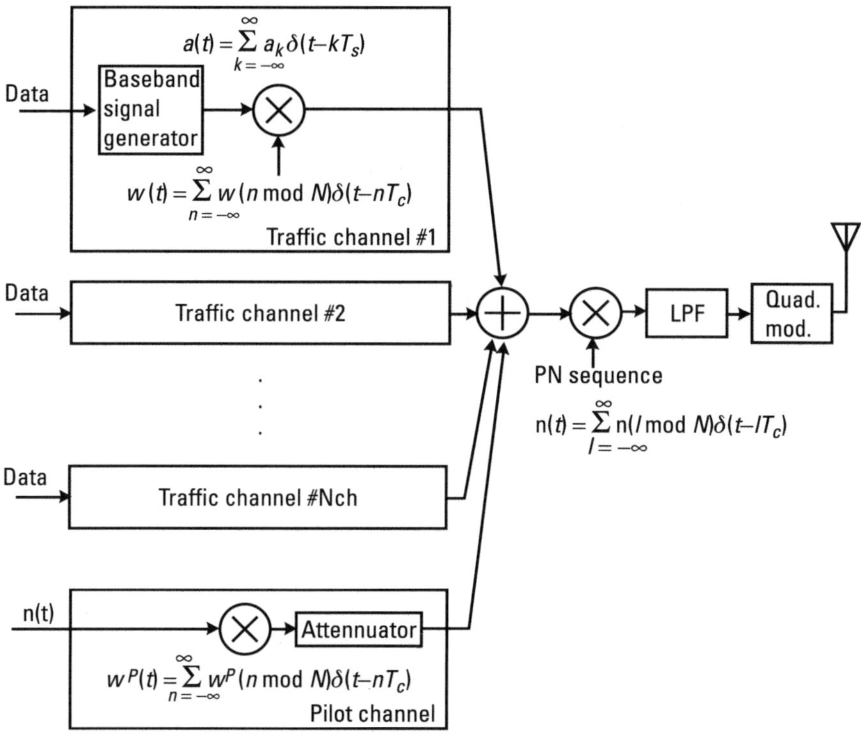

Figure 4.7 Pilot channel multiplexed CDMA. (Source: [2]. Reprinted with permission.)

where

$$n_{(l\bmod N_p)} = n_{I(l\bmod N_p)} + jn_{Q(l\bmod N_p)} \tag{4.42}$$

The modulated signals of the information channel and pilot channel, as shown in Figure 4.7, are given by

$$s(t) = \sum_{k=-\infty}^{\infty} \sum_{l=0}^{N-1} a_k w_l n_{(kN+l\bmod N_p)}\, c(t - k\tau_s - lT_c)\, \exp\left(j2\pi f_c t\right) \tag{4.43}$$

and the pilot channel

$$s^P(t) = \sum_{m=-\infty}^{\infty} n_{(m\bmod N_p)}\, c(t - mT_c)\, \delta(t - mT_c)\, \exp\left(j2\pi f_c t\right) \tag{4.44}$$

where N_p is the number of *PN* symbols in one period, its chip duration is the same as that of the Walsh code T_c, and $c(t)$ is the impulse response of the lowpass filter (LPF) of the transmitter.

The received signal of the information channel $s(t)$ and that of the pilot channel $s^p(t)$ transmitted via a multipath fading channel, with its impulse response of $c_{ch}(t)$, are given by [2]:

$$s(t) = \sum_{k=-\infty}^{\infty} \sum_{l=0}^{N-1} a_k w_{ln\,(kN+l\,\mathrm{mod}\,N_p)} c_1(t - kT_s - lT_c) \cdot \exp\left(j2\pi f_c t\right)$$

$$(4.45)$$

$$s^p(t) = \sum_{k=-\infty}^{\infty} \sum_{l=0}^{N-1} n_{(kN+l\,\mathrm{mod}\,N_p)} c_1(t - kT_s - lT_c) \cdot \exp\left(j2\pi f_c t\right)$$

$$(4.46)$$

where $c_1(t) = c(t) \otimes c_{ch}(t)$ and $\otimes$ is the convolution. The pilot channel can be assigned very high power. In some cases, the power of the pilot channel could be as much as 14 dB higher than that of the information channel. However, in the direction from the mobile station to the base station, because the received signal as the base station experiences different propagation path distortion, each information channel must be associated with its own pilot channel. In such a case, it is often considered that a DPSK system for the uplink will be applicable. Power suppression to a certain level of the pilot carrier is often used to solve this problem and avoid the usage of fading compensating techniques other than the pilot-aided method.

The receiver for such a case is shown in Figure 4.8.

We observe that first the signal is fed to a matched filter to obtain a delay profile of the received signal. Because the SNR of the delay profile at this point is low, we can improve it by coherently accumulating delay profiles at the complex delay profile estimator, as shown in Figure 4.9.

In [2], it is shown via experimental results that pilot signal–aided techniques provide an acceptable means for compensating flat fading and improving BER. The significance of these results cannot be appreciated if the reader of this book did not have a chance to review the parameters involved with the system, channel, and transmission levels in Chapters 1, 2, and 3. In the following sections of this chapter, we will study other compensating schemes for both flat and frequency selective fading. In Chapter 6, however, we will review, study, and evaluate the problem of interference and signal distortion reduction in a general context.

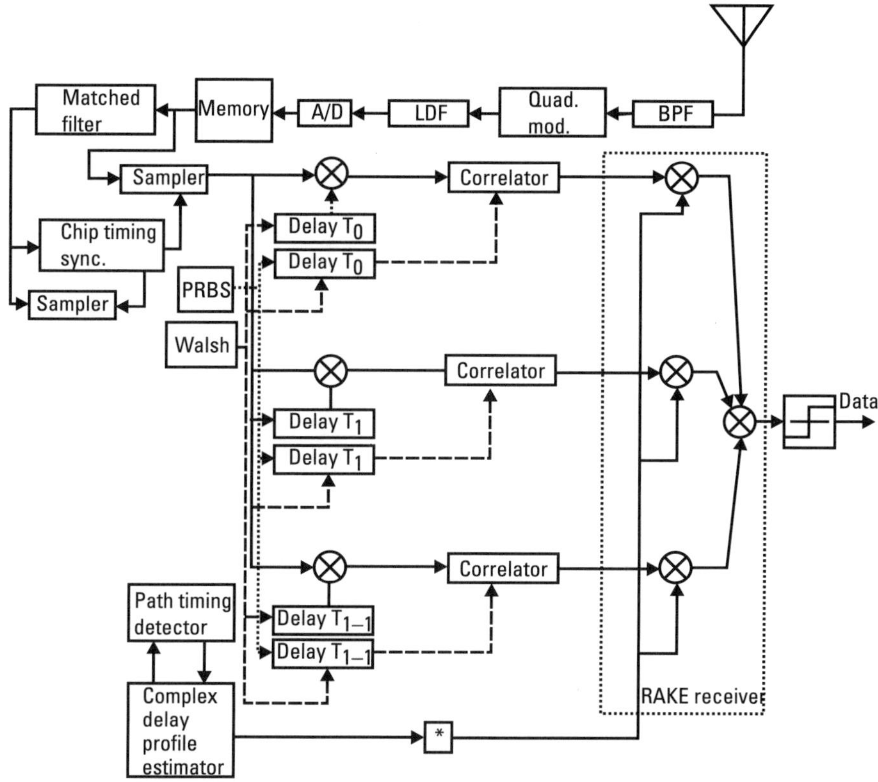

Figure 4.8 Receiver configuration of the suppressed pilot channel. (Source: [2]. Reprinted with permission.)

4.4.3 Diversity Techniques

In Chapter 2, we showed how the narrowband effects of the multipath channel cause very significant impairment of the quality of communication available from a mobile radio channel. Diversity is an important technique for overcoming these impairments and will be examined in this section. We shall also describe, in the section to follow, means of overcoming other impairments related to wideband fading and frequency selective fading. In some cases, these techniques work so successfully that communication quality is improved beyond the level, which would be achieved in the absence of the channel distortions [7–12].

The basic concept of diversity is that the receiver should have more than one version of the transmitted signal available, where each version is received through a distinct channel, as illustrated in Figure 4.10. In each

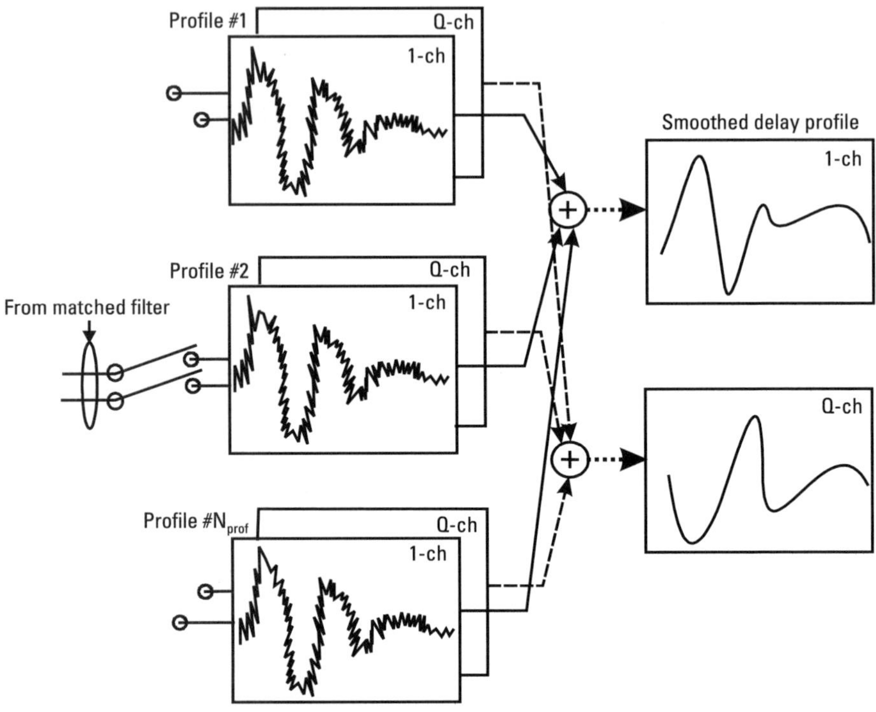

Figure 4.9 Concept of the complex delay profile estimation using coherent accumulation of the delay profiles. (Source: [2]. Reprinted with permission.)

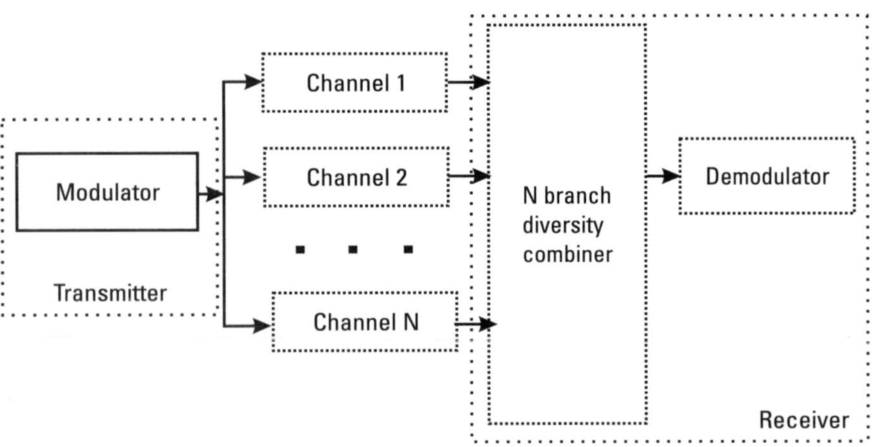

Figure 4.10 Channel diversity demodulator. (After: [12]. © 1999 John Wiley & Sons, Inc.)

channel, the fading is intended to be mostly independent, so the chance of a deep fade (and hence, loss of communication) occurring in all of the channels simultaneously is very much reduced. Each of the channels in Figure 4.10 (plus the corresponding receiver circuit) is called a branch, and the outputs of the channels are processed and routed to the demodulator by the diversity combiner.

Suppose the probability of experiencing a loss in communications due to a deep fade on one channel is p and this probability is independent on all of N channels. The probability of losing communications on all channels simultaneously is then p^N. Thus, a 10% chance of losing contact for one channel is reduced to $0.1^3 = 0.001 = 0.1\%$ with three independently fading channels.

This in illustrated in Figure 4.11, which shows two independent Rayleigh signals. The thick line shows the trajectory of the stronger of the two signals, which clearly experiences significantly fewer deep fades than either of the individual signals.

Two criteria are necessary to obtain a high degree of improvement from a diversity system. First, the fading in individual branches should have low cross-correlation. Second, the mean power available from each branch should be almost equal. If the correlation is too high, then deep fades in the branches will occur simultaneously. If, by contrast, the branches have low correlation but have very different mean power, then the signal in a weaker branch may not be useful even though it is less faded (below its mean) than the other branches.

Assuming that two branches numbered 1 and 2 can be represented by multiplicative narrowband channels a_1 and a_2, then the correlation between the two branches is expressed by the correlation coefficient ρ_{12} defined by

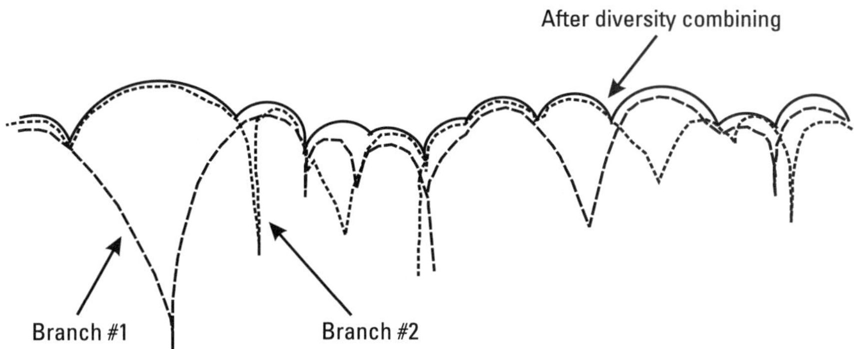

Figure 4.11 Diversity concept. (After: [12]. © 1999 John Wiley & Sons, Inc.)

$$\rho_{12} = \frac{E[(\alpha_1 - \mu_1)(\alpha_2 - \mu_2)^*]}{\sigma_1 \sigma_2} \tag{4.47}$$

* indicates complex conjugate

If both channels have zero mean (true for Rayleigh, but not for Rice fading), this reduces to

$$\rho_{12} = \frac{E[\alpha_1 \alpha_2^*]}{\sigma_1 \sigma_2} \tag{4.48}$$

The mean power in channel i is defined by

$$P_i = \frac{E\big[|\alpha_i|\big]}{2} \tag{4.49}$$

To design a good diversity system, therefore, we need to find methods of obtaining channels with low correlation coefficients and high mean power.

Among many diversity schemes, space diversity, polarization diversity, frequency diversity, time diversity, path diversity, directional diversity, and diversity-combining schemes are the most popular.

4.4.3.1 Space Diversity

The most fundamental way of obtaining diversity is to use two antennas, separated in space sufficiently that the relative phases of the multipath contributions are significantly different at the two antennas. The required spacing differs considerably at the mobile and the base station in a macrocell environment as follows [6–12].

Figure 4.12 shows two antennas separated by a distance, d; both receive waves from two scatterers, A and B. The phase differences between the total

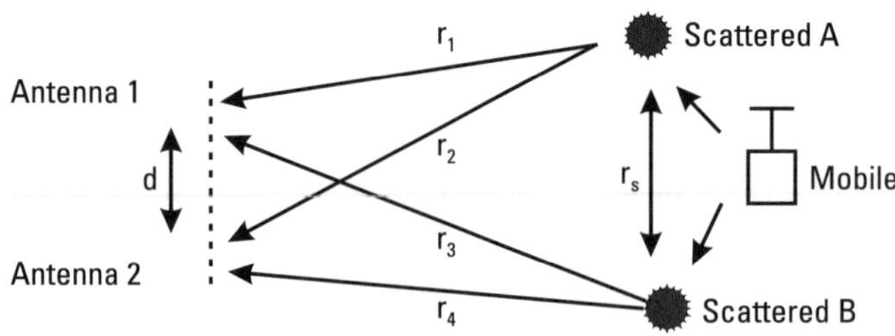

Figure 4.12 Space diversity antennas. (After: [12]. © 1999 John Wiley & Sons, Inc.)

signals received at each of the antennas is proportional to the differences in the path lengths from the scatterers to each antenna, namely $(r_1 - r_3)$ and $(r_2 - r_4)$. If the distance between the scatterers, r_s, or the distance between the antennas, d, increases, then these path length differences also increase. When large phase differences are averaged over a number of mobile positions, they give rise to a low correlation between the signals at the antennas. Hence, we expect the correlation to decrease with increases in either d or r_s.

Examining this effect more formally, Figure 4.13 shows the path to a single scatterer at an angle θ to the broadside direction (the normal to the line joining the antennas).

It is assumed that the distance to the scatterer is much greater than d, so both antennas view the scatterer from the same direction. The phase difference between the fields incident on the antennas is then

$$\phi = -kd \sin \theta \tag{4.50}$$

We can then represent the fields at the two antennas resulting from this scatterer as

$$\alpha_1 = r \text{ and } \alpha_2 = re^{j\phi} \tag{4.51}$$

If a large number of scatterers is present, the signals become a summation of the contributions from each of the scatterers:

$$\alpha_1 = \sum_{i=1}^{n_s} r_i \text{ and } \alpha_2 = \sum_{i=1}^{n_s} r_i e^{j\phi_1} \tag{4.52}$$

where r_i are the amplitudes associated with each of the scatterers. The correlation between α_1 and α_2 is then given by [12]:

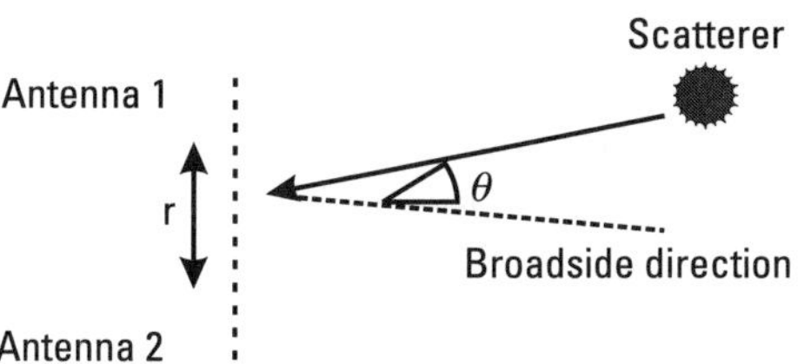

Figure 4.13 Geometry for prediction of space-diversity correlation. (After: [12]. © 1999 John Wiley & Sons, Inc.)

$$\rho_{12} = E\left[\sum_{i=1}^{n_s} \exp\left(-j\varphi_1\right)\right] = E\left[\sum_{i=1}^{n_s} \exp\left(jkd \sin \theta_i\right)\right] \qquad (4.53)$$

The scatterers are assumed uncorrelated and the expected value may be found by treating θ as a continuous random variable with PDF, $p(\theta)$ yielding

$$\rho_{12}(d) = \int_0^{2\pi} \exp\left(jkd \sin \theta\right) p(\theta)\, d\theta \qquad (4.54)$$

Equation (4.54) can be used in a wide range of situations, provided reasonable distributions for $p(\theta)$ can be found to be used in (4.54) to find the expected value. Note that (4.54) is essentially a Fourier transform relationship between $p(\theta)$ and $\rho(d)$. There is therefore an inverse relationship between the widths of the two functions. As a result, a narrow angular distribution will produce a slow decrease in the correlation with antenna spacing, which will limit the usefulness of space diversity, whereas environments with significant scatterers widely spread around the antenna will produce good space diversity for modest antenna spacings. It also implies that if the mobile is situated close to a line through antennas 1 and 2 (the endfire direction), the effective value of d becomes close to zero and the correlation will be higher.

4.4.3.2 Polarization Diversity

When a signal is transmitted by two polarized antennas and received by two polarized antennas, we can obtain two correlated fading variations. This is because the vertical and horizontal polarized components experience different fading variation due to different reflection coefficients of the building walls. There are two distinct features of this scheme: we can install two polarization antennas at the same place, and it does not require any extra spectrum. In other words, we do not have to be careful about the antenna separation, as in the case of space diversity. However, this scheme can achieve only two branch diversity schemes. One more drawback is that we have to transmit 3 dB more power because we have to feed signals to both polarization antennas at the transmitter.

In other words, due to the multiple reflections and scattering that the transmitted signal experiences during its propagation, and to the random mobile antenna orientation, a significant amount of the transmitted energy over the radio channel is usually transposed to a polarization state orthogonal to that of the transmitting antenna. The use of polarization diversity techniques thus allows the receiver to take advantage of both the copolarized and the cross-polarized states [13]. Additionally, it is also worth mentioning that polarization diversity is a simple means to improve performance of wireless communication systems without requiring the large antenna spacing that is necessary in spatial diversity techniques to obtain significant performance improvements. This implies that the use of polarization diversity has the potential to simplify the base station deployment. Moreover, if a microstrip antenna with two polarizations is used, only one antenna is sufficient to achieve diversity [13]. Polarization diversity may thus be a convenient and cheap way to exploit diversity benefits also in mobile transceivers that, due to space limitations, cannot support easily multiple antennas.

Finally, other advantages of polarization diversity include that this technique requires neither additional bandwidth with respect to a nondiversity system nor additional power to transmit the same information over two disjoint frequency bands. The beneficial impact of this technique on DS-CDMA systems has been investigated in [14]. It is shown that polarization diversity reception for nonorthogonal multipulse signals in a multiuser system operating over a single-path Rayleigh fading channel can be treated in a generalized way and achieve satisfactory results over existing techniques with or without polarization diversion. The treatment consists of the use of receivers that, as a first stage, implement a decorrelation filter to get rid of multiuser interference. The decoding process utilizes a generalized likelihood ratio first and uses the maximum likelihood estimates of each user-received waveform, as shown in [15].

4.4.3.3 Frequency Diversity

When a narrowband signal is transmitted over a frequency-selective fading channel, we can obtain independent fading variations if their frequency separation is larger than the coherence bandwidth. Although this scheme can easily obtain any number of diversity branches (L), it degrades system capacity because a channel occupies L times more bandwidth to achieve L-branch frequency diversity. Moreover, it requires L times more power. Therefore, this scheme is not applied much to land-mobile communications in which spectrum and power savings are the most important issues. However, fading variation independence between sufficiently separated frequency components is a very important effect for land mobile-communication technologies. This is called the frequency-diversity effect. For example, multicarrier transmission and frequency hopping techniques utilize this effect [16–18].

4.4.3.4   Time Diversity

As discussed in Chapter 2, the fading correlation coefficient $\rho(\tau)$ is low when $f_d\tau > 0.5$, where f_d maximum Doppler frequency, $\tau\alpha$ time delay. Therefore, when an identical message is transmitted over different time slots with a time slot interval of more than $0.5/f_d$, we can obtain diversity branch signals. Although this scheme requires L times more spectrum, it has the advantage that its hardware is very simple because the entire process is carried out at the baseband. Therefore, time diversity is effective for the CDMA systems in which bandwidth expansion of the source signal is not a problem. However, it is less effective when the terminal speed is very slow because a very long time slot interval is necessary to obtain sufficient diversity gain. Moreover, when the terminal is standing still, we cannot obtain diversity gain at all [19–20].

4.4.3.5   Directional Diversity

Because received signals at the terminal consist of reflection, diffraction, or scattered signals around the terminal, they come from incident angles. When we can resolve the received signal by using directive antennas, we can obtain independently faded signals because all of the paths coming from different angles are mutually independent. This scheme, however, is applicable only to the terminal because the received signal from a terminal comes from only limited directions at the base station. When we employ a directive antenna, we can reduce Doppler spread for each branch, as discussed in Chapter 2.

4.4.3.6   Path Diversity

Path diversity is a diversity-combining scheme that resolves direct and delayed components and coherently combines them. Therefore, this scheme is called implicit diversity because diversity branches are created after the signal reception. The adaptive equalizer and RAKE receiver are classified as path-diversity schemes. The most distinct feature of this method is that no extra antenna, power, or spectrum are necessary. To design such a diversity scheme, however, we must pay attention to the propagation path conditions because path diversity is less effective when the channel is under flat Rayleigh fading conditions. The advantages and disadvantages of each diversity scheme discussed so far will be summarized in Table 4.2.

4.4.3.7   Diversity Combining

Diversity techniques actively use the nature of propagation path characteristics to improve receiver sensitivity. The concept of diversity combining was shown in Figure 4.10.

Table 4.2
Features of Each Diversity Scheme

Diversity Scheme	Advantages	Disadvantages
Space diversity	Easy to design. Any number of diversity branches are (L) selectable. No extra power nor bandwidth is necessary. Applicable to macroscopic diversity.	Hardware size could be large (depends on device technologies). Large antenna spacing is necessary for microscopic diversity at the base station.
Polarization diversity	No space is necessary. No extra bandwidth is necessary.	Only two branch diversity schemes are possible. Three decibels more power is necessary.
Frequency diversity	Any number of diversity branches (L) are selectable.	L times more power and spectrum are necessary.
Time diversity	No space is necessary. Any number of diversity branches (L) are selectable. Hardware is very simple.	L times more spectrum are necessary. Large buffer memory is necessary when f_d is small.
Directional (angle) diversity	Doppler spread is reducible.	Diversity gain depends on the obstacles around the terminal. Applicable only to the terminal.
Path diversity	No space is necessary. No extra power or bandwidth is necessary.	Diversity gain depends on the delay profile.

(From: [2].)

We observe that if we select one of the antennas having the higher received signal level, we can reduce the probability of deep fading. This type of diversity is called microscopic because it intends to mitigate rapid fading variation by the microscopic configuration of the construction around the moving terminal. For the mitigation of shadowing (slow fading), we employ macrodiversity, which uses two or more base stations and sometimes it is called site diversity. Among the various diversity-combining schemes, we distinguish two: the pure-combining and hybrid-combining techniques.

Further, the pure-combining techniques are distinguished into four principal types, depending on the complexity restrictions put on the communication system and the amount of channel state information (CSI) available at the receiver. These are: selective combining (SC), maximal ratio combining (MRC), equal gain combining (EGC), and switch and stay combining (SSC). The hybrid-combining schemes are distinguished into generalized and multi-dimensional.

1. Selective combining.

SC is a category of pure combining for which the diversity branch with the strongest received signal is selected. If we assume that the instantaneous received signal level is subject to Rayleigh fading, we saw in Chapter 2 that the PDF received SNR level for the kth branch is given by

$$P(\gamma_k) = \frac{1}{\overline{\gamma}} \, e^{-(\gamma_k/\overline{\gamma})} \tag{4.55}$$

and the cumulative distribution function is given by

$$P(\gamma_k \leq x) = 1 - e^{-(x/\overline{\gamma})} \tag{4.56}$$

If we consider the contribution of the other L branches and use L branch selective combining, the probability that the signal levels of all the branches go below a certain level x is given next. Therefore,

$$P_{sel}(\gamma \leq x) = P(\gamma_1 \leq x) \cdot P(\gamma_2 \leq x) \ldots P(\gamma_L \leq x) = \prod_{k=1}^{L} [1 - e^{-(x/\overline{\gamma})}] \tag{4.57}$$

and the corresponding PDF is given by

$$P_{sel}(\gamma) = \frac{L}{\overline{\gamma}} \, e^{-(\gamma/\overline{\gamma})} \, [1 - e^{-(\gamma/\overline{\gamma})}]^{L-1} \tag{4.58}$$

The BER for various standards form of modulations as we saw in Chapter 2 are expressed in the form of $a\left(erfc\left(\sqrt{b\gamma}\right)\right)$ or $ae^{-b\gamma}$. In the case when, without fading, the BER is given by $a\left(erfc\left(\sqrt{b\gamma}\right)\right)$, the average BER for L branch selection diversity under Rayleigh fading conditions is given by

$$p(\overline{\gamma}) = \int_0^\infty a\left(erfc\left(\sqrt{b\gamma}\right)\right) \frac{L}{\overline{\gamma}} \cdot e^{-(\gamma/\overline{\gamma})} [1 - e^{-(\gamma/\overline{\gamma})}]^{L-1} \, d\gamma \qquad (4.59)$$

$$= aL \sum_{m=0}^{L-1} (-1)^m \binom{L-1}{m} \frac{1}{m+1} \left[1 - \frac{1}{\sqrt{1 + \dfrac{m+1}{b\overline{\gamma}}}}\right] \qquad (4.60)$$

where $\overline{\gamma}$ is the average E_b/N_0.

When the BER is given by $ae^{-b\gamma}$, the BER for L-branch selection diversity under Rayleigh fading conditions is given by

$$P(\overline{\gamma}) = \int_0^\infty ae^{-b\gamma} \frac{L}{\overline{\gamma}} \cdot e^{-(\gamma/\overline{\gamma})} [1 - e^{-(\gamma/\overline{\gamma})}]^{L-1} \, d\gamma \qquad (4.61)$$

$$= aL \sum_{m=0}^{L-1} (-1)^m \binom{L-1}{m} \frac{1}{m+1} \left[\frac{1}{1 + b\overline{\gamma}/(m+1)}\right]$$

2. Maximal ratio combining.

In the selections-combining scheme, only one of the diversity branch signals are discarded at the diversity combiner. Thus, if all the branch signals are coherently combined with an appropriate weighting coefficient for each branch signal, performance improvement could be expected. In the previous chapter, we saw that the RAKE receiver could be considered as an optimal ratio combining scheme in the absence of interference if all channel-fading parameters are known and regardless of fading statistics. For the case of Rayleigh fading, where each of the L independent identically distributed fading paths has a SNR per bit per pair, γ_l, of the form

$$P_{\gamma_l}(\gamma_l) = \frac{1}{\overline{\gamma}} \cdot e^{-(\gamma_l/\overline{\gamma})} \qquad (4.62)$$

and the SNR per bit of the combined SNR, γ_t, has a probability density function

$$P_{\gamma_t}(\gamma_t) = \frac{1}{(L-1)! \cdot \overline{\gamma}^L} \cdot \gamma_t^{L-1} e^{-(\gamma_t/\overline{\gamma})} \qquad (4.63)$$

where $\gamma_t = \sum\limits_{k=1}^{L} \gamma_k$ and the average probability of error under Rayleigh fading is given by

$$P(\gamma_t) = \int\limits_0^\infty a e^{-b\gamma} \frac{1}{(L-1)! \cdot \overline{\gamma}^L} \cdot \gamma_t^{L-1} e^{-(\gamma_t/\overline{\gamma})} \, d\gamma_t = a\left[\frac{(L-1)!}{(1 + b\overline{\gamma})^L}\right]$$

$$(4.64)$$

For this case, the BER under AWGN noise conditions is given by $ae^{-b\gamma}$.

In the case with the BER is given by $a\left(erfc\left(\sqrt{b\gamma}\right)\right)$

$$P(\gamma_t) = \int\limits_0^\infty a\,erfc\sqrt{b\gamma} \,\frac{1}{(L-1)! \cdot \overline{\gamma}^L} \cdot \gamma_t^{L-1} e^{-(\gamma_t/\overline{\gamma})} \, d\gamma_t \qquad (4.65)$$

This methodology leads to the derivation of these equations, which, when properly adjusted, will determine the average probability of error for a variety of modulation schemes [2].

(a) *MRC with pilot-aided techniques.* One reason that maximal ratio combining has hardly been used in land mobile communications systems, although it gives the maximum SNR of the combined signal, is that it requires at least an accurate estimate of channel parameters. This observation justifies the methodology we adapted in the Preface by which we had to include the material of Chapter 2. Before the emergence of sophisticated digital signal processing techniques, the estimation algorithms required resulted in the implementation of complicated hardware. As a result, MRC has become feasible using pilot signal–aided techniques with much simpler hardware. The pilot signal is used to estimate the fading variation and the maximum ratio-combining scheme in conjunction with the received baseband signal is implemented to detect the transmitted data.

3. Equal gain combining.

We saw that MRC provides the maximum performance improvement relative to all other diversity-combining techniques by maximizing the SNR at the combiner output. However, MRC has the highest complexity because it requires the knowledge of the fading amplitude in each signal branch. Equal gain combining (EGC), however, is much simpler and is considered a good alternative to MRC. It combines all of the branches with the same weighting factor. For equally likely transmitted symbols, it can be shown that the total conditional SNR per symbol γ_{EGC} at the output of the EGC combiner is given by

$$\gamma_{EGC} = \frac{\left(\sum_{l=1}^{L} (a_l)^2\right) E_s}{\sum_{l=1}^{L} N_l} \tag{4.66}$$

where E_s is the energy per symbol. N_l is the AWGN PSD on the lth path. For Rayleigh fading, we have

$$\overline{\gamma}_{EGC} = \overline{\gamma}\left(1 + (L-1)\,\frac{\pi}{4}\right) \tag{4.67}$$

For BPSK or binary FSK modulation over a multilink channel with L paths, the BER conditioned on the fading amplitudes a_l, is given by

$$P_b(E/a_l, l = 1, L) = Q\left(\sqrt{2g\,\frac{E_b\left(\sum_{l=4}^{L} a_l\right)^2}{\sum_{l=1}^{L} N_l}}\right) = Q\left(\sqrt{2g\gamma_{EGC}}\right) \tag{4.68}$$

where g depends on the modulation. $g = 1$ for BPSK and $g = 0.715$ for BFSK. The average BER, $P_b(E)$, is thus given by

$$P_b(E) = \int_0^{\infty} Q\left(\sqrt{\frac{2gE_b a_t^2}{\sum_{l=1}^{L} N_l}}\right) P_{a_t}(a_t)\, da_t \tag{4.69}$$

where $a_t = \sum_{l=1}^{L} a_l$

Equation (4.69) is evaluated in [1].

EGC can also be used in conjunction with differentially coherent and noncoherent cases. For differentially coherent detection, the receiver takes, at every branch l, the difference of two adjacent transmitted phases to arrive at the decision. For noncoherent detection, the decision is taken using a square law detector without estimating the phase. Using EGC, the L decision outputs are summed to form the final decision, and the receiver selects the symbol corresponding to the maximum decision variable. Again, the total conditional SNR per bit γ_t at the output of an EGC combiner is given by

$$\gamma_t = \sum_{l=1}^{L} \gamma_l \tag{4.70}$$

The average BER for these cases is analyzed in [1].

4. Switched and stay diversity.

Switched diversity refers to the case where the receiver switches to and stays with the other branch regardless of whether the SNR of that branch is above or below a predetermined threshold. Hence, this scheme is called SSC. The threshold is an additional design parameter for optimization. Figure 4.14 shows the operation of a dual branch SSC.

The philosophy of SSC can be translated statistically with the following cumulative distribution functions for the output

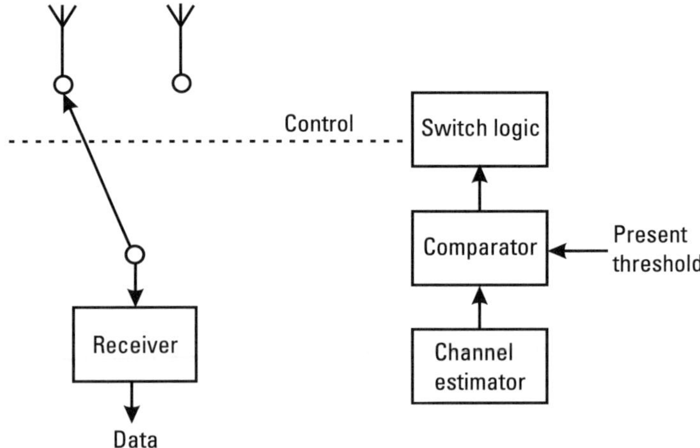

Figure 4.14 Dual-branch SSC diversity. (After: [1]. © 2000 John Wiley & Sons, Inc.)

$$P_{\gamma_{SSC}}(\gamma) = \begin{cases} P(\gamma_1 \leq \gamma_T) & \text{and} \quad \gamma_2 \leq \gamma \\ P(\gamma_T \leq \gamma_1 \leq \gamma) & \text{or} \quad \gamma_1 \leq \gamma_T \end{cases} \left. \begin{matrix} \gamma < \gamma_T \\ \gamma \geq \gamma_T \end{matrix} \right\} \quad (4.71)$$

and $\gamma_2 \leq \gamma$
or

$$P_{\gamma_{SSC}}(\gamma) = \begin{cases} P_\gamma(\gamma_T) P_\gamma(\gamma), & \gamma < \gamma_T \\ P_r(\gamma) - P_\gamma(\gamma_T) + P_\gamma(\gamma) P_\gamma(\gamma_T), & \gamma \geq \gamma_T \end{cases} \quad (4.72)$$

For the case of Rayleigh and Nakagami-m fading, the PDF and cumulative distribution functions (CDFs) for γ are given respectively by (4.73) and (4.74) for the PDFs and (4.75) and (4.76) for the CDFs.

Rayleigh fading

$$p_\gamma(\gamma) = \frac{1}{\overline{\gamma}} e^{-(\gamma/\overline{\gamma})}, \ P_\gamma(\gamma) = 1 - e^{-(\gamma/\overline{\gamma})} \quad (4.73)$$

Nakagami-m

$$p_\gamma(\gamma) = \frac{\left(\dfrac{m}{\overline{\gamma}}\right)^m \gamma^{m-1}}{\Gamma(m)} e^{-(m\gamma/\overline{\gamma})}, \ P_\gamma(\gamma) = 1 - \frac{\Gamma\left(m, \dfrac{m\gamma}{\overline{\gamma}}\right)}{\Gamma(m)} \quad (4.74)$$

Taking the derivative of the CDF, we obtain the PDFs, which become

$$p_{\gamma_{SSC}}(\gamma) = \begin{cases} P_\gamma(\gamma_T) p_\gamma(\gamma), & \gamma < \gamma_T \\ [1 + P_\gamma(\gamma_T)] p_\gamma(\gamma), & \gamma \geq \gamma_T \end{cases} \quad (4.75)$$

which can be written for Rayleigh fading as

$$P_{\gamma_{SSC}}(\gamma) = \begin{cases} \dfrac{1}{\overline{\gamma}} \left(1 - e^{-(\gamma_T/\overline{\gamma})}\right) e^{-(\gamma/\overline{\gamma})}, & \gamma < \gamma_T \\ \dfrac{1}{\overline{\gamma}} \left(2 - e^{-(\gamma_T/\overline{\gamma})}\right) e^{-(\gamma/\overline{\gamma})}, & \gamma \geq \gamma_T \end{cases} \quad (4.76)$$

and for Nakagami-m fading as

$$p_{\gamma_{SSC}}(\gamma) = \left\{ \left(1 - \frac{\Gamma\left(m, \frac{m}{\overline{\gamma}}\gamma_T\right)}{\Gamma(m)} \right) \frac{\left(\frac{m}{\overline{\gamma}}\right)^m \gamma^{m-1}}{\Gamma(m)} e^{-(m/\overline{\gamma})\gamma}, \quad \gamma < \gamma_T \right.$$

$$= \left\{ \left(2 - \frac{\Gamma\left(m, \frac{m}{\overline{\gamma}}\gamma_T\right)}{\Gamma(m)} \right) \frac{\left(\frac{m}{\overline{\gamma}}\right)^m \gamma^{m-1}}{\Gamma(m)} e^{-(m/\overline{\gamma})\gamma}, \quad \gamma \geq \gamma_T \right.$$

$$(4.77)$$

Similar expressions can be found for other types of fading by using Table 2.1 [1].

By averaging γ over $p_{\gamma_{SSC}}(\gamma)$ given by (4.76), we obtain the average of $\overline{\gamma}_{SSC}$ as

$$\overline{\gamma}_{SSC} = P_\gamma(\gamma_T) \int_0^\infty \gamma p_\gamma(\gamma)\, d\gamma + \int_{\gamma_T}^\infty \gamma p_\gamma(\gamma)\, d\gamma = P_\gamma(\gamma_T)\overline{\gamma} \quad (4.78)$$

$$+ \int_{\gamma_T}^\infty \gamma p_\gamma(\gamma)\, d\gamma$$

For Rayleigh fading

$$\overline{\gamma}_{SSC} = P_\gamma(\gamma_T)\overline{\gamma} + \int_{\gamma_T}^\infty \gamma \frac{1}{\overline{\gamma}} e^{-(\gamma/\overline{\gamma})}\, d\gamma = \overline{\gamma}\left(1 + \frac{\gamma_T}{\overline{\gamma}} e^{-(\gamma_T/\overline{\gamma})} \right)$$

$$(4.79)$$

whereas the average BER for BPSK is given by [1]:

$$P_b(E) = \left(1 - \frac{1}{2} e^{-(\gamma_T/\overline{\gamma})} \right)\left(1 - \sqrt{\frac{\overline{\gamma}}{1+\overline{\gamma}}} \right) \qquad (4.80)$$

$$+ \frac{\gamma_T}{2}\left[1 - 2e^{-1}Q(\sqrt{2\gamma_T}) - \sqrt{\frac{\gamma_T}{1+\gamma_T}}\left(1 - 2Q(\sqrt{2(1+\gamma_T)})\right) \right]$$

Figure 4.15 illustrates some of the important results obtained so far. We observe a consistent improvement using MRC.

We observe that the average BER of BPSK with MRC, SC, and SSC as a function of SNR per bit per branch $\overline{\gamma}$ for Nakagami-m fading channel changes drastically with m. For $m \geq 1$, the MRC case shows drastic improvement on BER over the other two cases.

So far, we have analyzed the performance of various multichannel receivers using various combining techniques under the assumptions that the diversity branches contribute independent and identically distributed signals. If this is not true, which is also what we practically expect in mobile communications, a more involved analysis is required. This problem has been studied by various researchers [1]. One can say that unbalanced branches $\gamma_1 \neq \gamma_2$ affects the performance of the receiver as far as BER significantly whereas correlated branches with as much as 0.6 correlation coefficients do not seriously degrade BER performance. In all cases, however, we assumed that the channel parameters were accurately estimated. The effects of channel estimations error or channel decorrelation on the performance of diversity systems has been studied in [21–29].

4.4.3.8 Hybrid Diversity Schemes

The diversity techniques associated with combining studied so far suffered either due to complexity or channel estimation errors such as MRC and

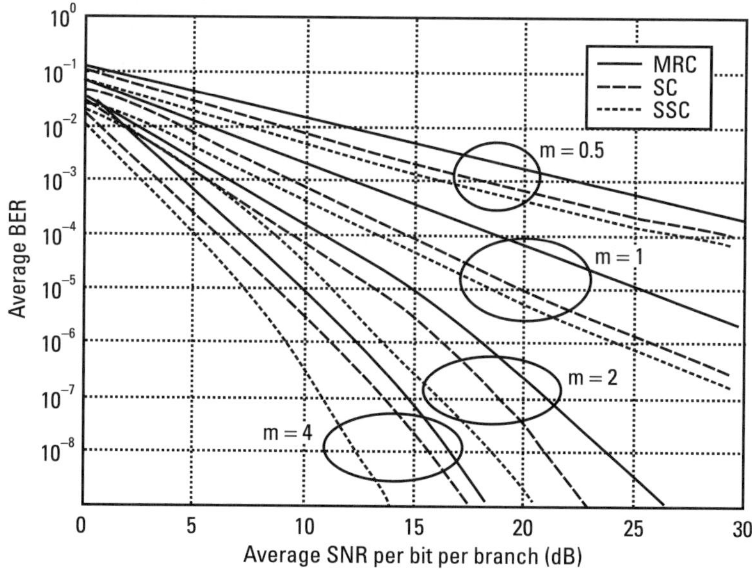

Figure 4.15 Average BER for MRC, SC, and SSC. (Source: [1]. © 2000 John Wiley & Sons, Inc.)

EGC. Moreover, the SC and SSC use only one path out of all the available multipaths and thus do not exploit the amount of diversity offered by the channel. In order to close the gap, generalized selection combining (GSC) techniques have been proposed [1], which combine adaptively under the scheme of MRC and EGC the L_c strongest paths among the L available. These types of receivers are also less complex because for the case of spread-spectrum systems, fewer fingers are used and are more robust towards channel estimations errors because the weakest SNR paths, which are more vulnerable to errors, are excluded.

1. *Generalized selection combining.* The procedure used to determine the *pdf* and average SNR as well as average BER for these diversity techniques have been presented in [1]. It is shown that for the case of BSPK signals with Nakagami-*m* and Rayleigh, fading is shown in Figure 4.16(a) and (b), respectively.

    It is observed that as the number of strongest paths increases, the performance is improved, but in a diminishing manner. Meanwhile, as the number of available diversity paths is increased, the performance is greatly improved.

2. *Generalized switched diversity.* Generalized switched selection combining (GSSC) involves the reception of even number $2L$ of diversity branches grouped in pairs. Every pair of signals is fed to a switching unit that operates according to the rules of SSC, and outputs from the L switching units are connected to MRC or EGC. GSSC is a decentralized scheme and can be viewed as a more practical implementation of GSC. More details are given in [1, 30, 31].

4.5 Frequency Selective Fading

This section will deal with the ways we can cope when we design high bit-rate wireless systems in a frequency selective fading environment. Figure 4.17 shows a general model of frequency selective fading.

From Figure 4.17, we can write

$$s_T(t) = \mathrm{Re}\left(c_T(t) \otimes u(t)\right) e^{j2\pi f_c t} \tag{4.81}$$

where

$\otimes$ = convolution operator

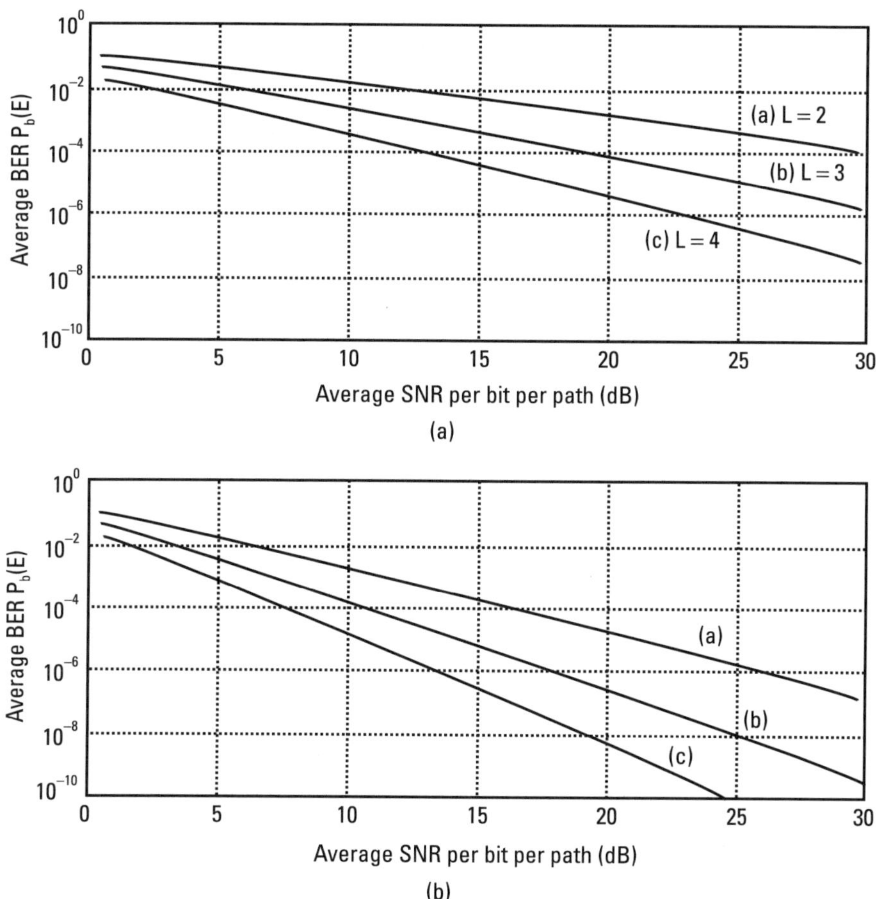

Figure 4.16 Average BER of BPSK signals with Rayleigh and Nakagami-*m*. (Source: [1]. © 2000 John Wiley & Sons, Inc.)

If we define

$$z_T(t) = c_T(t) \otimes u(t), \tag{4.82}$$

we can write

$$s_T(t) = \mathrm{Re}\left(z_T(t)\, e^{j2\pi f_c t}\right) \tag{4.83}$$

Assuming a discrete time operation (i.e., the output $y(t)$, is obtained by optimal sampling timing where $t_n = nT_s$), we obtain

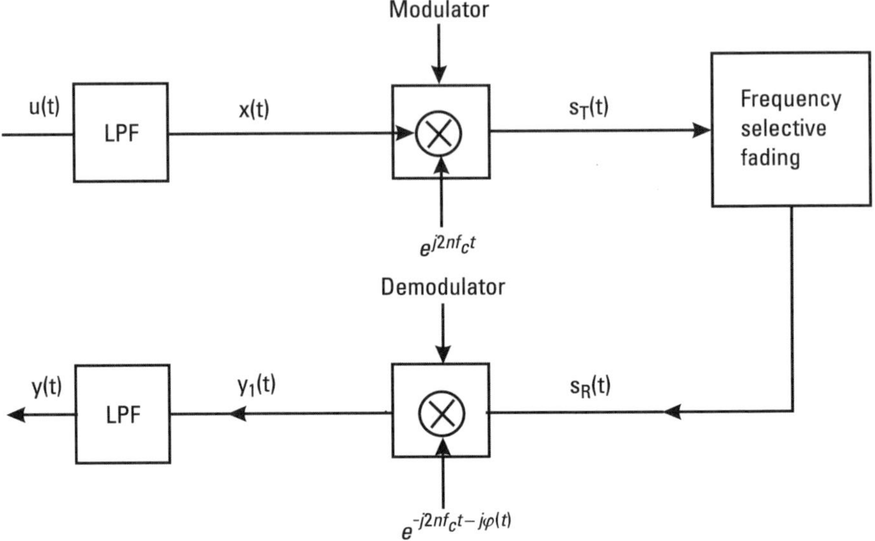

Figure 4.17 Frequency selective fading. (After: [2].)

$$y(nT_s) = \sum_{i=-\infty}^{\infty} u_i c((n-i)T_s)$$

$$= c(0)u_n + \sum_{\substack{i=-\infty \\ i\neq 0}}^{\infty} u_{n-i} c(iT_s) \qquad (4.84)$$

$$= c_0 u_n + \sum_{\substack{i\neq-\infty \\ i\neq 0}}^{\infty} u_{n-i} c_i$$

or where

$$c(t) = c_T(t) \otimes c_R(t) \otimes c_g(t)$$

$$c_R(t) = \text{Impulse response of received LPF} \qquad (4.85)$$

$$c_g(t) = c_b(t) e^{-j2\pi f_{off} t - j\phi(t)}$$

$c_b(t)$ is the baseband equivalent of the impulse response of the frequency selective fading channel $c_c(t)$, such that

$$c_c(t) = 2 \, \text{Re}\left(c_b(t)\, e^{j2\pi f_c t}\right) \tag{4.86}$$

f_{off} = offset frequency between carrier and local oscillator

$$u_n = u(nT_s), \quad c_i = c(iT_s) \tag{4.87}$$

We observe that the first term of the output $y(nT_s)$ at the instant $t = nT_s$ is the desired component of the input, and the rest is called intersymbol interference (ISI). If $c(iT_s) = 0$ for $i \neq 0$, then the propagation path characteristics can be treated as a flat Rayleigh fading. If, on the other hand $c(iT_s) \neq 0$ for $i = 0$ and 1 and for all other i, $c(iT_s)$ are negligible, we call the model a two-ray Rayleigh model.

In general, from the structure of (4.85), it seems that an equalizer of the form of a tranverse filter can provide the means of reducing the effects of ISI. In other words, a tranverse filter can compensate (reduce or eliminate) the ISI.

The problem of ISI is studied in Section 5.24 of Chapter 5 and Section 6.2.4.1 of Chapter 6. In this section, we shall outline some of the features of nonlinear schemes, which provide an alternative approach to ISI combating with fewer limitations.

4.5.1 Equalizers

We shall briefly describe the decision feedback equalizer (DFE) as shown in Figure 4.18. We shall also describe the maximum likelihood sequence estimator (MLSE) as an equalizer.

In Chapters 6 and 7, we shall show how these nonlinear equalizers are used to combat interference and/or signal distortion due to fading.

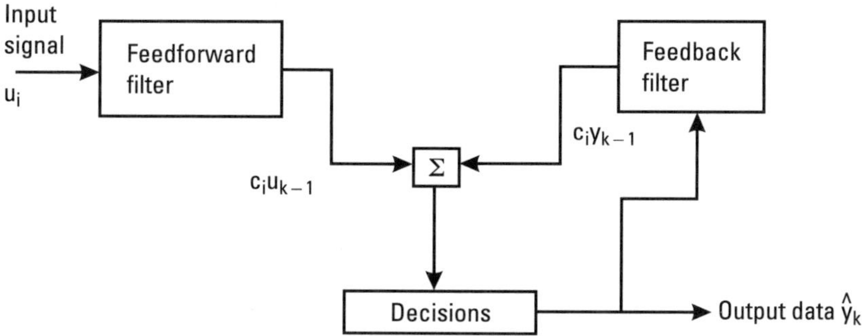

Figure 4.18 DFE.

4.5.1.1 DFE

It is almost obvious that in order to combat ISI, we must try to eliminate or somehow reduce the effect of the ISI term in (4.84). The output of the equalizer in Figure 4.18 can be written as (4.88), which follows.

$$\hat{y}(kT_s) = \sum_{i=-M_1}^{0} c_i u_{k-1} + \sum_{i=1}^{M_2} c_i y_{k-1} \tag{4.88}$$

where the feedforward filter contains $(M_i + 1)$ taps and the feedback filter contains M_2 taps. As long as the decision process is correct, the symbols fed back contain no noise and the resultant SNR at the equalizer output is higher than for a linear equalizer with the same total number of taps.

This process relies on the decision being correct; when a detection error is made, the subtraction process may give catastrophically wrong results, which may lead to further detection errors, and so on. This error propagation phenomenon is a significant disadvantage of the DFE.

It is seen from (4.88) that if we define the following vectors

$$\underline{u}_n^T \equiv [u_{n+M_1}, u_{n+M_{i-1}} \cdots u_{n-1}, y_{n-1}, y_{n-2}, \cdots y_{n-M_2}] \tag{4.89}$$

and the vector

$$\underline{b}^T = [c_{-M_1}, c_{-M_1+1}, \cdots c_{-1}, c_1, c_2, \cdots c_{M_2}] \tag{4.90}$$

and

$$\hat{y}(kT_s) \equiv \hat{y}_k \tag{4.91}$$

then we can write (4.88) as

$$\hat{y}_k = \underline{b}^T \underline{u}_n \tag{4.92}$$

If we knew exactly the parameters vector $\underline{b}$, we could choose the appropriate filter to reproduce them and thus eliminate ISI. This, however, is the problem. We must solve and thus choose various recursive algorithms to determine the optimal $\underline{b}$; we must, therefore, resort to recursive algorithms because these types of algorithms by their nature are adaptive, which means that they can be used on-line to estimate the channel. The observation makes it essential to review the material of Chapter 2. It also exhibits the relevance of Chapter 2 to the main theme of this book. The on-line operation is mandatory because of the time dependence of channel parameters, and thus any equalizer must implement a continually updating estimation procedure. In such a context (4.88), becomes

$$\hat{y}_k = \underline{h}_{n-1}^T \underline{u}_n \tag{4.93}$$

where $\underline{h}_{n-1}^T$ is the unknown vector to be determined in an adaptive manner. The best way to develop adaptive algorithms for a time-discrete process is to try to develop a process that leads to the minimization of the error between the desired and estimated output. It is shown in [2] that such a process, called recursive least squares (RLS), yields the following adaptive algorithms.

$$\underline{h}_n = \underline{h}_{n-1} + e_n \underline{k}_n \tag{4.94}$$

where

$$\underline{k}_n = P_{n-1}(\underline{u}_n^{*T} P_n \underline{u}_n + \lambda v_e)^{-1} \cdot \underline{u}_n \tag{4.95}$$

$$P_n = (P_{n-1} - \underline{k}_n \underline{u}_n^{*T} P_{n-1})\lambda^{-1} \tag{4.96}$$

$e_n = y_n - \hat{y}_n$;

v_e = variance of e_n (scalar);

$P_0 = I$;

λ = a weighting factor called forgetting factor.

$\underline{k}_n$ is the so-called Kalman gain. P_n is the covariance of $\underline{h}_n$ starting usually with $P_0 = I$ or multiplied by a constant smaller than one for convergence purposes.

Many modifications of RLS have been developed over the years [2], and one of them is the Kalman algorithm, which will be examined in detail in Chapter 6 and explained in Appendix B.

Minimum mean square estimation–based algorithms require a known training sequence to be transmitted to the receiver for the purpose of initially adjusting the equalizer coefficients. However, there are some applications, such as multipoint communication networks, where it is desirable for the receiver to synchronize to the received signal and to adjust the equalizer without requiring a known training sequence. Equalization techniques based on initial adjustment of the coefficients without the benefit of training sequence are said to be self-recovering or *blind*. A number of such blind equalizers exist, the most important of which [3] are:

1. LMS-based blind equalization;

2. Stochastic gradient–based blind equalization;

3. Blind equalization based on second and higher order statistics.

4.5.1.2 Maximum Likelihood Sequence Estimation

The maximum likelihood sequence estimation (MLSE) algorithm operates on the estimated data sequence directly, and for this reason it is associated with an appropriate decoder, which in many cases is a Viterbi decoder. The criterion usually adopted to define the optimum receiver is the maximum likelihood criterion, and it can work equally well for slow and fast variation of the channel and can be treated equally well as a time-continuous or discrete case. The typical MLSE algorithm with a Viterbi decoder is shown in Figure 4.19.

The mathematical model can be given in the form

$$J(b) = E\left[\left| u(t) - \hat{u}(t,\, b) \right|^2 \right] \tag{4.97}$$

$$= E\left[\left| u(t) \right|^2 + \left| \hat{u}(t,\, b) \right|^2 - 2\,\mathrm{Re}\left(u(t)\,\hat{u}^{\,*}(t,\, b) \right) \right]$$

where

$E(\cdot)$ is the expected value operator, b is the bit sequence.

Because the first term of (4.97) is not dependent on b, minimization of $J(b)$ is equivalent to minimizing only the term of equation $E\,[\mathrm{Re}\ u(t)\,\hat{u}$ $^*(t,\, b)]$. Even the second term of (4.97) is independent of the data sequence, as it represents the energy of the sequence, which is standard. Assuming that we can replace the expected value operator by the integral, which is equivalent to assuming that the statistics of the disturbance are constant over time, we obtain

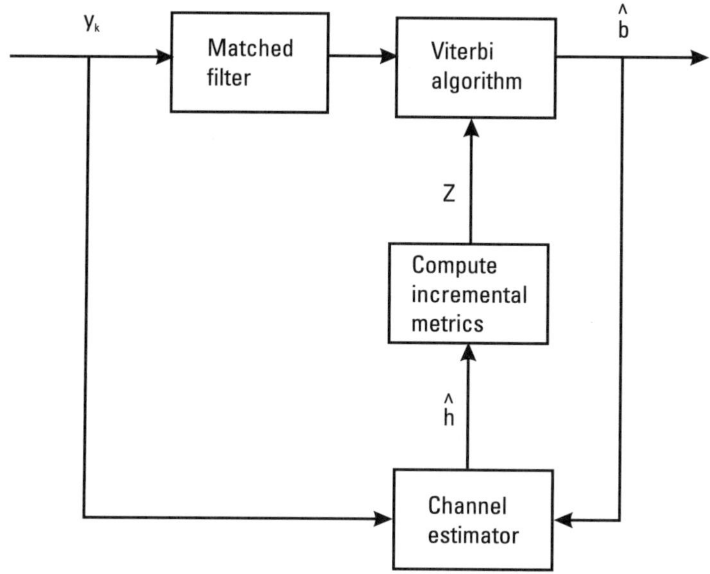

Figure 4.19 MLSE with Viterbi decoder. (After: [12]. © 1999 John Wiley & Sons, Inc.)

$$J_n(b) = \int_0^{NT_s} \mathrm{Re}\left(u(t)\,\hat{u}^*(t, b)\right)\, dt \tag{4.98}$$

Minimization of the integral means finding the sequence of bits most likely used during transmission, as they estimate that signal $u(t)$ that was most likely to have been transmitted. In order to avoid too many computations, we break up the optimization process into two parts.

$$J_n(b) = \int_0^{(N-1)T} \mathrm{Re}\left(u(t)u^*(t, b)\right)\, dt + \int_{(N-1)T}^{NT} \mathrm{Re}\left(u(t)u^*(t, b)\right)\, dt$$

$$= J_{N-1}(b) + Z_N(b) \tag{4.99}$$

where $Z_N(b)$ is known as incremental metric or branch metric.

It is shown in [31] that the implementation of a Viterbi algorithm, which follows a Trellis diagram depicting the various states that ISI gives rise to in each bit interval, is nearly optimal.

The question many times is raised as to which of these two major algorithms (i.e., the DFE or the MLSE/Viterbi) should be used and when.

As a general rule, because the Viterbi algorithm uses implicitly MRC whereas DFE uses SC, the performance of the Viterbi algorithm is better, as we saw before. This, of course, depends on the modulation used. On the other hand, DFE is less complex, especially when we need to employ a higher modulation level. The general rule is shown graphically in Figure 4.20. The popular GSM system is on the borderline.

4.5.1.3 Subband Diversity

An important issue when implementing CDMA systems is to choose appropriate orthogonal code families for both bit-spreading and user-separating purposes.

Unfortunately, the ultimate goal of all of those traditional orthogonal codes was to achieve time-domain orthogonality. No attention was given to enabling the inherent capability candidate codes against frequency-selective fading, which commonly exists in mobile channels and poses a great danger

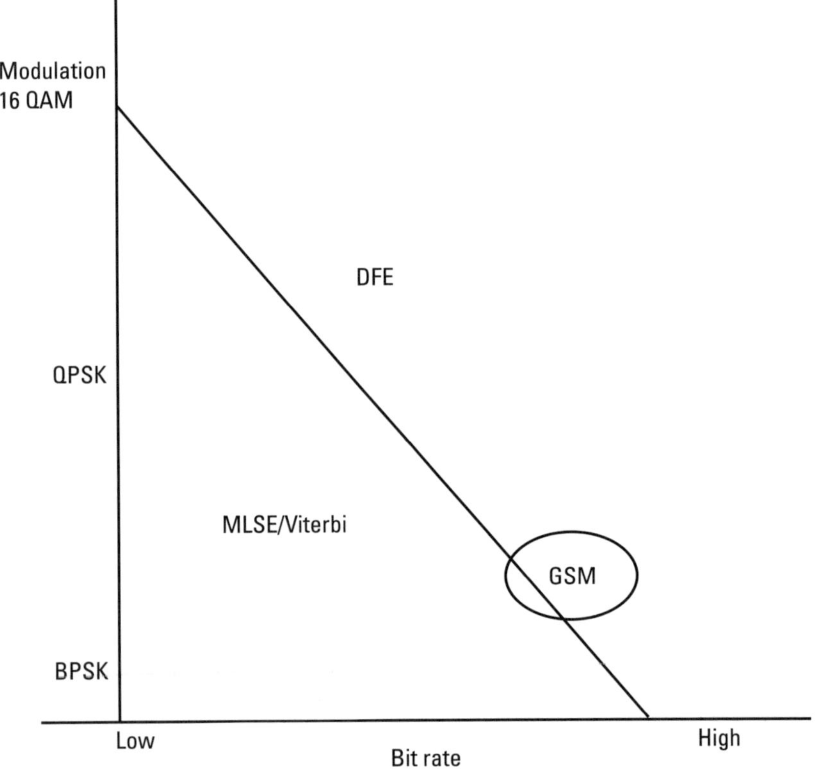

Figure 4.20 Comparison of DFE and MLSE. (After: [2].)

to successful signal reception. In fact, of all traditional spreading codes, time-domain orthogonality is based on uniform chip duration across the code period in order to achieve a specific spreading gain. Fixed-chip duration entails some advantages for signal processing at receivers, such as constant sampling rate, and can be applied to all chips of received signals. However, it gives no frequency diversity advantage for a receiver to mitigate frequency-selective fading.

A new class of CDMA code (wavelet-packet orthogonal codes) is capable of retaining time-domain orthogonality as well as providing intracode sub-band diversity to mitigate frequency-selective fading [32]. The new codes are constructed by congregating several wavelet waveforms with various dilations and shifts. The combination of the wavelet waveforms in different nodes in a wavelet packet full binary tree enable frequency diversity capability. Due to the even code length, they can be readily used in mobile communication systems for multirate streaming and multibit spreading. Wavelet-packet codes, combined with a RAKE receiver, perform much better than traditional time-domain orthogonal codes in frequency-selective fading channels.

To facilitate cross-correlation dependent performance study on wavelet packet spreading codes, a new methodology is introduced, the correlation statistics distribution convolution (CSDC) algorithm. The CSDC algorithm is universally applicable to study spreading codes–dependent performance of a CDMA system with various receiver structures, including correlator and RAKE receiver. The CSDC algorithm can provide a framework, under which extensive studies on both wavelet packet and traditional spreading codes could be carried out [33–36].

4.5.1.4 Optimum Combining

For cellular mobile systems in general, the capacity is limited by the interference from cochannel mobiles in neighboring cells. A popular means of combating this type of interference is the use of an intelligent combination of the signals at multiple antenna elements, as shown in Figure 4.21.

The intelligence is required due to the time dependence of the interference effect on the mobile, which suffers, and thus adaptive techniques must be used, as we saw in the case of DFEs. In this case, we need to find an adaptive way to choose the weights of the antenna.

If a base station in a cellular system uses an adaptive array to direct a radiation pattern towards the mobile with which it is communicating, then several benefits are produced. Depending on the direction of the mobile, the probability of a base station causing interference to cochannel mobiles in surrounding cells is reduced. This situation represents a kind of spatial

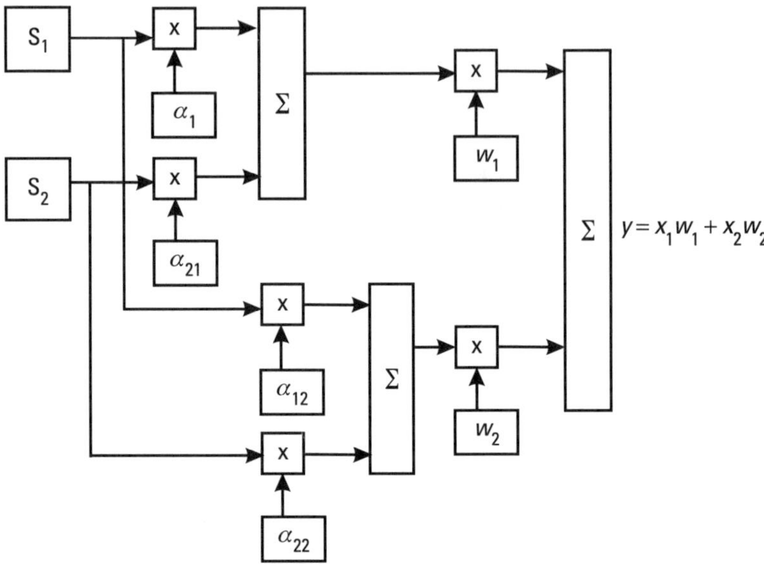

Figure 4.21 Optimum combiner structure/combination of multiple antenna elements. (After: [12]. © 1999 John Wiley & Sons, Inc.)

filtering of interference reduction (SFIR), which in some sense is also an extreme form of sectorization. If, in a system that implements adaptive antennas for interference reduction, as shown in Figure 4.2, we add another combiner that eliminates the signal of one of the mobiles, the system can operate with two mobiles in the same cell on the same channel. This is called space division multiple access (SDMA).

SDMA

Clearly, there will be times when multiple beams will be produced by the base stations that overlap, making it impossible to separate mobiles completely. In order to be effective, this technique must be used only if it is foreseen during the original design of the cellular system. For a simple model of a two-mobile system, the output of the combiner can be give as

$$\underline{x} = s_1 \underline{u}_1 + s_2 \underline{u}_2 + \underline{n} \tag{4.100}$$

where

$$\underline{x} = \begin{bmatrix} x_1 \\ x_2 \end{bmatrix}, \text{ received signals}$$

$$\underline{u}_1 = \begin{bmatrix} a_{11} \\ a_{12} \end{bmatrix}, \underline{u}_2 = \begin{bmatrix} a_{21} \\ a_{22} \end{bmatrix}, \text{ fading factors}$$

$$\underline{n} = \begin{bmatrix} n_1 \\ n_2 \end{bmatrix}, \text{ noise components}$$

Equation (4.100), which gives the output in Figure 4.21, can be written

$$\underline{y} = \underline{w}^T \cdot \underline{x} \tag{4.101}$$

where

$\underline{y}$ is the output of the combiner;

$$\underline{w} = \begin{bmatrix} w_1 \\ w_2 \end{bmatrix} \text{ are the weights.}$$

The Wiener solution, which provides the optimal value for the weight, is given by [33]:

$$\underline{w}_{op} = R_{xx}^{-1} \underline{u}_d \tag{4.102}$$

where $\underline{u}_d$ is the received signals from the desired mobile.

The entire implementation process for the optimum combiner is shown in Figure 4.22.

In practice, it is not easy to directly implement the optimum combiner for the following reasons:

1. Exact knowledge of the channels is required to form the correlation matrix. These channels can only be estimated in the presence of noise and interference.

2. The channel may be different between uplink and downlink because these may be separated in time, frequency, or space.

3. The channel may change rapidly in time, so only a limited amount of data is available for estimation.

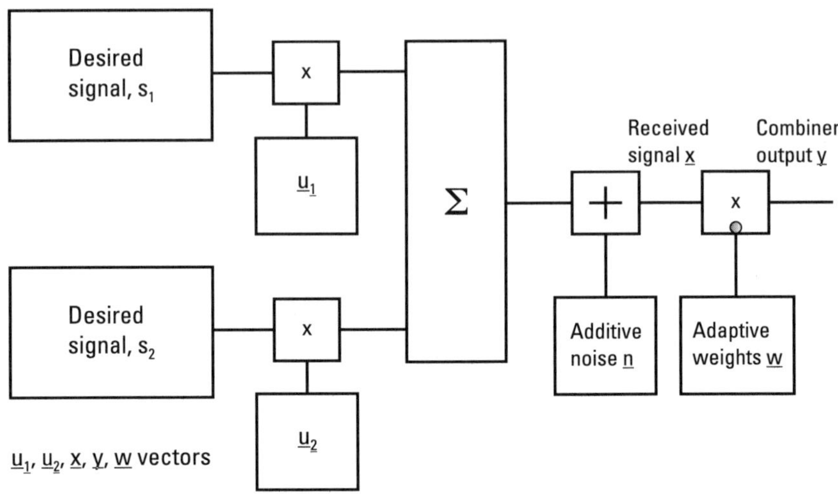

Figure 4.22 Vectorial form of optimum combiner. (After: [12]. © 1999 John Wiley & Sons, Inc.)

These issues may be partially dealt with as follows:

1. Weights must be recalculated with a frequency of around 10 times the maximum Doppler frequency of the channel.

2. Instead of implementing the optimum combiner on the reverse link, the fast fading may be averaged in time to produce estimates of the angles of arrival of the signal sources. Because these angles change more slowly, more reliable results may be obtained at the expense of suboptimal performance. Having thus dealt with these difficulties, the application of adaptive antennas to mobile systems presents significant advantages. However, it is necessary to have a good understanding of the propagation channel and to use this understanding to design systems that have performance benefits outweighing the extra costs involved. Currently, few operational mobile systems actually use adaptive antennas in a standard operation. It is expected, however, that in the next few years, such antennas will form a standard feature of virtually all systems.

Optimum Combining in a Fading and Interference Environment

We have seen so far how fading can be incorporated to determine the average BER. When operating in the scenario that includes interference, the appropriate diversity scheme to employ is one that combines the branch outputs in such a way as to maximize the signal to interference plus noise ratio (SINR) at the combiner output [1].

In such a case, we can write

$$\underline{x}(t) = \sqrt{A_d}\,\underline{a}_d s_d(t) + \sqrt{A_I}\,\underline{a}_I s_I(t) + \underline{n}(t) \tag{4.103}$$

where

$s_d(t)$, $s_I(t)$ are the desired and interfering signals;

A_d and A_I are the respective powers;

$\underline{a}_d$ and $\underline{a}_I$ are channel propagation (fading) vector components.

We saw in Section 4.4.3.7 that when an MRC is used in conjunction with a RAKE receiver, the objective is to select the weights of the receiver to maximize the SNR. For optimum combining (OC), the weights are chosen to maximize SINR. It is shown [1] that for this formulation

$$\gamma_t = A_d\,\underline{a}_d^H R_{ni}^{-1}\,\underline{a}_d \tag{4.104}$$

where

$\gamma_t =$ SINR;

$R_{ni} =$ covariance matrix between interference and noise defined by

$$R_{ni} = E\left[\left(\sqrt{A_I}\,\underline{a}_I s_I(t) + \underline{n}(t)\right)\left(\sqrt{A_I}\,\underline{a}_I s_I(t) + n(t)\right)^H\right] = A_I \underline{a}_I \underline{a}_I^H + \sigma^2 I$$

where I is an $L \times L$ identity matrix.

In order to directly relate SINR with the desired vector, we choose the unitary matrix u of the eigenvectors of the matrix R_{ni} corresponding to the eigenvalues of $\lambda_1, \lambda_2, \ldots, \lambda_L$ such that (4.104) becomes

$$\gamma_t = A_d\,\underline{a}_d^H U\Lambda^{-1}U^H \underline{a}_d = A_d \underline{s}^H \Lambda^{-1}\underline{s} = A_d \sum_{l=1}^{L} \frac{|s_l|^2}{\gamma_l} \tag{4.105}$$

since $R_{ni}^{-1} = U\Lambda^{-1}U^H$

where Λ is the diagonal matrix with elements the eigenvalues of R_{ni} and

$$\underline{s} = U\underline{a}_d$$

which represents the transformed desired signal propagation vector with components

$$s_i, \; i = 1, 2, \ldots L$$

In (4.104) if there is no interference, then (4.104) becomes

$$\gamma_t = \frac{A_d}{\sigma^2} \, \underline{a}_d^H \, \underline{a}_d = \frac{A_d}{\sigma^2} \cdot \sum_{l=1}^{L} \alpha_{dl}^2 = \sum_{l=1}^{L} \gamma_l$$

as it should where α_{dl} is the l element of the vector $\underline{a}_d$.

It can be shown [1] that for the simple case of BPSK in conjunction with an OC to find the average BEP, we must average the conditional (on fading) BEP over the fading distribution of the combiner output statistic. In other words,

$$P_b(E) = \int_0^{\infty} Q\left(\sqrt{2\gamma_t}\right) p_{\gamma_t}(\gamma_t) \, d\gamma_t \tag{4.106}$$

$$= \int_{\sigma^2}^{\infty} \int_0^{\infty} Q\left(\sqrt{2\gamma_t}\right) p_{\gamma_t}(\gamma_t/\lambda_1) \, d\gamma_t p_{\lambda_1}(\lambda_1) \, d\lambda_1$$

From the covariance matrix R_{ni} we can determine

$$\lambda_l = A_I \sum_{n=1}^{L} a_{In}^2 + \sigma^2, \qquad l = 1 \tag{4.107}$$

$$\lambda_l = \sigma^2, \qquad l \geq 2$$

In this integral, only λ_1 is taken as a random variable, because all other λ_l are constants equal to σ^2. The determination of $p_{\gamma_t}(\gamma_t/\lambda_1)$ and $p(\lambda_1)$ in closed form is a very complex and difficult mathematical exercise [1], and many times the closed-form expression involves functions not readily found in simulation packages. The problem is compounded if the fading environment is more involved than Rayleigh fading.

4.5.2 A Comparison of Frequency Selective Fading Compensation Algorithms

DFEs, MLSE, and hard output Viterbi algorithms as forward-backward algorithms are compared on the basis of a multiple accessing scheme, namely, code time division multiple access (CTDMA). The comparison is made in terms of a spectral efficiency criterion (bits/hertz/seconds/user) [34].

CTDMA resembles that of a CDMA system, and this aspect constitutes its advantage, except that the users are assigned unique time shifts (or time slots) and that they all use the same spreading sequence. The benefits of this structure are obtained at the receiver, as we shall see later. One may use the same dispreading filter for all users, and then "equalize" the channel and separate the users by using TDMA equalization techniques. At the same time, the advantages over CDMA are due to the fact that only one linear filter common to all users in one cell is needed to transform the CDMA signal into a TDMA one, thus eliminating the need for a costly CDMA joint detection.

A block diagram of the multiple accessing system CTDMA in a multipath environment is shown in Figure 4.23. In this figure, K users send the coded bit streams $b_k[\cdot]$ through the multipath channels $h_k[\cdot]$, $k = 1 \ldots K$. First, however, their bit streams are spread by the spreading sequence $s[\cdot]$ and delayed by $k \cdot \Delta$ chips. At the receiver, the noise $Z[\cdot]$

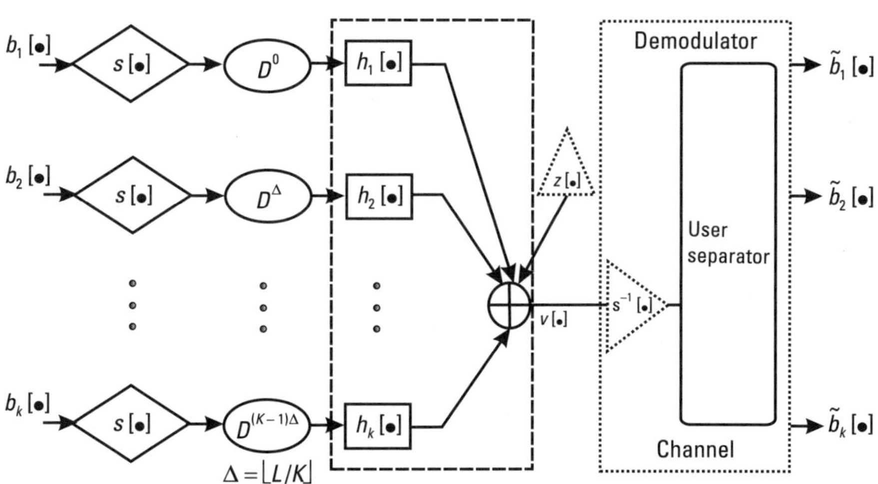

Figure 4.23 CTDMA system. (After: [34].)

and the sum of all the users' symbol streams are accumulated for detection. For reasons of simplicity, we split the demodulation operation into two parts, a linear despreading operation and a user separation/equalization operation. Finally, the demodulator outputs soft decisions $b_k[\cdot]$ of the coded bit streams.

With this set up, any user in a CTDMA system does not affect the other users (i.e., there is no interuser interference). However, for long multipath channels, this is not possible without large degradations in capacity. We must therefore allow interuser interference and consider schemes to separate interfering CTDMA users.

In [33] it is shown, however, that the spectral efficiency of a conventional asynchronous CDMA system is drastically reduced in an environment where typical urban/suburban multipath and fading phenomena occur and no power control ameliorates them. CTDMA loses virtually nothing, even in propagation environments with long channel responses, by shifting the users appropriately and by using user separation algorithms of modest complexity. This issue enters in the comparison process between competing systems, and we arrive at an acceptable measure of quality.

Receiver Complexity Versus Performance

Though capacity and call quality may be of primary concern, other key requirements of a receiver are cost, power consumption, and size. All of the latter features are dominated by the computational complexity of the implemented algorithms (i.e., the number of operations per unit time). Thus, we define the relevant criterion of comparison as

$$\gamma = \frac{\text{Performance}}{\text{Complexity}}$$

where the performance will be measured in terms of cut-off rate and the complexity in terms of floating point operations. This parameter, referred to as cut-off rate given in bits/user, indicates the number, which gives the region of rates where it is possible to operate with an acceptable probability of error.

In [34], it is stated that whichever multiple access technique is employed, the ultimate performance limitation is the system's susceptibility to interference. The different multiple access system designs differ in the possibility to resolve both interuser and intersymbol interference. But that is where CTDMA excels over other multiple access techniques by employing demodulation techniques with modest complexity. Moreover, by considering the *inner receiver,* (i.e., the user separator) as part of the channel, the coding for the multiple access system simplifies to coding for an AWGN channel. Doubtlessly, this is a much simpler task than coding for the multiple access channel, as shown in Figure 4.24.

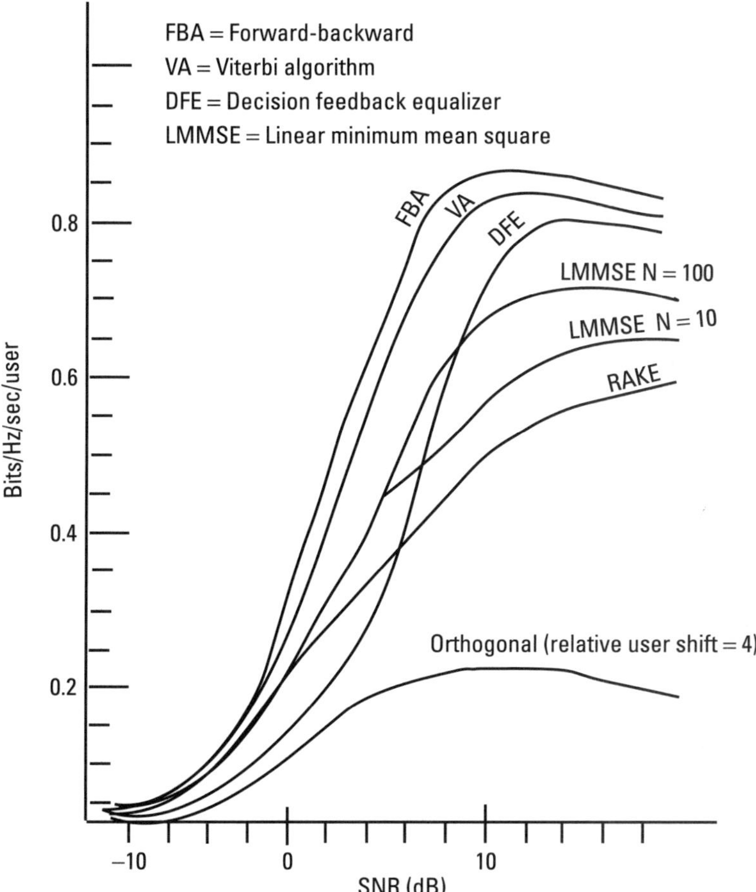

Figure 4.24 Special efficiency for various CTDMA systems. (After: [34].)

References

[1]   Simon, M. K., and M. S. Alouini, *Digital Communications over Fading Channels,* New York: John Wiley, 2000.

[2]   Sampei, Seiichi, *Applications of Digital Wireless Technologies to Global Wireless Communications,* Upper Saddle River, NJ: Prentice-Hall, 1997.

[3] Simon, M. K., Hinedi, M. S., and W. C. Lindsey, *Digital Communication Techniques: Signal Design and Detection,* Upper Saddle River, NJ: Prentice-Hall, 1995.

[4] Lindsey, W. C., and M. K. Simon, *Telecommunication Systems,* Upper Saddle River, NJ: Prentice-Hall, 1973.

[5] Gardiner, F. M., *Phase Lock Techniques,* New York: John Wiley, 1979.

[6] Saito, S., and H. Suzuki, "Fast Carrier Tracking Coherent Detection with Dual Mode Carrier Recovery Circuit for Land Mobile Radio Communication," *IEEE Journal of Select. Areas of Comm.,* Vol. 7, No. 1, January 1989, pp. 130–139.

[7] Salmasi, A., and K. S. Gilhousen, "On the System Design Aspects of Code Division Multiple Access CDMA Applied to Digital Cellular and Personal Communicaton Network," *Proceedings of VTC,* May 1991, pp. 57–62.

[8] Abeta, S., S. Sampei, and N. Morinaga, "A DS/CDMA Coherent Detection System with a Suppressed Pilot Channel," *Globecom 94,* November 1994, pp. 1622–1626.

[9] Brennan, D., "Linear Diversity Combining Techniques," *Proc. IRE,* Vol. 47, June 1959, pp. 1075–1102.

[10] Rappaport, T. S., *Wireless Communications: Principles and Practice,* Upper Saddle River, NJ: Prentice Hall, 1996.

[11] Buzzi, S., et al., "Diversity Reception of Nonorthogonal Multipulse Signals in Multiuser Nakagami Fading Channels," *IEEE Communications Letters,* Vol. 5, May 2001, pp. 188–190.

[12] Saunders, S. R., *Antennas and Propagation for Wireless Communication Systems,* New York: John Wiley, 1999.

[13] Shin, E., and S. Safari-Naeini, "A Simple Theoretical Model for Polarization Diversity Reception in Wireless Mobile Environments," *IEEE Int. Symp. Antennas and Propagation Society,* Vol. 2, 1999, pp. 1332–1335.

[14] Sapienza, F., M. Nilsson, and C. Beckmann, "Polarization Diversity in CDMA," *Proc. of the 1998 IEEE Aerospace Conference,* Vol. 3, 1988, pp. 317–322.

[15] Varanasi, M. K. A. and Russ, "Noncoherent Decorrelative Detection for Nonorthogonal Multipulse Modulations Over the Multiuser Gaussian Channel," *IEEE Trans. Comm.,* Vol. 44, Dec. 1998.

[16] Kamio, Y., "Performance of Trellis Coded Modulation Using Multi-Frequency Channels in Land Mobile Communications," *IEEE, VTC,* May 1990.

[17] Hara, S., et al., "Multicarrier Modulation Techniques for Broadband Indoor Wireless Communications," *PIMRC 93,* Japan 1993.

[18] Kamio, Y., S. and Sampei, "Performance of a Trellis Coded 16 QAM/TDMA System for Land Mobile Communications," *IEEE Trans. Veh. Technology,* Vol. 43, Aug. 1994.

[19] Kubota, S., S. Kato, S., and K. Feher, "A Time Diversity CDMA Scheme Employing Orthogonal Modulation for Time Variant Channels," *IEEE, VTC,* May 1993.

[20] Meyer, M., "Improvement of DS-CDMA Mobile Communications Systems by Symbol Splitting," *IEEE, VTC,* July 1995.

[21] Proakis, J. G., "Probabilities of Error for Adaptive Reception of M-Phase Signals," *IEEE Trans. Comm. Technology,* Vol. Com-16, February 1968.

[22] Biglieri, Ezio, J. G. Proakis, and Shlomo Shamai, "Fading Channels: Information Theoretic and Communication Aspects," *IEEE Transactions on Information Theory*, Vol. 44, No. 6, October 1998.

[23] Cavers, J. K., "An Analysis of Pilot Symbol Assisted Modulation for Rayleigh Fading Channels," *IEEE Trans. Veh. Technol.*, Vol. VT-40, November 1991, pp. 686–693.

[24] Webb, W. T., and L. Hanzo, *Modern Quadrature Amplitude Modulation*, New York: IEEE Press, 1994.

[25] Bello, P. A., and B. D. Nelin, "Predetection Diversity Combining with Selectively Fading Channels," *IEEE Trans. Commun. Syst.*, Vol. CS-10, 1962, pp. 32–42.

[26] Gans, M. J., *"The Effect of Gaussian Error in Maximal Ratio Combiners,"* *IEEE Trans. Commun. Technol.*, Vol. COM-19, August 1971, pp. 492–500.

[27] Alouini, M. S., S. W. Kim, and A. Goldsmith, "RAKE Reception with Maximal-Ratio and Equal-Gain Combining for CDMA Systems in Nakagami Fading," *Proc. IEEE Int. Conf. Univ. Personal Commun. (ICUPC '97)*, San Diego, CA, October 1997, pp. 708–712.

[28] Tomiuk, B. R., N. C. Beaulieu, and A. A. Abu-Dayya, "General Forms for Maximal Ratio Diversity with Weighting Errors," *IEEE Trans. Commun.*, Vol. COM-47, April 1999, pp. 488–492. See also *Proc. IEEE Pacific Rim Conf. Commun. Comput. Signal Process. (PACRIM '95)*, Victoria, British Columbia, Canada, May 1995, pp. 363–368.

[29] Kong, N., T. Emy, and B. L. Milstein, "A Selection Combining Scheme for RAKE Receivers," *Proc. IEEE Int. Conf. Univ. Personal Comm. (ICUPC 95)*, Tokyo, 1995.

[30] Ko, Y. C., M. S. Alouini, and M. K. Simon, "Performance Analysis and Optimization of Switched Diversity," *IEEE Trans. Veh. Technol.*, Vol. 149, Sept. 2000.

[31] Viterbi, A. J., "Error Bounds for Convolutional Codes and an Asymptotically Optimum Decoding Algorithm," *IEEE Trans. Info. Technology*, Vol. 13, 1967.

[32] Chen, Hsiao-Hua, "On Multi-Band Wavelet Packet Spreading Codes with Intra-Code Subband Diversity to Mitigate Frequency Selective Fading in Mobile Communications to Appear," *IEEE Proceeding in Communications*.

[33] Haykin, S., *Adaptive Filter Theory*, third edition, Upper Saddle River, NJ: Prentice Hall, 1996.

[34] Kramer, G., et al. "A Comparison of Demodulation Techniques for Code Time Division Multiple Access," *IEEE, VTC*, 1996.

[35] Trun, G. L., "The Effects of Multipath and Fading on the Performance of Direct-Sequence CDMA Systems," *IEEE J. Sel. Areas in Comm.*, Vol. 2, July 1984.

[36] Vaseghi, Saeed V., *Advanced Digital Signal Processing and Noise Reduction*, New York: John Wiley, second edition, 2000.

5

Interference Analysis

5.1 Introduction

Throughout time, people have and will continue to use communications at an ever-increasing pace in an interference environment [1]. In addition to the widespread use of satellite systems during the decades of the 1970s and the 1980s, we are now living through the mobile revolution. A large percentage of the communications needs can be carried out satisfactorily, even in a bad interference situation, and people are willing to show moderation. For example, hearing a distant cochannel repeater when your local repeater is not active, while annoying, is not "unacceptable interference." Hearing adjacent channel splatter while carrying on a conversation on simplex or your local repeater, while affecting the quality of the conversation, is not truly unacceptable interference. If it makes communication completely impossible, then it should be considered interference, although it still may not necessarily be harmful or willful. Take note at this point that many of the noise sources to be defined here do not affect FM/PM type radio operation except to cause desensing of the radio, possibly masking the desired signal. This is the reason we strive to define and derive the qualitative measures by which we can design modern wireless system in an ever-increasing interference background.

Up to this point, we have examined and analyzed distortion mainly in the form of fading that is caused to information signals by the wireless channel for the types of wireless systems currently being used. In this and the following chapters, we shall analyze and study interference and include additive effects. We shall define signal to interference ratio (S/I, or SIR) as a quality measure and relate probability of error to carrier-to-interference ratio (C/I, or CIR) or S/I. In general terms, however, interference is considered in this book as any distortion agent to the desired signal. With the expected increase in congestion of frequency spectrum by the use of satellites, mobile systems, and wireless local loops (WLLs) in conjunction with various frequency reuse mitigation techniques in order to allow usage of higher bands, the role played by interference is likely to increase in the future. In the first three chapters, we introduced the design parameters of wireless systems in general, discussed the basic characteristics of the channel, analyzed coding, and defined the quality measure for various modulation techniques as the wireless system operates in an interference environment. We also defined the interference environment and recognized that we could categorize in characteristics into two groups. One is referred to the additive types of interference, which include cochannel, adjacent channel, intersystem intermodulation, and intersymbol. The other is referred to the multiplicative type, which is mainly the effect of multipath reflections, diffraction, and dispersion of transmitted signals as they enter the receiver of wireless systems, especially mobile. The effects of this type of interference are analyzed in Chapter 4. In this chapter, we shall analyze and discuss the additive type of interference. We shall also point out the parameters that will be incorporated in Chapter 6 to develop realistic interference-reduction techniques.

5.2 Types of Interference

The interference signals in wireless communication systems can be placed in two categories for the purposes of this chapter: those caused by natural phenomena, which are not within our capability to eliminate, and those manmade signals that, by and large, can be attenuated or controlled. Our objective here is to define interference as a signal that affects communications, define its sources, and then point out those methods that can be used in the design of modern wireless systems, in order to have acceptable communications in this type of setting.

5.2.1 Cochannel Interference

Cochannel interference is defined as the interfering signal that has the same carrier frequency as the useful information signal. For analysis purposes, we utilize the conditional cochannel interference probability (CCIP) measure [2].

Blocking probability or CCCIP is defined as the probability that the undesired signal local mean power (LMP) exceeds the desired LMP by a protection ratio denoted as β. Amplitude fading in a multipath pico- or microcellular environment may follow different distributions depending on the area covered, presence or absence of a dominating strong component, and some other conditions. For example, the motion of people within a building causes Rician fading in LOS paths, while Rayleigh fading still dominates in non-LOS paths. The Rician distribution contains the Rayleigh distribution as a special case and simultaneously is well approximated by a Gaussian distribution.

The calculation of the CCIP in a Nakagami mobile environment is particularly important, as Nakagami fading is one of the most appropriate models in many mobile communication practical applications. Nakagami distribution (also called m-distribution) contains a set of other distributions for special cases and provides the optimum in analyzing data from outdoor and indoor environments.

The CCIP P_c can be expressed as [2]:

$$P_c = \text{Prob}\left(\frac{\dfrac{s}{k}}{\sum\limits_{i=1}^{k} I_i} < \beta \right) \tag{5.1}$$

where

$s = $ the LMP of the desired signal;

$I_i = $ the LMP of the ith interferer;

$\beta = $ the protection ratio;

$k = $ the number of interferers.

If $w = s - \beta \cdot \sum\limits_{i=1}^{k} I_i$ then (5.1) can be written as follows

$$P_c = \text{Prob}\,(w > 0) \tag{5.2}$$

Considering log-normal PDFs for the I_i and s, the following expressions are given

$$p_i(y_i) = \frac{1}{\sqrt{2} \cdot \pi \cdot \sigma_i \cdot y_i} \exp\left(\frac{-(\ln y_i - m_i)^2}{2\sigma_i^2}\right), \qquad y_i \geq 0 \quad (5.3)$$

where

y_i = the LMP of the ith interferer;

σ_i = the standard deviation of the LMP of the ith interferer.

$$p_S(y) = \frac{1}{\sqrt{2} \cdot \pi \cdot \sigma_S \cdot y} \exp\left(\frac{-(\ln y - m_S)^2}{2\sigma_S^2}\right), \qquad y \geq 0 \quad (5.4)$$

where

y = the LMP of the desired signal;

σ_S = the standard deviation of the LMP of the desired signal.

It can be proven that the PDF of the βI_i is given [2]:

$$p_{\beta I_i}(y) = \frac{1}{\beta} p_{I_i}\left(\frac{y}{\beta}\right) \tag{5.5}$$

Let $\Phi_W(r)$, $\Phi_S(r)$, $\Phi_{\beta I_i}(r)$ be the characteristic functions of the variables w, s, βI_i, respectively. By taking into account that s, I_i are statistically independent, the following can be written [2]:

$$\Phi_W(r) = \Phi_S(r) \cdot \prod_{i=1}^{k} \Phi_{\beta I_i}(-r) \tag{5.6}$$

Using the definition of the characteristic function, (5.6) assumes the form

$$\Phi_W(r) = \Phi_S(r) \cdot \prod_{i=0}^{k} \int_0^\infty \exp(-irx_i) \cdot f_{\beta I_i}(x_i) \, dx_i \tag{5.7}$$

where

$$f_{\beta I_i}(x) = \frac{1}{\beta} \cdot f_{I_i}(x) = \frac{1}{\beta} \cdot \left(\frac{m_k}{\Omega_k}\right)^{m_k} \cdot \frac{x^{m_k-1}}{\beta^{m_k-1} \cdot \Gamma(m_k)} \cdot \exp\left(-\frac{m_k}{\Omega_k} \cdot \frac{x}{\beta}\right)$$

(5.8)

where $f_{\beta I_i}(x)$ could be the log-normal or other well-known PDFs as m-Nakagami

m_k = an arbitrary fading parameter;

Ω_k = the average power.

$$\Gamma(\cdot) \equiv \int_0^\infty \rho^{(\cdot)-1} e^{-\rho} \, d\rho$$

where

$$\Gamma(x) = \text{the Gamma function}$$

Setting $\ln x_i - \ln \beta = m_i + \sigma_i \cdot r_i$ in (5.5), from (5.6) and (5.7) and making certain simplifications, we obtain:

$$\Phi_w(r) = \frac{k}{2} \cdot 2\pi \cdot \Phi_S(r) \cdot \int_{-\infty}^{\infty} \dots \int_{-\infty}^{\infty} \exp\left(-jr\left(\tau + \sum_{i=1}^{k} \beta \cdot e^{(m_i + \sigma_i \cdot r_i)}\right)\right)$$

$$\cdot \exp\left(-\sum_{i=1}^{k} \frac{r_i^2}{2}\right) dr_1 \dots dr_k \tag{5.9}$$

The random variable r_i, which represents the amplitude of the ith cochannel interferer, follows log-normal of Nakagami distribution. All of the r_i are statistically independent with $r_i \geq 0$.

But, using (5.2) and by definition, we have

$$P_C = \int_{-\infty}^{0} f_W(\tau) \, d\tau = \frac{1}{2\pi} \cdot \int_{-\infty}^{0} \int_{-\infty}^{\infty} \Phi_W(r) \cdot \exp(-jr\tau) \, dr d\tau \tag{5.10}$$

Now, using (5.5) in (5.10) and taking into account that by definition

$$\int_{-\infty}^{\infty} \Phi_S(r) \cdot \exp\left[-jr\left(\tau + \sum_{i=1}^{k} \beta \cdot e^{(m_i + \sigma_i \cdot r_i)}\right)\right] dr \qquad (5.11)$$

$$= 2\pi f\left(\tau + \sum_{i=1}^{k} \beta \cdot e^{(m_i + \sigma_i \cdot r_i)^2}\right)$$

Then, the expression for P_C when $f(x)$ is the log-normal PDF of the desired signal, then the P_C can be written as

$$P_C = \frac{1}{(2\pi)^{k/2}} \cdot \int_{-\infty}^{\infty} \cdots \int_{-\infty}^{\infty} \exp\left(-\sum_{i=1}^{k} \frac{r_i^2}{2}\right) \qquad (5.12)$$

$$\cdot F\left(\sum_{i=1}^{k} \beta \cdot e^{(m_i + \sigma_i \cdot r_i)}\right) dr_1 \ldots dr_k$$

where $F(x)$ is the CDF given by

$$F(x) = G_{NORMAL}\left(\frac{\ln x - m_S}{2\sigma_S}\right) \qquad (5.13)$$

with G_{NORMAL} being the CDF of the normal distribution. Hence, the final form for the P_C is

$$P_C = \frac{1}{(2\pi)^{k/2}} \cdot \int_{-\infty}^{\infty} \cdots \int_{-\infty}^{\infty} \exp\left(-\sum_{i=1}^{k} \frac{r_i^2}{2}\right) \qquad (5.14)$$

$$\cdot G_{NORMAL}\left(\frac{\ln \beta - m_S + \sum_{i=1}^{k} e^{(m_i + \sigma_i \cdot r_i)}}{\sigma_S}\right) dr_1 \ldots dr_k$$

where

β = the protection ratio in natural units;

γ = the path loss propagation factor;

σ_i = the standard deviation of the LMP of the interferers in natural units;

σ_s = the standard deviation of the LMP of the desired signal in natural units.

The second part of (5.14) can be calculated using the following Gauss-Hermite formula

$$\int_{-\infty}^{\infty} \exp\left[-x^2\right] \cdot g(x)\, dx = \sum_{i=0}^{\nu} \alpha_i \cdot g(x_i) \tag{5.15}$$

where

α_i, x_i = constants given by special tables;

ν = a constant that denotes the accuracy at the νth decade digit.

With the Gauss-Hermite formula, we can control the error in desired levels, but in a real cellular mobile radio environment, the shadow-fading parameter σ has different values in different regions of the system area. In this case, our formula for the CCIP is modified to

$$P_c = \frac{1}{(2 \cdot \pi)^{k/2}} \int_{-\infty}^{\infty} \cdots \int_{-\infty}^{\infty} \exp\left[-\sum_{i=1}^{k} \frac{r_i^2}{2}\right] \tag{5.16}$$

$$\cdot\, G\left[\frac{ln\left(\beta \cdot (3 \cdot n_g)^{(-\gamma/2)} \cdot (e^{\sigma_1 \cdot r_1} + e^{\sigma_2 \cdot r_2} + \ldots)\right)}{\sigma_s}\right] dr_1 \ldots dr_k$$

with σ_1, σ_2, $\ldots$, σ_k as the standard deviations of the logarithm of the LMPs of the k interferers.

Because

$$(3 \cdot n_g)^{1/2} = \frac{D}{R} \tag{5.17}$$

with

R = the radius of the cell and;

D = the distance from the first tier;

n_g = the cluster size.

P_C can be written as

$$P_C = \frac{1}{(2 \cdot \pi)^{k/2}} \int_{-\infty}^{\infty} \cdots \int_{-\infty}^{\infty} \exp\left[-\sum_{i=1}^{k} \frac{r_i^2}{2}\right] \tag{5.18}$$

$$\cdot F\left[\frac{ln\left(\beta \cdot \left(\frac{D}{R}\right)^{-\gamma} \cdot (e^{\sigma_1 \cdot r_1} + e^{\sigma_2 \cdot r_2} + \ldots)\right)}{\sigma_s}\right] dr_1 \ldots dr_k$$

Equation (5.18) gives a general form for the CCIP in terms of the critical (for the cellular system) cochannel interference reduction factor D/R. This is very important for the system designer because there is a direct connection between CCIP and this factor. Hence, giving a desired value for P_c in (5.18) and using an approximate mathematical method to solve this equation, the factor D/R can be calculated for several shadow and path loss environments of the system.

Equation (5.18) is true as long as the cell size is fixed and the cochannel interference is thus independent from the transmitted power of each cell. But, in the case where cell size is not fixed, the distances from the first tier are not the same for all the k interferers and (5.18) must be modified to

$$P_c = \frac{1}{(2 \cdot \pi)^{k/2}} \int_{-\infty}^{\infty} \cdots \int_{-\infty}^{\infty} \exp\left[-\sum_{i=1}^{k} \frac{r_i^2}{2}\right] \tag{5.19}$$

$$\cdot F\left[\frac{ln\,\beta - m_s + ln\left(\sum_{i=1}^{k} e^{m_s \cdot \left(\frac{D_i}{R_i}\right)^{-\gamma} \cdot (\sigma_i \cdot r_i)}\right)}{\sigma_s}\right] dr_1 \ldots dr_k$$

with

m_s = the area mean power of the desired signal;

R = the radius of the cell contained the desired transmission;

R_i = the radius of the cell contained the ith interferer and;

D_i = the distance of the ith interferer from this cell.

Blocking probability should be kept below 2%. As for the transmission aspect, the aim is to provide good quality service for 90% of the time. The analysis so far resulted in a simple criterion of relating design parameters such as D/R with quality of service in an interference environment.

5.2.2 Adjacent Channel Interference

The adjacent channel interference can be classified as either *inband* or *out-of-band* interference. The term *inband* is applied when the center of the interfering signal bandwidth falls within the bandwidth of the desired signal. The term *out of band* is applied when the center of the interfering signal bandwidth falls outside the bandwidth of the desired signal.

In the mobile radio environment, the desired signal and the adjacent channel signal may be partially correlated with their fades. Then the probability exists that $r_2 \geq \alpha r_1$, where r_1 and r_2 are the two envelopes of the desired and the interfering signals, respectively. In that case, the probability can be obtained from the joint density function, assuming that $E\left[r_1^2\right] = E\left[r_2^2\right] = 2\sigma^2$ and that α is a constant

$$P(r_2 \geq \alpha r_1) = \int_0^\infty dr_1 \int_{\alpha r_1}^\infty p(r_1, r_2)\, dr_2$$

$$= \int_0^\infty dr_1 \int_{\alpha r_1}^\infty r_1 r_2 \exp\left[-\frac{r_1^2 + r_2^2}{2\sigma^2(1 - \rho_r)}\right] I_0\left[\frac{r_1 r_2}{\sigma^2} \cdot \frac{\sqrt{\rho_r}}{(1 - \rho_r)}\right] dr_2$$

$$= \frac{1}{2} + \frac{1}{2} \cdot \frac{1 - \alpha^2}{\sqrt{(1 + \alpha^2)^2 - 4\rho_r \alpha^2}} \tag{5.20}$$

where

ρ_r = is the correlation coefficient between r_1 and r_2.

The probability density function $p_r(y)$ of $r = r_2/r_1$ can be obtained as follows

$$p_r(y)\big|_{y=\alpha} = -\frac{d}{d\alpha}\, P\!\left(\frac{r_2}{r_1} \geq \alpha\right) \tag{5.21}$$

We determine the term $R = \sqrt{Gr}$, where G is the power gain at the intermediate frequency filter output for the desired signal relative to the adjacent channel interferer. Then, we have

$$p_{R^2}(x) = p_r(y)\,\frac{1}{2yG}\bigg|_{y=\sqrt{x/G}} = \frac{(1-\rho_r)\left(1 + \dfrac{x}{G}\right)}{G\left[\left(1 + \dfrac{x}{G}\right)^2 - 4\rho_r\dfrac{x}{G}\right]^{3/2}} \tag{5.22}$$

where ρ_r is given by the formula

$$\rho_r(\Delta\omega,\, \tau) = \frac{J_0^2(\beta V\tau)}{1 + (\Delta\omega)^2\Delta^2}$$

and with $\tau = 0$, it is simplified in the following form

$$\rho_r = \frac{1}{1 + (\Delta\omega)^2\Delta^2} \tag{5.23}$$

where the term $\Delta\omega/2\pi$ is the difference in frequency between the desired signal and the interferer. The term Δ is the time delay spread. The ρ_r decrease, which will vary in value depending on the different types of mobile environments proportionately as either Δ or $\Delta\omega$ increases. As ρ_r decreases, the adjacent channel interference also decreases. The same procedure used to find the cochannel baseband SNR can also be used to find the baseband SNR, due to an adjacent channel interferer in a fading environment, by substituting the PDF of (5.19) in place of the PDF in a Rayleigh fading environment.

As a final consideration, when adjacent channel interference is compared with cochannel interference at the same level of interfering power, the effects of the adjacent channel interference are always less.

5.2.3 Intermodulation Interference

Nonlinear system components, especially in analog signal transmission, cause spurious signals, which may play the role of interference in adjacent channels. When a nonlinear device (amplifier) is used simultaneously by a number of carriers, intermodulation products are generated, which cause distortion in the signals. The nonlinearities in such cases are of two types: amplitude nonlinearities and amplitude to phase conversions (AM/PM), by which the change in the envelope of multicarrier input causes a change in the output phase of each signal component. In many instances, especially when the nonlinear element operates below saturation level, the AM/PM effects dominate the instantaneous amplitude nonlinearity.

In this section, we shall follow a procedure similar to the one described in previous sections and try to relate quality of communication with design system parameters in an environment, which operates in relation to phase intermodulation interference. Both nonlinearities will be treated jointly and the AM/PM conversion is modeled as follows.

Assuming an input signal of the following form [3]:

$$s_1(t) = R_e(Ae^{j\omega_0 t}) \tag{5.24}$$

is used as an input to a nonlinear device, then the output of the particular nonlinear device with AM/PM characteristics is given by

$$s_0(t) = R_e g(A) \cdot e^{j(\omega_0 t + f(A))} \tag{5.25}$$

where $g(A)$ and $f(A)$ are the amplitude and phase functions, respectively. For the rest of the analysis, (5.25) will represent the reference model for the nonlinearities we are going to consider.

In order to facilitate calculations, it is customary to use the approximation suggested in [4], which is given here:

$$g(\rho) e^{jf(\rho)} \approx \sum_{\ell=1}^{L} b_\ell J_1(a\ell\rho) \tag{5.26}$$

Intermodulation effects caused by this type of nonlinearity are important in multicarrier signals, which will be considered next.

Assume that the input signal to a nonlinear device of the type described earlier is given by:

$$s_1(t) = R_e\left[\sum_{i=1}^{M-1} A_i e^{(j\omega_0 t + j\vartheta_i(t))} + (N_c(t) + jN_s(t)) e^{j(\omega_0 + \omega_m)t}\right]$$

(5.27)

or

$$s_1(t) = R_e\left[\sum_{i=1}^{m-1} A_i e^{j(\omega_0 t + \vartheta_i(t))} + A_m(t) e^{j(\omega_0 t + \vartheta_m(t))}\right]$$

(5.28)

where

$$\vartheta_m(t) = \omega_m t + \tan^{-1}\frac{N_s(t)}{N_o(t)}$$

$$A_i = \text{constant}$$

(5.29)

$$A_m(t) = \sqrt{N_c^2(t) + N_s^2(t)}$$

$$\vartheta_i(t) = \text{phase input carrier}$$

If this input multicarrier signal goes through a nonlinear device of the type described earlier, the output is given by [1–5].

$$s_0(t) = R_e\left[e^{j\omega_0 t}\sum_{\substack{k_1,k_2,\ldots k_m=-\infty \\ k_1+k_2+\ldots+k_m=1}}^{\infty} e^{j\sum_{j=1}^{m-1} k_i \vartheta_i(t)} \cdot M_{(k_1,k_2,\ldots k_m)} \cdot e^{jk_m \vartheta_m(t)}\right]$$

(5.30)

where

$$M(k_1, k_2, \ldots k_m) = \int_0^{\infty} \gamma \prod_{i=1}^{m} J_{k_i}(\gamma A_i)\, d\gamma \cdot \int_0^{\infty} \rho g(\rho) e^{jf(\rho)} \cdot J_1(\gamma\rho)\, d\rho$$

(5.31)

In the absence of a noise signal at the input of the device, the output, $s_o(t)$, consists of the angle-modulated carriers and intermodulation products, which also have properties of angle-modulated carriers. With the introduction of noise at the input, the output may be divided into two categories:

1. The original output components with modified complex amplitudes;

2. Additional intermodulation components caused by the introduction of noise.

$$s_o(t) = s_s(t) + s_N(t) \tag{5.32}$$

These two classes can be represented by s_s and s_N, respectively. For the particular case of Gaussian noise whose *rms* power is $R(0)$, this yields

$$s_s(t) = R_e \left[e^{j\omega_0 t} \sum_{\substack{k_1, k_2, \ldots k_{m-1}=-\infty \\ k_1+k_2+\ldots+k_{m-1}=1}}^{\infty} e^{j\sum_{j=1}^{m-1} k_i \vartheta_i(t)} \cdot M_s(k_1, k_2, \ldots k_{m-1}) \right] \tag{5.33}$$

where

$$M(k_1, k_2, \ldots k_{m-1}) = \int_0^{\infty} \gamma \prod_{i=1}^{m-1} J_{k_i}(\gamma A_i) e^{\frac{-\gamma^2}{2} R(o)} \, d\gamma \tag{5.34}$$

$$\cdot \int_0^{\infty} \rho g(\rho) e^{jf(\rho)} J_1(\gamma \rho) \, d\rho$$

This output signal can further be categorized into three types:

1. The main carrier to be demodulated;

2. The intermodulation products and noise falling within the band of the receiver filter of this main carrier;

3. The other carriers, intermodulation products, and noise falling away from the main carrier, which can be filtered out.

Categories 1 and 2 are important in the process of demodulating the main carrier. Before it passes the demodulator, and if we further assume for simplicity that $k_1 = 1$ and all other $k_i = 0$, the output signal can be represented as follows

$$s_o(t) = R_e e^{j(\omega_0 t + \vartheta_1(t))} M_o(1 + R(t) + jI(t)) \tag{5.35}$$

where

$$M_o = \int_0^\infty \gamma \left[\prod_{i=2}^{m-1} J_o(\gamma A_i) \right] J_1(\gamma A_1) C(\gamma)\, d\gamma \cdot \int_0^\infty \rho g(\rho) e^{jf(\rho)} J_1(\gamma \rho)\, d\rho \tag{5.36}$$

$$C(\gamma) = \int_0^\infty J_o(\xi \gamma) p(\xi)\, d\xi \tag{5.37}$$

$p(\xi)$ = probability density function of the noise amplitude ξ usually Rayleigh distributed;

$R(t)$ = Real part of the expression in (5.38);

$I(t)$ = Imaginary part of the expression in (5.38).

Hence,

$$I(t) = I_m M_0^{-1} [M(1, 0, \ldots, 0; t) - M_0]$$

$$+ I_M \left\{ M_0^{-1} \cdot \sum_{\substack{k_1, k_2, \ldots k_M = -\infty \\ k_1 + k_2 + \ldots + k_m = 1}}^{\infty} M(k_1, k_2, \ldots k_m; t) \right. \tag{5.38}$$

$$\left. \cdot e^{jk_m\left(\omega_m t + \tan^{-1}\frac{N_s(t)}{N_c(t)}\right)} \cdot e^{\left(j(k_1 - 1)\vartheta_1(t) + j\sum_{i=2}^{m-1} k_i \vartheta_i(t)\right)} \right\}$$

where $M(1, 0, \ldots; 0, t)$ and $M(k_1, k_2, \ldots k_M; t)$ are given by expressions similar to that given in 5.31 and 5.34 [4], the output signal $s_o(t)$ is then passed through an ideal angle demodulator.

The output of an ideal demodulator therefore based on (5.35) is given

$$S_0^1 = \phi_1(t) + \tan^{-1}\frac{I(t)}{1 + R(t)} \tag{5.39}$$

Because in normal situations, $I(t)$ and $R(t)$ are small, the (5.39) can be approximated by

$$S_0^1 \approx \phi_1(t) + I(t) \tag{5.40}$$

In order to determine the effect of $I(t)$ on the desired angle modulated signal, we need to calculate and determine the power spectrum of $I(t)$. The power spectrum of $I(t)$ in (5.40) as a function of the frequency is given in [3].

Having determined the power spectral density of I, we can then form the ratio of signal power to noise power in the specified frequency band, as we shall see in Section 5.3.1.

For example, if the case under consideration is frequency modulation with multichannel telephony signals, the ratio is given by [5]

$$\text{NPR}(f) = \frac{S}{N} = \frac{P(f) \cdot f_r \cdot f_{rms}^2}{(1 - \epsilon)r^2 f^2 S_I(f)} \tag{5.41}$$

$$S_f = f^2 S_I(f) \tag{5.42}$$

where

NPR = noise power ratio as a function of frequency;

$S_I(f)$ = assumed constant over a telephone channel;

f_{rms} = *rms* frequency deviation;

ϵ = ratio of minimum to maximum baseband frequencies;

f_r = top-based frequency of wanted signal;

$P(f)$ = the pre-emphasis weighting factor;

$r^2 = [C/I]^{-1}$ carrier to interference ratio as a function of frequency.

If we deal with the digital carriers, the impairment is measured in terms of the bit error probability, P_e, as we shall see later. For special cases, 4-phase (phase shift keying modulation) this parameter is given by

$$P_e = \frac{1}{2}\, erfc\left(\sqrt{\gamma}\right) \tag{5.43}$$

where it is assumed that the interference is a close approximation to Gaussian noise and that γ is the S/N at the filter output at the sampling instant. We see that we have been able to relate S/N and error probability with crucial design parameters of the wireless systems under consideration.

5.2.4 Intersymbol Interference

For several types of digital modulation, the equivalent lowpass transmitted signal has the following form [4–8]:

$$s_m(t) = \sum_{n=0}^{\infty} I_n u(t - nT) \tag{5.44}$$

where I_n represents the discrete information bearing sequence of symbols and $u(t)$ represents a pulse that, for simplicity, is assumed to have a bandlimited frequency characteristic $U(f)$ (i.e., $U(f) = 0$ for $|f| > W$).

We assume that the channel frequency response $C(f)$ is also bandlimited such as $C(f) = 0$ for $|f| > W.$

The received signal has the form

$$s_0(t) = \sum_{n=0}^{\infty} I_n h(t - nT) + n(t) \tag{5.45}$$

where

$$h(t) \equiv \int_{-\infty}^{\infty} u(t)\, c(t - r)\, dr \tag{5.46}$$

$n(t)$ = represents additive Gaussian noise

The received signal is usually first passed through a filter and then sampled at the rate of $1/T$ samples per second.

We denote the output of the receiving filter as

$$y(t) = \sum_{n=0}^{\infty} I_n x(t - nT) + V(t) \tag{5.47}$$

where $V(t)$ is the response of the receiving filter to the noise $n(t)$. Sampling $y(t)$ at sampling instants T seconds apart, we should be able to obtain the transmitted information symbol. The sampling gives

$$y(\kappa T + \tau_0) \equiv y_k = \sum_{\kappa=0}^{\infty} I_n x(kT - nT + \tau_0) + V(\kappa T + \tau_0) \tag{5.48}$$

$$y_\kappa = \sum_{n=0}^{\infty} I_n x_{\kappa-n} + V_k \tag{5.49}$$

where

$$x_{k-n} = x(kT - nT + \tau_0)$$

$$V_k = V(kT + \tau_0) \tag{5.50}$$

$$y_k = x_0 I_\kappa + \sum_{\substack{n=0 \\ n \neq k}}^{\infty} I_n x_{k-n} + V_\kappa$$

Because x_0 is a scaling factor, we can set it arbitrarily to unity and thus the previous equation becomes

$$y_\kappa = I_\kappa + \sum_{\substack{n=0 \\ n \neq k}}^{\infty} I_n x_{k-n} + V_\kappa \tag{5.51}$$

In (5.51), the first term is the transmitted information symbol at the kth sampling instant and the second term

$$\sum_{\substack{n=0 \\ n \neq \kappa}}^{\infty} I_n x_{\kappa-n} \tag{5.52}$$

is the unwanted signal (*intersymbol interference*), which is the interference contribution of other symbols to the symbol under consideration. v_κ is the contribution of the additive Gaussian noise. This unwanted interference, depending on the type of modulation used, could be viewed on an oscilloscope as an *eye pattern,* or as a two-dimensional scatter diagram, as shown in Figures 5.1 and 5.2 [6].

It will be shown next that in order to eliminate this interference, the received signal must pass through a filter, which *is matched to the received pulse.* That is, the frequency response of the receiving filter should be $H^*(f)$.

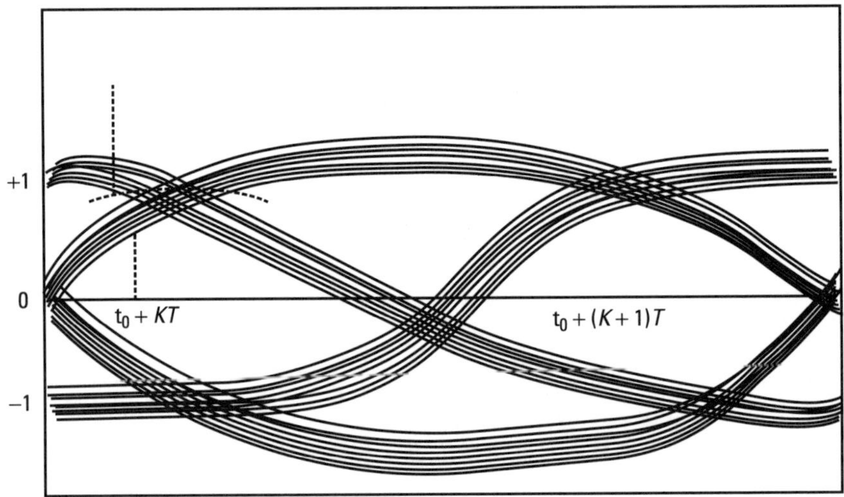

Figure 5.1 Eye pattern for binary pulse amplitude modulation (PAM).

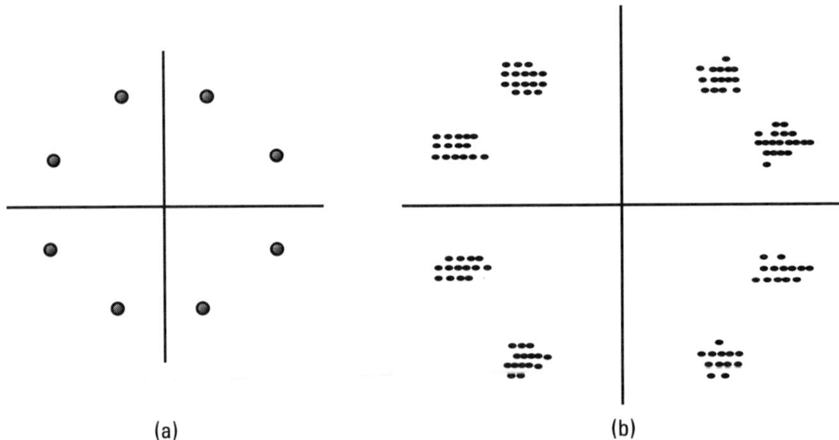

Figure 5.2 Two-dimensional digital eye patterns: (a) transmitted eight-phase signal and (b) received signal samples at the output of demodulator.

$H^*(f)$ is the complex conjugate of the frequency characteristic $H(f)$ of the input pulse $h(t)$.

If for simplicity we assume $C(f) = 1$ for all $|f| \leq W$, then $x(t)$ shown in (5.47) can be given by

$$x(t) = \int_{-W}^{W} X(f) e^{j2\pi ft} \, df \qquad (5.53)$$

where

$$X(f) = U(f) U^*(f) = |U(f)|^2 \qquad (5.54)$$

For no intersymbol interference to exist, it is necessary that

$$x(t = \kappa T) = 1 \qquad \text{for} \qquad \kappa = 0 \qquad (5.55)$$
$$x(t = \kappa T) = 0 \qquad \text{for} \qquad \kappa \neq 0$$

Because $x(t)$ is a bandlimited signal, use of the sampling theorem gives [6]:

$$x(t) = \sum_{n=-\infty}^{\infty} x\left(\frac{n}{2W}\right) \frac{\sin 2\pi W\left(t - \frac{n}{2W}\right)}{2\pi W\left(t - \frac{n}{2W}\right)} \qquad (5.56)$$

where

$$x\left(\frac{n}{2W}\right) = \int_{-W}^{W} X(f) e^{j2\pi f \frac{n}{2W}} \, df \qquad (5.57)$$

If, moreover, we assume:

$$T = \frac{1}{2W} \qquad (5.58)$$

and the symbol rate is the Nyquist rate, (5.56) becomes:

$$x(t) = \sum_{n=-\infty}^{\infty} x(nT) \frac{\sin n\pi \dfrac{(t-nT)}{T}}{\pi \dfrac{(t-nT)}{T}} \qquad (5.59)$$

For zero intersymbol interference, it is required that all $x(nT)$ terms be zero except $x(0)$. The previous equation then becomes

$$x(t) = \frac{\sin\left(\dfrac{\pi t}{T}\right)}{\dfrac{\pi t}{T}} \qquad (5.60)$$

Three major problems are raised with this type of pulse in addition to the conditions set in order to eliminate ISI by (5.55).

1. This type of pulse is not physically realizable.
2. The tails of $x(t)$ decay as $1/t$, and a mistiming error in sampling results in an infinite series of ISI components.
3. There is absolutely no flexibility in the symbol rate, but it must be precisely defined and restricted $T = 1/2W$.

In practical situations, it is not possible to satisfy all three conditions simultaneously. If we impose the condition that the symbol rate be $2W$ symbols per second and remove the constraint that there is zero ISI, we obtain a class of physically realizable pulses called partial response signals. The compensation for the interference is then obtained through equalization and/or various optimization techniques. By these optimization techniques, we seek to obtain optimal receiver filter parameters with which, in turn, we obtain the best estimate of the received symbols. This, in effect, results in minimizing interference. The concept contained in this paragraph will be the cornerstone of some of the various methodologies, which will be developed to combat interference.

As far as equalization is concerned, the main thrust of the procedure lies in the fact that we seek to design discrete-time linear receiver filters to eliminate or reduce ISI— see (5.52)—which have impulse responses of the form

$$q_n = \sum_{j=-\infty}^{\infty} c_j f_{n-j} \qquad (5.61)$$

where q_n is simply the convolution of c_n and f_n. Moreover, c_n is the impulse response of the equalizer, and f_n is the impulse response of the filter. In such a case, the estimate of the kth symbol is given by

$$\hat{I}_k = q_0 I_k + \sum_{n \neq k} I_n q_{k-n} + V_k \tag{5.62}$$

The first term represents the scaled version of the desired symbol, which can be normalized to unity. The second term is the ISI, and the third term represents the noise. A standard procedure that leads to acceptable filter designs is to find the tap weight coefficients c_j of the equalizer, as shown in Figure 5.3, which minimize the mean square error (MSE) value of the error $e_k = I_k - \hat{I}_k$.

In most cases, we use optimization techniques, which seek to minimize the MSE, $E\left[e_\kappa^2\right]$ having as a starting point an assumed receive filter structure of the form of (5.61), whose optimal design parameters are determined by the optimization algorithm that is developed.

In order to present how the optimization techniques would work in the design of receiver filters, we can consider a binary PAM system. The transmitted continuous-time signal can be expressed as [7]

$$s(t) = \sum_{i=-\infty}^{\infty} a_i^o h_\tau(t - iT) + i_T(t) \tag{5.63}$$

where $a_i^o \in [-A, A]$ are the transmitted PAM symbols of the desired channel, which are assumed to be statistically independent.

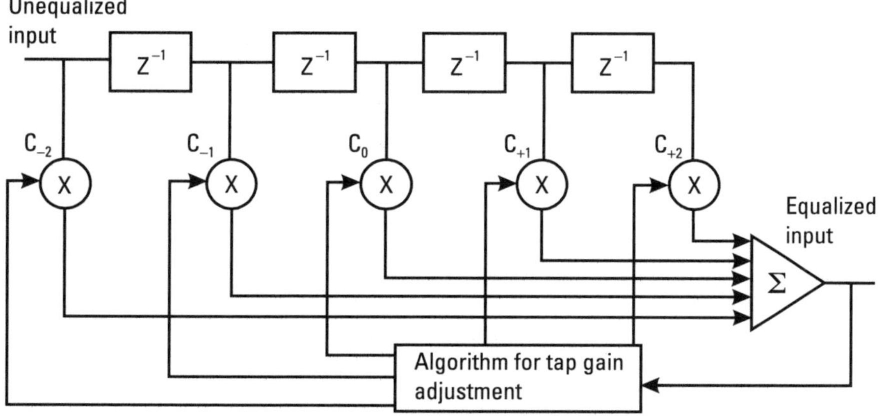

Figure 5.3 Linear filter equalizer. (After: [7].)

$i_T(t)$ is the interference from $2N$ adjacent channels, with a spacing of B_C Hz expressed as follows

$$i_T(t) = \sum_{\substack{\ell=-N \\ \ell \neq o}}^{N} e^{j(2\pi\ell B_c t+\varphi_\ell)} \sum_{i=-\infty}^{\infty} a_{i\ell} h_T(t - iT - \tau_\ell) \qquad (5.64)$$

where $a_{i\ell}$, φ_ℓ, τ_ℓ are the ith symbol, the phase shift, and the delay of the ℓth adjacent channel, respectively.

The transmitted signal, $s(t)$, goes through a linear channel that has impulse response, $c(t)$, and it is also corrupted by Gaussian noise, $n(t)$. We assume that the receiver filter has impulse response $h_R(t)$, shown in Figure 5.4.

The output of the receiver filter is then given by

$$s_o(t) = h_R(t) * \left[c(t) * \left[\sum_{i=-\infty}^{\infty} a_i^o h_T(t - iT) + i_T(t) \right] + n(t) \right] \qquad (5.65)$$

$$= \sum_{i=-\infty}^{\infty} a_i^o g(t - iT) + i_R(t) + n_R(t)$$

where

$$g(t) = h_T(t) * c(t) * h_R(t) \qquad (5.66)$$

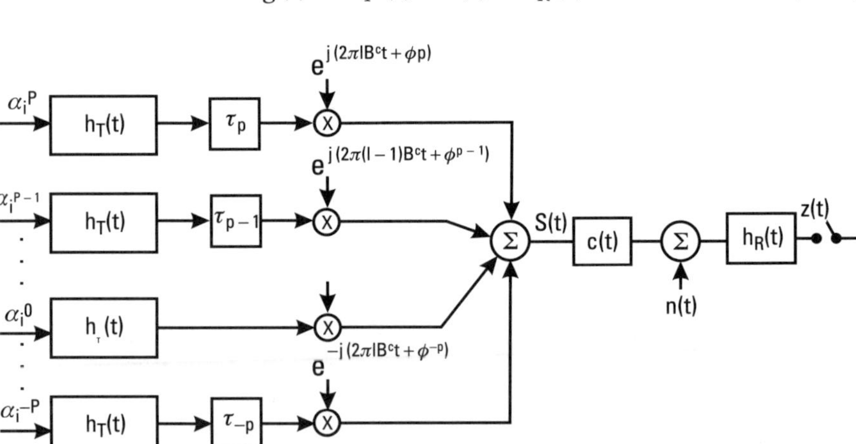

Figure 5.4 Linear filter optimization. (After: [7].)

* denotes the convolution operation

$$n_R = h_R(t) * n(t) \tag{5.67}$$

$$i_R(t) = i_T(t) * c(t) * h_R(t)$$

If we assume that the length of the pulse $g(t)$ is at most $N = 2M + 1$ symbols, the signal sampled at $t = 0$ can be expressed as

$$s_0(0) = \sum_{i=-M}^{M} a_i^o g(-iT) + i_R(0) + n_R(0) = \underline{a}^T \underline{g} + i_R(0) + n_R(0) \tag{5.68}$$

where the vectors $\underline{a}$ and $\underline{g}$ are given in (5.69).

$$\underline{a}^T = \left[a_{-M}^o \dots a_{M-1}^o \cdot a_M^o \right]^T \tag{5.69}$$

$$\underline{g}^T = \left[g(MT) \dots g(-(M-1)T) \cdot g(-MT) \right]^T$$

The error between transmitted symbol a_0^0 and sampled symbol $s_o(0)$ using (5.68) is given by:

$$e_o = a_0^0 - s(0) = a_0^0 - \underline{a}^T \underline{g} - i_R(0) - n_R(0) \tag{5.70}$$

Assuming uncorrelated signal and noise samples, as well as uncorrelated adjacent channel interfering signals, the mean square value of e_o, MSE, is given by squaring (5.70) and finding its mean. This process results in [7]:

$$E\left[e_0^2 \right] = A^2 (1 - g(o))^2 + A^2 \sum_{\substack{i=-M \\ i \neq 0}}^{M} (g(iT))^2 + \sigma_{ACI}^2 + \sigma_N^2 \tag{5.71}$$

where

$$A^2 = E\{ a_0^2 \};$$

σ_{ACI}^2 = average power of adjacent channel interference (ACI);

σ_N^2 = noise variance.

Our objective is to design a discrete-time receive filter to minimize MSE. We shall further assume that L samples are taken per symbol interval $(1/f_s = T/L)$.

The receive filter coefficients will be defined as

$$\underline{h}_R = [h_R(-M_R) \ldots h_R(+M_R - 1) h_R(M_R)]^T \tag{5.72}$$

where the receiver filter coefficients can be expressed as a length $N_R = 2M_R + 1$, whereas similarly the coefficients of the combined transmit filter and channel response will be given by

$$\underline{h}_{TC}(\kappa) = h_T(\kappa) * c(\kappa) \tag{5.73}$$

We shall assume that $\underline{h}_{TC}$ has the following form.

$$\underline{h}_{TC} = [h_{TC}(-M_{TC}) \ldots h_{TC}(+M_{TC} - 1) h_{TC}(M_{TC})]^T$$

where again $N_{TC} = 2M_{TC} + 1$

The kth sample of $g(t)$ then will be given by

$$g(\kappa) = \underline{h}_R^T J(\kappa) \underline{h}_{TC} \tag{5.74}$$

$J(\kappa)$ is an $N_R X N_{TC}$ swapping matrix performing the discrete convolution and $M_R + M_{TC} \leq ML$.

The second term in (5.71) gives [7]:

$$\sigma_{ISI}^2 \equiv A^2 \sum_{\substack{i=-M \\ i \neq 0}}^{M} g^2(iL) = A^2 \sum_{\substack{i=-M \\ i \neq 0}}^{M} \left(\underline{h}_R^T \underline{w}_i\right)^2 = A^2 \underline{h}_R^T W \underline{h}_R \tag{5.75}$$

where

$$W = \sum_{\substack{i=-M \\ i \neq 0}}^{M} \underline{w}_i \cdot \underline{w}_i^T, \quad \underline{w}_i = J(iL) \underline{h}_{TC}$$

Similarly

$$\sigma_n^2 = \underline{h}_R^T R_n \underline{h}_R$$

where R_n = covariance matrix of the noise whose elements R_{ni}^m are given by:

$$R_{lm}^n = \int_{-1/2}^{1/2} S_n(\rho) \cos(\ell - m)\, 2\pi\rho\, d\rho, \text{ where } \rho = f/f_s \quad (5.76)$$

$S_n(\rho)$ = power spectrum of noise

The variance of ACI is given by

$$\sigma_{ACI}^2 = \int_{-1/2}^{1/2} S_I(\rho) |C(\rho) H_R(\rho)|^2\, d\rho \quad (5.77)$$

where

$S_I(\rho)$ is the power spectrum of ACI;

$C(\rho)$ is the Fourier transform of the channel response, $c(t)$;

$$H_R(\rho) = \sum_{i=-M_R}^{M_R} \underline{h}_R(i)\, e^{-j2\pi i} \text{ (Fourier transform of receive filter).}$$

If we define

$$S_I(\rho) = \sum_{k=-\infty}^{\infty} r_I(\rho)\, e^{-j2\pi\kappa\rho}, \text{ where} \quad (5.78)$$

$$r_I(\kappa) = E\left[i_T(nT)\, i_T^*(n + \kappa)\, T \right] \quad (5.79)$$

Using (5.64) and (5.78) and assuming $E\{\varphi_\ell\} = E[\tau_\ell] = 0$, $E = \{\alpha_i^\ell\, \alpha_j^\nu\} = A^2$ if $\ell = \nu$ and $i = j$, otherwise zero, we obtain the power spectrum of ACI as

$$S_I(f) = A^2 \sum_{\substack{\ell=-P \\ \ell \neq 0}}^{P} |H_T(f + \ell B_c T)|^2 \quad (5.80)$$

From (5.77) we obtain

$$\sigma^2_{ACI} = A^2 \underline{h}_R^T R_{ACI} \underline{h}_R \tag{5.81}$$

where R_{ACI} is the covariance matrix of ACI with elements

$$R_{ACI}(k, \ell) = \int_{-1/2}^{1/2} \left[\sum_{\substack{n=-P \\ n \neq 0}}^{P} |H_T(\rho + nB_c T)|^2 \right] |C(\rho)|^2 \cos\left[2\pi(k-\ell)\rho\right] d\rho \tag{5.82}$$

Having expressed all of the terms of $E\left[e_0^2\right]$ in term of $\underline{h}_R$, we can form the following cost functional

$$Q(\underline{h}_R, \lambda) = \beta_{ISI} \underline{h}_R^T W \underline{h}_R + \beta_{ACI} A^2 H_R^T R_{ACI} \underline{h}_R + \underline{h}_R^T R_n \underline{h}_R \tag{5.83}$$
$$+ \lambda\left[\underline{h}_R^T \underline{w}_o - 1\right]$$

where λ is a Lagrange multiplier, β_{ISI} and β_{ACI}, are weight parameters, depending on what emphasis we want to place on ISI and/or ACI.

Taking the derivative of (5.83) with respect to $\underline{h}_R$ and setting it to zero, we obtain the optimal value of the design receive filter parameters, $\underline{h}_R^*$.

$$\underline{h}_R^* = \frac{P^{-1} \underline{w}_o}{\underline{w}_o^T P^{-1} \underline{w}_o} \tag{5.84}$$

where

$$P = \beta_{ISI} A^2 W + \beta_{ACI} A^2 R_{ACI} + R_n \tag{5.85}$$

Having determined $\underline{h}_R^*$, which is the vector that contains the filter design parameters, we can now proceed to construct the receive filter, which in turn will minimize the error between transmit and receive symbols. In other words, the design parameters of this filter have been obtained, which in turn minimize simultaneously intersymbol and adjacent channel interference. The great advantage of this formulation is that it leads to an optimal design that takes into consideration simultenously intersymbol, adjacent channel interference, and noise factors. The procedure described earlier, if it is seen from a different angle, it is equivalent to a having led to an optimization of carrier to interference ration, *C/I*. Over the years, *C/I* has been used as a measure of performance of wireless systems and as such has also been used lately for cases which deal with resource allocation in wireless communication systems.

5.2.5 Near End to Far End Ratio Interference

One type of interference, which occurs only in mobile communication systems, is the near end to far end type of interference [9]. That kind of interference appears when the distance between a mobile unit and the base station transmitter becomes critical with respect to another mobile transmission that is close enough to override the desired base station signal. This phenomenon occurs when a mobile unit is relatively far from its desired base station transmitter at a distance d_1, but close enough to its undesired nearby mobile transmitter at a distance d_2 and $d_1 > d_2$. The problem in that situation is whether the two transmitters will transmit simultaneously at the same power and frequency, thus masking the signals received by the mobile unit from the desired source by the signals received from the undesired source. Also, this type of interference can take place at the base station when signals are received simultaneously from two mobile units that are at unequal distances from the base station. The power difference due to the path loss between the receiving location and the two transmitters is called the *near end to far end ratio interference* and is expressed by the ratio of path loss at distance d_1 to the path loss at distance d_2.

This form of interference is unique to the mobile radio systems. It may occur both within one cell or within cells of two systems.

In One Cell

When mobile station A is located close to the base station, and at the same time mobile station B is located far away from the same base station (e.g., at the cell boundaries), mobile station A causes adjacent-channel interference to the base station and mobile station B (Figure 5.5). The *C/I* at mobile station B is expressed by the following equation [9]:

$$\frac{C}{I} = \left(\frac{d_0}{d_1}\right)^{-\gamma} \tag{5.86}$$

where γ is the path loss slope.

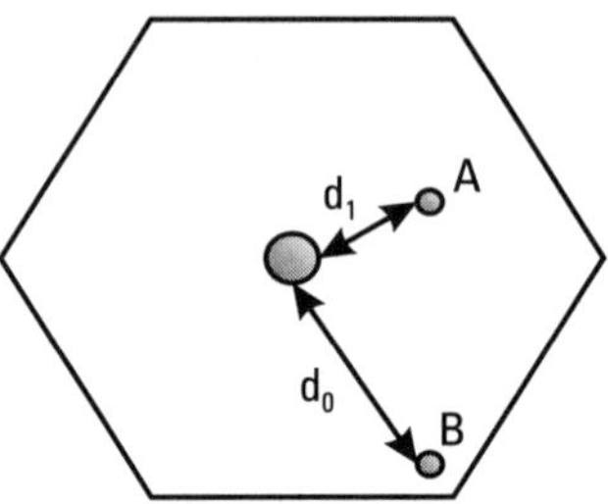

Figure 5.5 Near-far interference in one cell. (After: [9].)

Because $d_0 > d_1$, from (5.86) we obtain $C/I < 1$. This means that the interfering signal is stronger than the desired signal.

This problem can be rectified if the filters used for frequency separation have sharp cut-off slopes. The frequency separation can be expressed as follows [9]:

$$\text{frequency band separation} = 2^{G-1} B$$

where

$$G = \frac{\gamma \log_{10}\left(\dfrac{d_0}{d_1}\right)}{L}$$

B = the channel bandwidth;

L = the filter cut-off slope.

In Cells of Two Systems

If two different mobile operators cover an area, adjacent-channel interference may occur if the frequency channels of the two systems are not properly coordinated.

In Figure 5.6, two different mobile radio systems are depicted. Mobile station A is located at the cell boundaries of system A, but very close to base station B. Also, mobile station B is located at the cell boundaries of system B, but very close to base station A. Interference may occur at base station A from mobile station B and at mobile station B from base station A. The same interference will be introduced at base station B and at mobile station A.

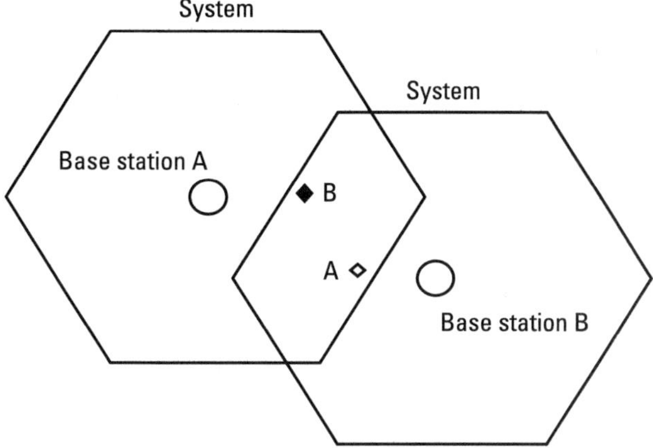

Figure 5.6 Near-far interference in cells of two systems (After: [9].)

This form of interference can be eliminated if the frequency channels of the two systems are properly coordinated, as mentioned earlier. If such a case occurs, two different systems operating in the same area may have colocated base stations.

5.3 Interference Analysis Methodology

One of the main design goals in the cellular mobile terrestrial and satellite communication systems is to provide high capacity in combination with the required quality of service. Due to the architectural structure of these systems, a very crucial issue is the determination methodologies for analyzing the nature and the influence of any kind of interference. Up to now, the system designer almost always assumed that the limiting corrupting signal has Gaussian characteristics, such as the characteristics of thermal noise. With the advent of low-noise receivers and congestion in the radio frequency bands, this assumption can no longer be justified, and interference of non-Gaussian nature into our present and future communication systems is an important issue. The method of analysis used to determine the effect of thermal noise on communication systems cannot, therefore, be used blindly to determine the effect of interference of non-Gaussian nature on the new and evolving wireless systems and thus to design system components. Various analysis tools have been developed, which take into consideration interference not only as an additive distorting agent but also as a multiplicative agent, as in fading, as we saw in Chapter 4. The main objective is then to analyze how the interference as a general distortion agent affects well-accepted criteria of performance of wireless systems, such as C/I or S/I and BER, and then proceed to develop optimal or suboptimal design tools that lead to practical system implementation and that satisfy predetermined minimum performance levels. Chapter 6 will do just that, and Chapter 7 will show how the results of Chapters 5 and 6 will be used to design practical implementations, which satisfy set goals of performance. The analysis methodology that is involved in order to achieve our objective is presented in the following sections. It takes, as a basic analysis tool, the determination of the C/I, S/I, or BER as functions of critical design parameters. The reader is encouraged to refer to the Preface of this book to appreciate the importance of the methodology developed here as a part of the overall methodology developed in the beginning of this book to cover and justify the relevance and interrelationship of all chapters.

This methodology consists of the following steps:

1. Calculation or estimation of interference power density;

2. Calculation of C/I power ratio;

3. Determination of relationship between C/I and S/I or error probability (P_e);

4. Determination of relationship between S/I or P_e and system performance;

5. Determination of relationship between system performance and acceptable level of system parameter changes for improving system performance;

6. Use of C/I as a measure for the optimization of resource allocation and quality;

7. Develop mechanisms and criteria for interference reduction. In any case, develop methods that calibrate the affect of interference by manipulating design parameters.

The two parameters C/I and S/I as a quality measure are intimately related with the grade of service of the wireless systems and for the case of cellular systems with the following parameters:

- Carrier to cochannel interference ratio;
- Blocking probability.

Over the years, practical values for these parameters have been obtained, which set the quality criteria for specific practical wireless systems in use.

5.3.1 Analog Signals

Analog signals are those signals that are produced by the information source (voice or image) and are used for transmission in analog form (i.e., continuous in time). Even though most of the information signals used nowadays for transmission are either digitized (digital) or are produced by the source in data form, we still need to discuss and analyze their interference aspects because the development of interference reduction techniques of digital signals are mostly based on these classical schemes, as we shall see in Chapter 6.

Essential for computing the baseband interference is knowledge of the RF power spectral densities of both the desired and interfering signals. Let the desired angle-modulated signal $s_1(t)$ and an arbitrary narrowband interfering signal $s_2(t)$ be given by [1–9]:

$$s_1(t) = \mathrm{Re}\,[z_1(t)] = \mathrm{Re}\,[A_1 \exp\{j[\omega_1 t + x_1(t) + \mu]\}] \qquad (5.87)$$

$$\mathrm{Re}\,\{A_1 u_1(t)\,[\exp j\omega_1 t]\}$$

$$s_2(t) = \mathrm{Re}\,[z_2(t)] = \mathrm{Re}\,\{V_2(t) \exp[j\omega_2 t]\} \qquad (5.88)$$

respectively.

It is assumed that $s_1(t)$ and $s_2(t)$ are both wide-sense stationary and are generated from separate sources; thus, they are statistically independent of each other. Furthermore, $x_1(t)$ and μ are assumed to be independent and μ is assumed to be uniformly distributed in $\{0\text{–}2\pi\}$. At the input of a demodulator, both signals are added and go through a phase detector, assuming we're dealing with phase modulated signals. The sum of these two signals is given by

$$s(t) = s_1(t) + s_2(t) = \mathrm{Re}\,(a(t))\,\exp\,(j\omega_1 t + x_1(t) + \lambda(t))$$

where

$$a(t)\,e^{j\lambda(t)} = 1 + z(t)\,\exp\,(j(\omega_2 - \omega_1)t - x_1(t) + \mu)$$

and

$$z(t) = \frac{z_2(t)}{z_1(t)}$$

In [1], it is shown that under certain mild conditions, the output of the demodulator will contain the desired signal plus the excess phase cause by the interfering signal. The excess phase angle (caused by the presence of the interference) at the output of an ideal demodulator is given by

$$\lambda(t) = \text{Im} \ln\left[1 + \frac{z_2(t)}{z_1(t)}\right] \tag{5.89}$$

For $\left|z_2(t)/z_1(t)\right| < 1$, $\lambda(t)$ can be expanded as

$$\lambda(t) = \text{Im} \sum_{m=1}^{\infty} \frac{(-1)^{m+1}}{m} \left(\frac{z_2(t)}{z_1(t)}\right)^m = \sum_{m=1}^{\infty} \lambda_m(t) \tag{5.90}$$

The baseband power spectrum of the demodulated interference is obtained from the autocorrelation function of the total detected phase $\phi(t)$ where

$$\varphi(t) = x_1(t) + \lambda(t) \tag{5.91}$$

the autocorrelation function is thereby given

$$R_\phi(\tau) = \langle [x_1(t) + \lambda(t)] \cdot [x_1(t + \tau) + \lambda(t + \tau)] \rangle \tag{5.92}$$
$$= R_{x_1}(\tau) + R_\lambda(\tau)$$

Because the cross terms vanish when averaged over μ, the mth term of $\lambda(t)$ can be written as

$$\lambda_m(t) = \text{Im}\left\{V_2^m(t) \exp(jm\omega_2 t) \exp[jx_m(t)]\right\} K_m$$
$$\lambda_M(t) = \frac{K_m}{2j}\left\{V_2^m(t) \exp[jm\omega_2 t] \exp[jx_m(t)]\right. \tag{5.93}$$
$$\left. - V_2^m(t)^* \exp[-jm\omega_2] \exp[-jx_m(t)]\right\}$$

where

$$x_m(t) = -m[\omega_1 t + x_1(t) + \mu] \tag{5.94}$$

and

$$K_m = \frac{-1^{m+1}}{mA_1^m} \tag{5.95}$$

The term A_1 represents the wanted carrier amplitude.

Equation (5.93) can be used to find the PSD of $\lambda_m(t)$ and the autocorrelation function of $\lambda_m(t)$, $R_\lambda^{mn}(\tau)$. It is shown that [5]:

$$R_\lambda^{mn}(\tau) = \left[\frac{1}{4m^2 A_1^{2m}} \, R_{V_2^m}(\tau) R_{u_1^m}^*(\tau) \, \exp\left[jm(\omega_2 - \omega_1)\tau \right] \right. \tag{5.96}$$

$$\left. + \, R_{V_2^m}^*(\tau) R_{u_1^m}(\tau) \, \exp\left[-jm(\omega_2 - \omega_1)\tau \right] \right]$$

where the $R_{V_2^m}(\tau)$ is the autocorrelation function of $V_2^m(t)$, and the $R_{u_1^m}^*(\tau)$ is the complex conjugation of the autocorrelation function of $u_1^m(t)$.

The power spectrum of the baseband interference is then given by

$$I(f) = \sum_{m=1}^{\infty} \frac{1}{4m^2 A_1^{2m}} \left[T_m(f - mf_s) + T_m(-f - mf_s) \right] \tag{5.97}$$

where

$$T_m(f) = S_{V_2^m}(f) \otimes S_{u_1^m}(f) \tag{5.98}$$

with

$$S_{V_2^m}(f) = F[R_{V_2^m}(\tau)] = \text{power spectral density of } V_2^m(t);$$

$$S_{u_1^m}(f) = F[R_{u_1^m}(\tau)] = \text{power spectral density of } u_1^m(t).$$

where

* denotes complex conjugate;

$\otimes$ denotes convolution.

The solution to the problem of interference into an angle-modulated system in its most general form therefore comprises two convolution terms. Convolving the power spectral of the mth power of the complex envelopes can generate each term. These spectral densities will be used to calculate C/I.

5.3.1.1 Calculation of C/I

For simplicity, we shall assume that the transmitted modulated analog signal is given by the following equation:

$$s(t) = A \cos(\omega_1 t + \varphi(t)) \tag{5.99}$$

and the interference is expressed by

$$i(t) = R(t) \cos(\omega_2 t + \psi(t) + \mu) = \mathrm{Re}\{u(t) \exp(j\omega_2 t + \mu)\} \tag{5.100}$$

where

$\varphi(t)$ includes the information signal;

$\psi(t)$ includes the interference signal;

μ is assumed to be uniformly distributed on $[0, 2\pi]$.

$$u(t) = R(t) e^{j\psi(t)}$$

It is assumed that at the receiver, we obtain the sum of these two signals, which is indicated as $s_0(t)$ and is given by the following formula

$$s_0(t) = s(t) + i(t) \tag{5.101}$$

Equation (5.101), using (5.99) and (5.100), can be written as shown in (5.102) using simple trigonometric identities

$$s_0(t) = \mathrm{Re}\left(A a(t) e^{j(\omega_1 t + \varphi(t) + \lambda(t))}\right) \tag{5.102}$$

where

$$\lambda(t) = \mathrm{Im} \ln\left(1 + z(t) e^{j(2\pi f_\Delta t - \varphi(t) + \mu)}\right) \tag{5.103}$$

and

$$a(t) e^{j\lambda(t)} = 1 + z(t) e^{j(2\pi f_\Delta t - \varphi(t) + \mu)} \tag{5.104}$$

$$z(t) = \frac{u(t)}{A}, \quad f_\Delta = f_2 - f_1 \tag{5.105}$$

$$f = \frac{\omega}{2\pi} \tag{5.106}$$

If we assume $|z(t)| \ll 1$, thus (5.91) can be expanded in a series given by

$$\lambda(t) = \sum_{m-1}^{\infty} \frac{(-1)^{m+1}}{mA^m} (R(t))^m \sin(m2\pi f_\Delta t - \varphi(t) + \psi(t) + \mu)$$

If the receiver we use for detection is an ideal phase detector, as shown in Figure 5.7, we obtain as output $\phi(t) + \lambda(t)$.

We observe that the contribution of the interference signal $i(t)$ to the transmitted information signal $s(t)$ is the signal $\lambda(t)$.

In order to determine the level of performance deterioration for analog signal transmission, a criterion of performance measured in decibels has been developed that is given by

$$20 \log\left(\frac{S}{I}\right) \tag{5.107}$$

This is 20 times the logarithm of the ratio of the signal power to interference power. We need, therefore, to calculate the power ratio S/I.

For any given $\phi(t)$ and $\psi(t)$, the calculation of this power ratio is very difficult, and we usually use some approximation. In most cases of analog transmission, this yields acceptable results.

For example, for typical multichannel frequency division multiplex frequency modulated telephony signals, both $\phi(t)$ and $\psi(t)$ can be assumed to be Gaussian independent and stationary. With these assumptions, it can be shown [5] that

$$\frac{S}{I} = \frac{(2\pi)^2 M_1^2 f_{m1} b}{r^2 (1 - \epsilon_1)} \left[\int_{f_c - b/2}^{f_c + b/2} (2\pi f)^2 S_\lambda(f) \, df \right]^{-1} \tag{5.108}$$

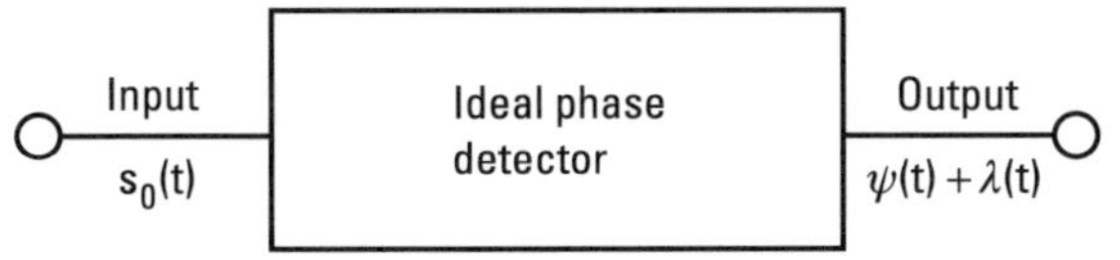

Figure 5.7 Model of an ideal phase detector.

where

f_c = center frequency of channel under construction;

b = telephone channel bandwidth;

f_{m_1} = top baseband frequency of want signal;

$M_I = r_{MS}$ modulation index of wanted multichannel baseband;

ϵ_1 = ratio of lowest to highest frequency of multichannel baseband.

We observe that if this ratio is not acceptable, it can be changed by changing the appropriate parameter in (5.108). The reader can realize that for cases when the assumptions taken for the calculation of (5.108) don't hold, the derivation of an equation equivalent to (5.108) may not be possible in closed form. In such cases, we have to resort to various computational methods and simulation techniques. In any case, for any type of analog modulation used and for any type of service, such as voice or TV implemented, the relationship between performance deterioration with S/I is given to the system designer beforehand. The current approach for very complex analog systems is to use heuristic methods, such as neural networks for the determination of S/I and other design parameters of wireless systems [10].

If we go back to (5.108), we observe that S/I depends directly on $1/r^2$, which is the carrier power $A^2/2$ to interference $r^2 A^2/2$ ratio denoted by C/I. Hence,

$$\frac{C}{I} = \left(\frac{1}{r^2}\right) = \frac{\dfrac{A^2}{2}}{\dfrac{r^2 A^2}{2}} \tag{5.109}$$

In other words, in all cases for any modulation system under consideration, we have a priori:

$$\frac{S}{I} = R\frac{C}{I} \tag{5.110}$$

where R is a constant. For our case, under the assumptions made, this constant is given by (5.108) if we factor out $1/r^2$.

For the case of satellite systems, it can be shown [1] that the C/I is a function of intersatellite spacing, $\Delta\theta$. For certain modulation systems used and for certain services provided by a satellite system, setting a specific level of quality for the service provided, we determine the value of the required S/I and thus C/I. This specific value of C/I sets a limit on the intersatellite spacing and thus on the orbit utilization. It is therefore important to realize that for satellite systems, the interference plays a major role in the orbit utilization. Coupled with thermal noise and for a specified limit of total noise into a certain channel, interference is one of the major of factors of orbit utilization in satellite systems.

5.3.2 Digital Signals

For the case of digital systems, we shall take a standard PSK signal of the form [3–8]:

$$s(t) = A \cos(\omega_1 t + \varphi_1(t)) \tag{5.111}$$

For simplicity, we choose $A = 1$. The digital modulation is carried in the angle of $s(t)$ by $\phi_1(t)$, which assumes discrete values from a set of M equally spaced points in $[0, 2\pi]$ at the sample times T seconds apart. Thus the Nth message or baud is modulated by

$$\varphi_1(NT) = \frac{2\pi k}{M}, \; k = 0, 1, 2, \ldots, M-1$$

where each of M values of k is equally probable. For a coherent receiver, which compares the received wave with the unmodulated carrier, $A \cos \omega_1 t$, and produces instantly the signed phase difference between the two points, an M-ary symbol is transmitted in one baud by the value of k.

The external mainly thermal noise is modeled in the usual fashion by a stationary zero mean Gaussian random process with uniform spectral density, as mentioned in the previous section. Hence,

$$n(t) = n_1(t) \cos \omega_1 t - n_2(t) \sin \omega_2 t \tag{5.112}$$

where $n_1(t)$ and $n_2(t)$ are stationary independent, zero mean Gaussian random processes with power σ^2.

The interference signal shall be modeled by

$$i(t) = rA \cos(\omega_2 t + \varphi_2(t) + \mu) \tag{5.113}$$

At a certain instant, the combined input signal at the detector is given by

$$s'(t) = s(t) + n(t) + i(t)$$

and is presented in Figure 5.8.

The detector examines the difference between the phase of the received signal and the reference phase and decides which symbol was transmitted. Assuming equal a priori symbol probabilities, for a proper decision we need to define decision thresholds by dividing the circle into regions, as shown in Figure 5.9 for the case of $M = 8$.

$$\frac{\pi}{M}, \frac{3\pi}{M}, \dots, \frac{(2M - 1)\pi}{M}$$

Therefore, at the instant of detection, if the phase of the received signal lies within the region, $0 \le \theta \le \pi/4$, we make the decision that the symbol, having been transmitted, corresponds to the value $k = 1$.

5.3.2.1 *C/I* or *S/I* as a Performance Measure—Digital Signals

In the previous section, we defined the *C/I*s and SNRs and we said that, depending on the service to be offered by the system designed, certain values

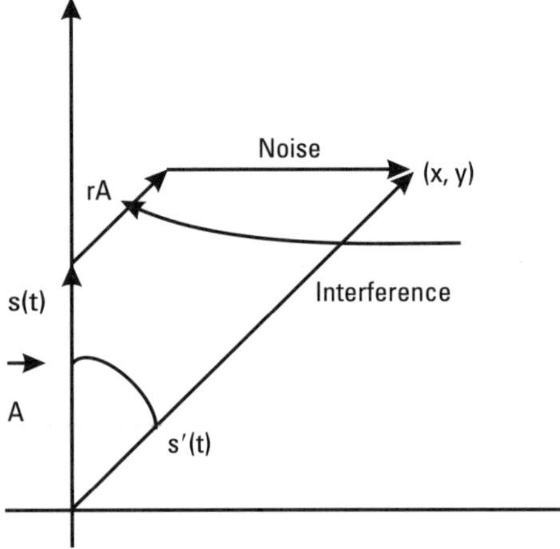

Figure 5.8 Phasor diagram of the signal, noise, and interference.

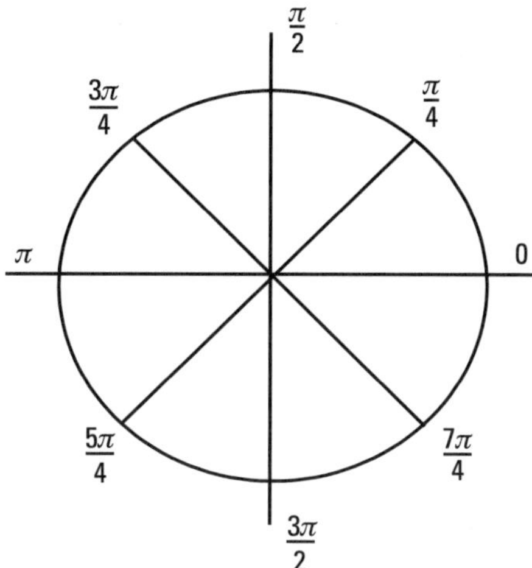

Figure 5.9 Signal-space diagram.

for these ratios are specified beforehand. These values are usually determined by qualitative evaluation of the service offered. For digital signals, the qualitative evaluation depends on the number of errors the system causes to the received data operating in a particular interference environment. Having thus set the values of these ratios for satisfactory quality of service, we can use them as references and thus can consider these ratios as quality measures.

1) PSK Systems

In order to develop a measure of performance for digital systems, as we saw in the previous section for analog systems, it is customary to seek to establish a relationship between the ratio of the transmitted signal carrier power to interference power and the probability of error. By probability of error, we understand the probability that the angle θ is outside of the decision region.

The coordinates of θ $(x; y)$, which are random variables, have means given next.

$$\bar{x} = r \sin \varphi \tag{5.114}$$

$$\bar{y} = 1 + r \cos \varphi$$

conditioned, of course, on the angle ϕ. We also see $A = 1$ for simplicity. The conditional joint PDF of x, y is given by

$$f_{XY}(x, y \mid \varphi) = \frac{1}{2\pi\sigma^2} e^{-\frac{1}{2\sigma^2}[(x - r\sin\varphi)^2 + (y - 1 - r\cos\varphi)^2]} \tag{5.115}$$

If we multiply (5.115) by the PDF of ϕ, which is $1/2\pi$ integrate over the interval $[0, 2\pi]$ we obtain [1–3]:

$$f_{XY}(x, y) = \frac{e^{-\frac{1}{2\sigma^2}(x^2+(y-1)^2+r^2)}}{(2\pi\sigma)^2} \int_0^{2\pi} e^{\frac{r}{\sigma^2}(x^2+(y-1)^2)^{1/2}\cos(\varphi+\eta)} \, d\varphi$$

(5.116)

where

$$\eta = \tan^{-1}\frac{y-1}{x}$$

Equation (5.116) gives

$$f_{XY}(x, y) = \frac{e^{-\frac{1}{2\sigma^2}(x^2+(y-1)^2+r^2)}}{(2\pi\sigma)^2} \, I_0\left(\frac{r}{\sigma^2}(x^2 + (y-1)^2)^{1/2}\right)$$

(5.117)

where

I_0 is the zero modified Bessel function of the first kind.

Equation (5.117) gives the joint PDF of the components of the received signal. We need, however, the PDF of the received signal phase. We achieve our objective if we change the variable x, y into polar coordinates and integrate over the phasor's length. If we set

$$x = \delta \sin a$$

(5.118)

$$y = \delta \cos a$$

Equation (5.117) becomes

$$f_\Theta(\vartheta) = \frac{1}{(2\pi\sigma)^2} \int_0^\infty e^{-\frac{1}{2\sigma^2}(\delta^2+r^2+1-2\delta\cos a)}$$

(5.119)

$$I_0\left(\frac{r}{\sigma^2}(\delta^2 + 1 - 2\delta \cos a)^{1/2}\right) \delta \, d\delta$$

If we now integrate equation (5.119) over the region, which lies outside the boundaries from $-\pi/M$ to π/M, we obtain

$$\text{Probability of error} = P_e = 2 \int_{(\pi/M)}^{\pi} f_\Theta(\vartheta)\, d\vartheta \qquad (5.120)$$

We use the factor 2 because $f_\Theta(\theta)$ is symmetric with respect to θ. A graphical representation of (5.120) is given in the Figure 5.10 for $M = 4$ [3].

We observe from (5.120) and Figure 5.10 that P_e depends directly on the parameters $1/r$ and $1/\sigma$, which are the C/Is and CNRs. Having related probability of error, which is a quality measure, to the C/I with design parameters, our analysis has led to our original objective to relate quality measures to design objectives in any type of interference environment. The remaining sections will be devoted to calculating C/I for other types of applications of wireless systems. In the following sections we shall apply this analysis to other cases of interference.

2) Terrestrial Mobile Cellular Communications Systems

In this section we shall present a methodology used to calculate the C/I for cellular and mobile systems. This methodology will then be applied to calculate C/I for standard mobile systems currently in use.

The C/I of a cellular system can be approximated by [9, 11].

$$\frac{C}{I} = \frac{1}{M} \cdot \left(\frac{D}{R}\right)^n \qquad (5.121)$$

where

M = the number of cochannel interfering cells;

n = a path loss exponent that ranges between two and four in urban cellular systems;

D = distance between two cochannel cells;

R = the radius of a cell.

a) TDMA Cellular

For TDMA cellular networks, the mean C/I at any given location is given by [12–14]:

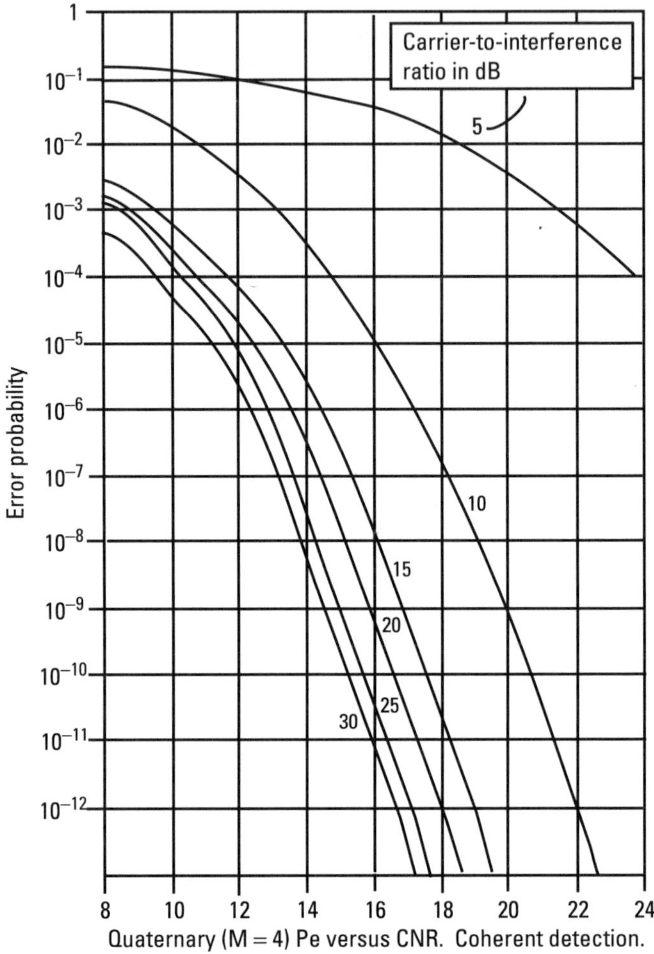

Figure 5.10 Curves of error probability versus CNR. (After: [6].)

$$C/I = 10 \log \left[S_d \bigg/ \sum_{i=1}^{n} I_i \right] \tag{5.122}$$

where

S_d = the desired signal strength;

I_i = the interference from the ith cochannel base station.

Calculation of Signal to Interference Plus Noise Ratio for TDMA

Many contemporary cellular radio resource management algorithms for hand-offs, channel assignment, and power control assume fast and accurate measurements for the signal to interference plus noise ratio $S/(I + N)$.

Several methods have been recently developed to generate *real-time* estimates of the $S/(I + N)$ in TDMA cellular systems:

1. Interference projection (IP), which uses the training and/or color code sequences that are typically present within cellular TDMA slots to obtain an unbiased estimate of $S/(I + N)$.

2. Use of the autocorrelation sequence of the received signal samples over a short time scale.

3. Subspace-based (SB) estimates of $S/(I + N)$ obtained by the use of the eigenvalues of the co-variance matrix of the received signal sequence.

4. Use of signal to variation power (SVP) estimator. This method uses the autocorrelation sequence of the received signal samples for a short time scale. However, numerical results (in DECT SYS) reveal that the estimator suffers from a large bias for interesting values of the S/I.

5. Signal projection (SP) methods have a computational complexity comparable to the IP methods and an average absolute $S/(I + N)$ prediction error comparable to the SB methods.

b) OFDM/CDMA

The basic equation for the C/I of a user and a carrier for an OFDM/CDMA system in the case of synchronously arriving signals is [11]:

$$(C/I)_i = \frac{P_{iR}\, G_p}{\displaystyle\sum_{\substack{j=0 \\ j \neq 1}}^{N} a_j \cdot P_{jR} + \sum_{k=0}^{M} \beta_{IC} \cdot P_{kR}^{tot} + N_0} \tag{5.123}$$

where

P_{iR} = the receiver power of the carrier i;

P_{kR}^{tot} = the total received power from base transceiver station (BTS) k;

G_p = the processing gain;

a_j = the orthogonality factor for intracell interference;

β_{IC} = models the orthogonality loss due to nonideal channel estimation and due fading multipath channel;

N_0 = models the thermal noise.

The equation for single carrier is given as

$$(C/I)_i = \frac{P_{iR} \cdot G_p}{\left[\sum_{k=0}^{M} \beta_{IC} \cdot P_{kR}^{tot} \right] \cdot \gamma} \tag{5.124}$$

The equation for a single user will now be

$$C/I = \frac{P_R \cdot G_p}{\left[\sum_{k=0}^{M} \beta_{IC} \cdot P_{kR}^{tot} \right] \cdot \gamma} \quad \text{where } P_R = \sum_{i=0}^{N} \frac{P_{iR}}{N} \tag{5.125}$$

The parameter γ models the orthogonality between the signals from different BTS.

c) CDMA Cellular Systems

To maintain the communications quality in CDMA cellular systems at the target level, S/I-based power control methods have been proposed [12–15]. In the uplink, all MSs in a cell control their transmission power so that the received power attains the desired power level at the connecting base station. In the downlink, a base station allocates its transmission power so that the MSs in the cell have the same S/I. Therefore, all MSs in a cell have the same uplink S/I and the same downlink S/I, as shown in Figure 5.11.

Uplink communication quality at BS_0, SIR_{0_up}, is expressed as

$$SIR_{0\_up} = \frac{P_{R0}}{(N_0 - 1) \cdot P_{R0} + B_0} = \frac{1}{(N_0 - 1) \cdot + B_0/P_{R0}} \tag{5.126}$$

Here P_{R0} represents the desired power level at BS_0 and becomes the target for transmission power control when MSs connect to BS_0. The first term in the denominator is the interference from other MSs in the same cell. The second term expresses the interference from other cells, and it's denoted as B_0.

Downlink communications quality of $MS(0, j)$ connected to BS_0, $SIR(0, j)_{down}$ is expressed as

$$SIR(0, j)_{down} = \frac{P_A(0, j) \cdot L_0(0, j)}{(1 - F_0) \cdot P_{BS_0} \cdot L_0(0, j) + C(0, j)} \qquad (5.127)$$

The signal sent from BS_0 to $MS(0, j)$ is transmitted with a power of $P_A(0, j)$. The propagation loss between $MS(0, j)$ and BS_0 is represented as $L_0(0, j)$ [15].

The denominator at the right-hand side in (5.127) represents the total interference at $MS(0, j)$. The total transmission power at BS_0 is expressed as P_{BS0}. $C(0, j)$ is the total interference from the other cells at $MS(0, j)$. We define an orthogonality factor F_0 in the downlink, and thus $(1 - F_0)$ represent the degree of loss in orthogonality. The orthogonality factor depends on such characteristics as the number of propagation paths, the power ratio between paths, and the number of fingers in the RAKE receiver (see Figure 5.11).

The downlink SIR at BS_0, SIR_{0_down} is expressed as

$$SIR_{0\_down} = \frac{P_{BS_0} - P_{pl}}{N_0 \cdot (1 - F_0) \cdot P_{BS_0} + \sum_{j=1}^{N_0} \frac{C(0, j)}{L_0(0, j)}} \qquad (5.128)$$

where P_{pl} indicates the pilot-signal transmission power.

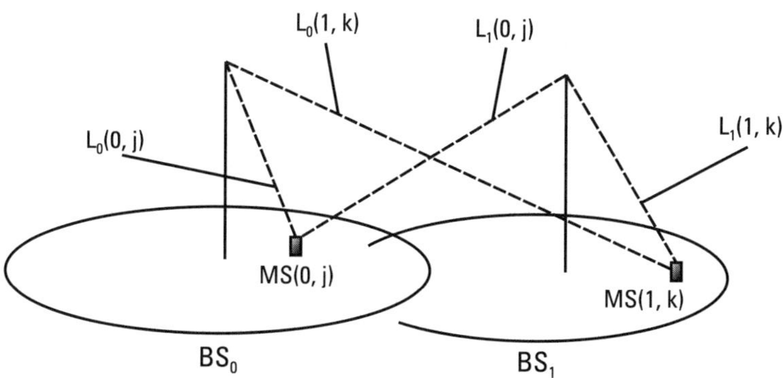

Figure 5.11 Example of possible links in a cellular system. (After: [9].)

From (5.128), the communication quality in the downlink is affected by the following factors:

- Number of MSs in the target cell N;
- The total transmission power P_{BS};
- The orthogonal factor F_0;
- The interference from other cells $C(i, j)$;
- The propagation loss $L(i, j)$.

d) Macrocell and Microcell Systems

Without power control for downlink, the transmitted power of the BS to MS located anywhere, is the same. The C/I experienced by a mobile in the central macrocell and in the microcell can be derived as [15, 16]

$$\left[\frac{C}{I}\right]_l = \frac{\dfrac{p_{tl}(1 - a_l)}{N} \cdot L_p}{\left[1 - \dfrac{(1 - a_l)}{N}\right] p_{tl} \cdot L_P + p'_{ts} \cdot L'_P + \sum_{i=1}^{6} p_{tl} \cdot L'_{Pi}} \tag{5.129}$$

$$\left[\frac{C}{I}\right]_S = \frac{\dfrac{p'_{ts}(1 - a_S)}{M} \cdot L'_P}{\left[1 - \dfrac{(1 - a_s)}{M}\right] p'_{ts} \cdot L'_P + p_{tl} \cdot L_P + \sum_{i=1}^{6} p_{tl} \cdot L'_{Pi}} \tag{5.130}$$

where

p_{tl} and p_{ts} = the transmitted power from the macrocell BS and microcell BS;

L_p, L'_P and L_{pi} are the path loss for macrocell, microcell, and adjacent macrocell, respectively, and a_l and a_s parameters set to certain values (around 0.1) in order to maximize the capacity of macrocell and microcell and

$$\sum_{i=1}^{6} p_{tl} \cdot L'_{Pi} = \text{the interference from the six adjacent macrocell}$$

$$\tag{5.131}$$

These formulas are the decision rules for downlink to accept a newly active *MS*.

The total transmission power of macrocell or microcell for users should be less than $(1 - a)$ of total power. That is,

$$\sum_{i=1}^{N} P(y) \le p_t (1 - a) \tag{5.132}$$

where

N = the number of users in a dedicated cell;

a = the pilot power fraction.

With downlink power control applied to a macrocell or a microcell, the (C/I) for mobile i is modified as

$$\left[\frac{C}{I} \right]_d = \frac{f(y_i) \cdot P_R \cdot L_P}{(p_t - f(y_i) P_R) \cdot L_P + p_t' \cdot L_P' + \sum_{i=1}^{6} p_t' \cdot L_{Pi}'} \ge \left(\frac{C}{I} \right)_{td} = -16 \text{ db} \tag{5.133}$$

where

p_t = total transmitted power from the *BS* belonging to the other kind of cell with corresponding (path loss);

p_t' = transmitted power from each adjacent macrocell *BS* with corresponding (path loss).

Equation (5.133) represents the decision rule for accepting a newly active *MS* in the downlink.

e) Carrier to Cochannel Interference Ratio in Mobile (C/I)

In this section, we shall use simplified models of standard cellular mobile systems currently implemented in order to determine C/I due to cochannel interference. The mobile unit at location M in Figure 5.12 receives the desired signal on frequency F_1 from the nearest base station. Simultaneously, the mobile unit at M also receives independent undesirable interfering signals from other base stations on the same frequency. The same receiver receives

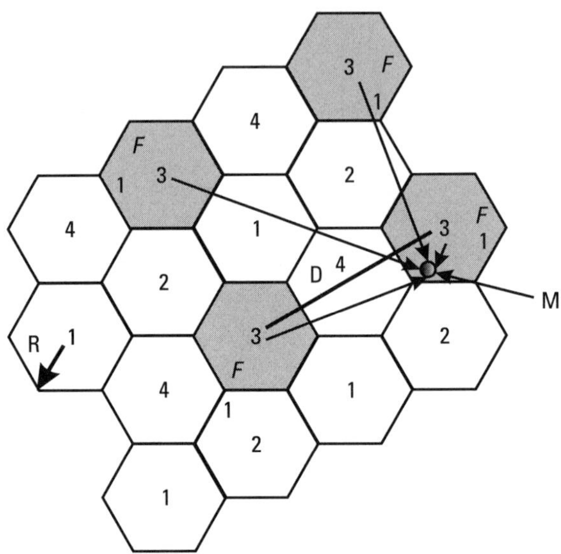

Figure 5.12 Cochannel interference. (After: [9].)

these independent signals, all of which are on the same frequency, simultaneously. This results in the presence of cochannel interference.

The frequency reuse distance D is a function of the number K_0 of the interfering cells, as well as the C/I ratio at the mobile receiver. This ratio is defined using the following equation [9]:

$$\frac{C}{I} = \frac{C}{\sum_{k=1}^{K_0} I_K} \tag{5.134}$$

where I_k is the power of the interfering signal originating from the Kth cochannel cell. The interfering signals originating from base stations other than those belonging in the first tier are considered to be negligible.

It is known that

$$C \propto R^{-\gamma} \tag{5.135}$$

and

$$I \propto D^{-\gamma} \tag{5.136}$$

It can be proved that

$$\frac{C}{I} = \frac{R^{-\gamma}}{\displaystyle\sum_{K=1}^{K_0} D_K^{-\gamma}}$$

(5.137)

where R is the cell radius.

In the following paragraphs certain cases, where cochannel interference occurs, are presented. Cases using omnidirectional base station antennas, as well as cases using directional antennas of different directivities are described.

f) System Using Omnidirectional Antennas

First we will examine the case with a seven-cell cluster, Figure 5.13.

In a cluster with seven cells, in the presence of six interfering cells in the first tier, the C/I becomes

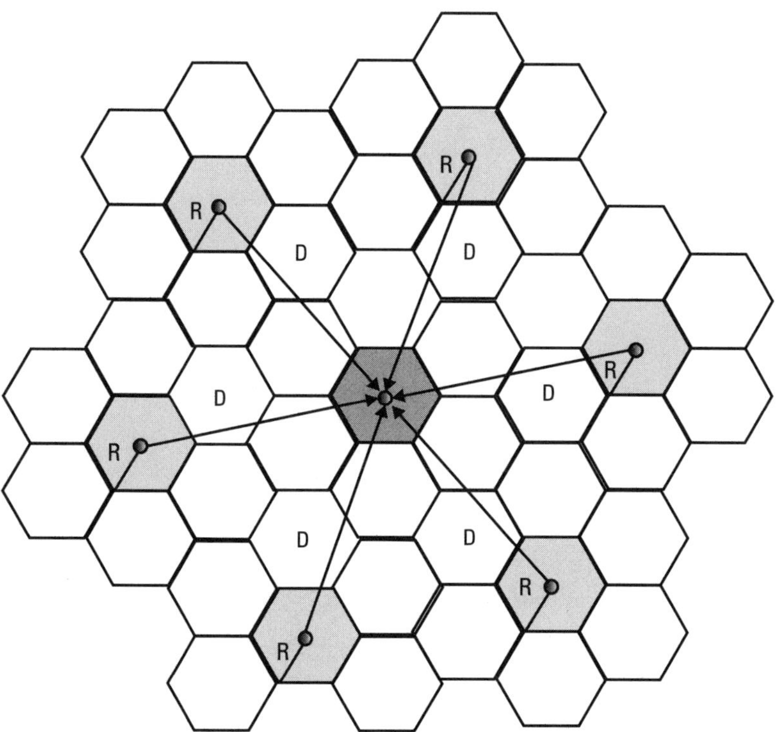

Figure 5.13 Seven-cell cluster with omnidirectional antennas. (After: [9].)

$$\frac{C}{I} = \frac{R^{-\gamma}}{\sum_{k=1}^{6} D_k^{-\gamma}} = \frac{1}{\sum_{k=1}^{6} \left(\frac{D_k}{R}\right)^{-\gamma}} = \frac{1}{\sum_{k=1}^{6} q_k^{-\gamma}} \tag{5.138}$$

where q_k is the cochannel interfering reduction factor at the kth cell.

Assuming that the propagation path loss slope, γ, is equal to 4, and all distances D_k are equal to D, (5.138) yields

$$\frac{C}{I} = \frac{1}{6\left(\dfrac{D}{R}\right)^{-4}} \tag{5.139}$$

In a cluster with seven cells, in the presence of six interferers and when the mobile unit is located at the cell boundaries (worst case), the C/I becomes

$$\frac{C}{I} = \frac{R^{-\gamma}}{2(D-R)^{-\gamma} + 2(D)^{-\gamma} + 2(D+R)^{-\gamma}} \tag{5.140}$$

$$= \frac{1}{2(q-1)^{-\gamma} + 2(q)^{-\gamma} + 2(q+1)^{-\gamma}}$$

For $\gamma = 4$, (5.140) becomes

$$\frac{C}{I} = \frac{R^{-4}}{6(D-R)^{-4}} = \frac{1}{6(q-1)^{-4}} \tag{5.141}$$

This is the case shown in Figure 5.14.

g) System Using Directional Antennas

Using directional antennas on the base stations in the architecture of seven-cell cluster, we have the following possible cases.

Three-Sector Case

In the case depicted in Figure 5.15, directional antennas of 120° directivity are used.

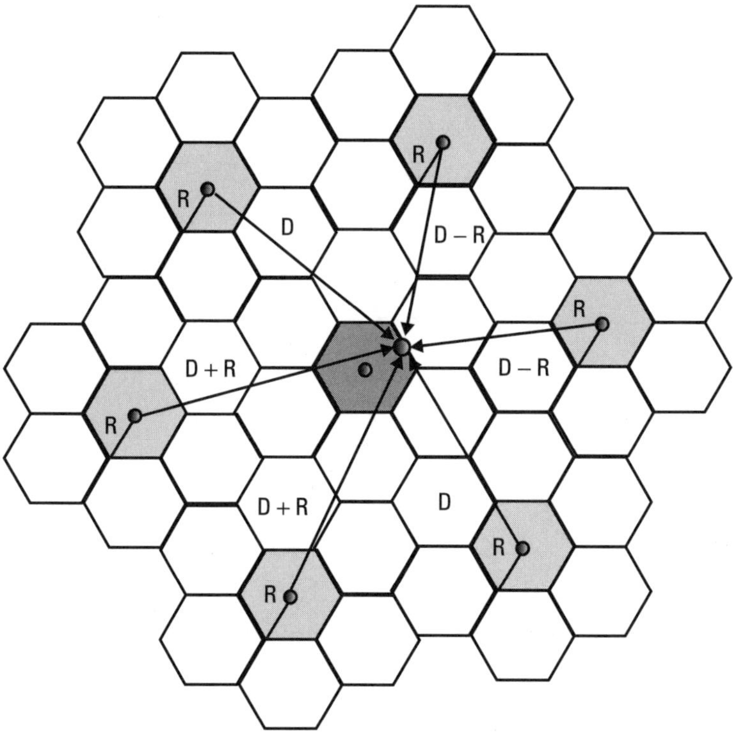

Figure 5.14 Worst case in a seven-cell cluster with omnidirectional antennas.

In the worst case described earlier, C/I becomes:

$$\frac{C}{I} = \frac{R^{-4}}{(D + 0.7R)^{-4} + D^{-4}} = \frac{1}{\left(\frac{D}{R} + 0.7\right)^{-4} + \left(\frac{D}{R}\right)^{-4}} = \frac{1}{(q + 0.7)^{-4} + q^{-4}}$$

$$(5.142)$$

Six-Sector Case

In this case directional antennas of 60° directivity are used. In the worst case, as shown in Figure 5.16, C/I becomes

$$\frac{C}{I} = \frac{R^{-4}}{(D + 0.7R)^{-4}} = \frac{1}{\left(\frac{D}{R} + 0.7\right)^{-4}} = \frac{1}{(q + 0.7)^{-4}} \qquad (5.143)$$

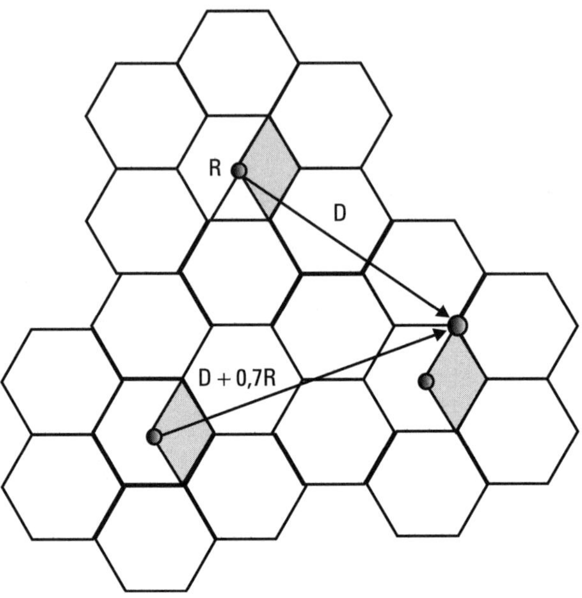

Figure 5.15 Worst case in a seven-cell cluster with antennas of 120° directivity.

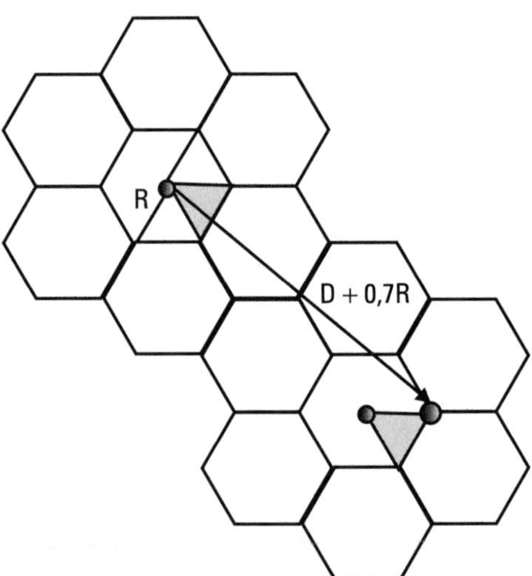

Figure 5.16 Worst case in a seven-cell cluster with antennas of 60° directivity (After: [9].)

3) Mobile Satellite Systems

The assumption for the C/I calculations in nongeostationary satellite systems are [17, 18]:

- The selected satellite for the communication link is the one from which maximum power is received.

- Any mobile will have the same carrier power at the receiver input. Hence, power control is required to compensate the variation in losses, which are dependent on the mobile location relative to the satellite.

- The mobile antenna is omnidirectional.

- The interferer causes maximum interference when the frequency spectrum is totally overlapped.

Using these assumptions, the C/I equations are as follows:
The C is the carrier power received at the mobile terminal.

$$C = \frac{P_{Tw}\, G_{TW}(\vartheta)\, G_{RW}(\alpha)}{L(d)\, P_e\, am} \tag{5.144}$$

The PN density I_{01}, is the multiple access interference resulting from $(m-1)$ interferers (i.e., m users are communicating simultaneously per carrier in each spotbeam) in the same spotbeam and can be written as:

$$I_{01} = \frac{Ca\,(m-1)\,F_1\,(10)^{\left(\frac{\Delta}{10}\right)}}{B} \tag{5.145}$$

The PN density I_{02}, is the beam-to-beam interference resulting from m interferers in all adjacent spotbeams, assuming frequency reuse of the neighbor cell, and can be written as:

$$C_1 = \frac{G_{TI}(\theta)\, G_{RW}(\alpha)}{L(d)\, P_e\, am} \tag{5.146}$$

$$I_{02} = \frac{C_1\, am F_2 \cdot (10)^{\frac{\Delta}{10}}}{B} \tag{5.147}$$

The C/I levels are added as thermal noise. The total interfering signal power is the sum of the powers from all 'N' visible satellite spotbeams in the interfering system.

$$\left[\frac{C}{I_0}\right]_T = \left[\frac{C}{I_{01} + \sum_{k=1}^{N} I_{02,k} + I_{03}}\right] \tag{5.148}$$

where

$$P_{TW} = \text{wanted satellite spotbeam power;}$$

$$P_{TI} = \text{interfering satellite spotbeam power;}$$

$$G_{TW}(\theta) = \text{wanted spotbeam gain in direction } \theta;$$

$$G_{TI}(\theta) = \text{interfering spotbeam gain in direction } \theta;$$

$$G_{RW}(\alpha) = \text{mobile terminal antenna gain in direction } \alpha;$$

$$L(d) = \text{free space path loss;}$$

$$P_e = \text{propagation effects which takes into account the shadowing (fading loss) for the link—a function of elevation angle and environment;}$$

$$a = \text{voice activity ratio;}$$

$$m = \text{number of users per carrier;}$$

$$\Delta = \text{power control error;}$$

$$F_i = \text{correlation factor;}$$

$$B = \text{subband bandwidth;}$$

$$I_{03} = \text{external interference inband-shared scenario.}$$

4) WLL Communications Systems

For the purpose of the analysis here, WLL indicates a system that connects subscribers to the public switched telephone network using radio signals as a substitute for copper for the entire connection between the subscriber and the switch. The physical layout of such a system is shown in Figure 5.17.

We shall assume that the WLL system operates in an area where there exists a regular microwave link of fixed service-frequency division multiplex/

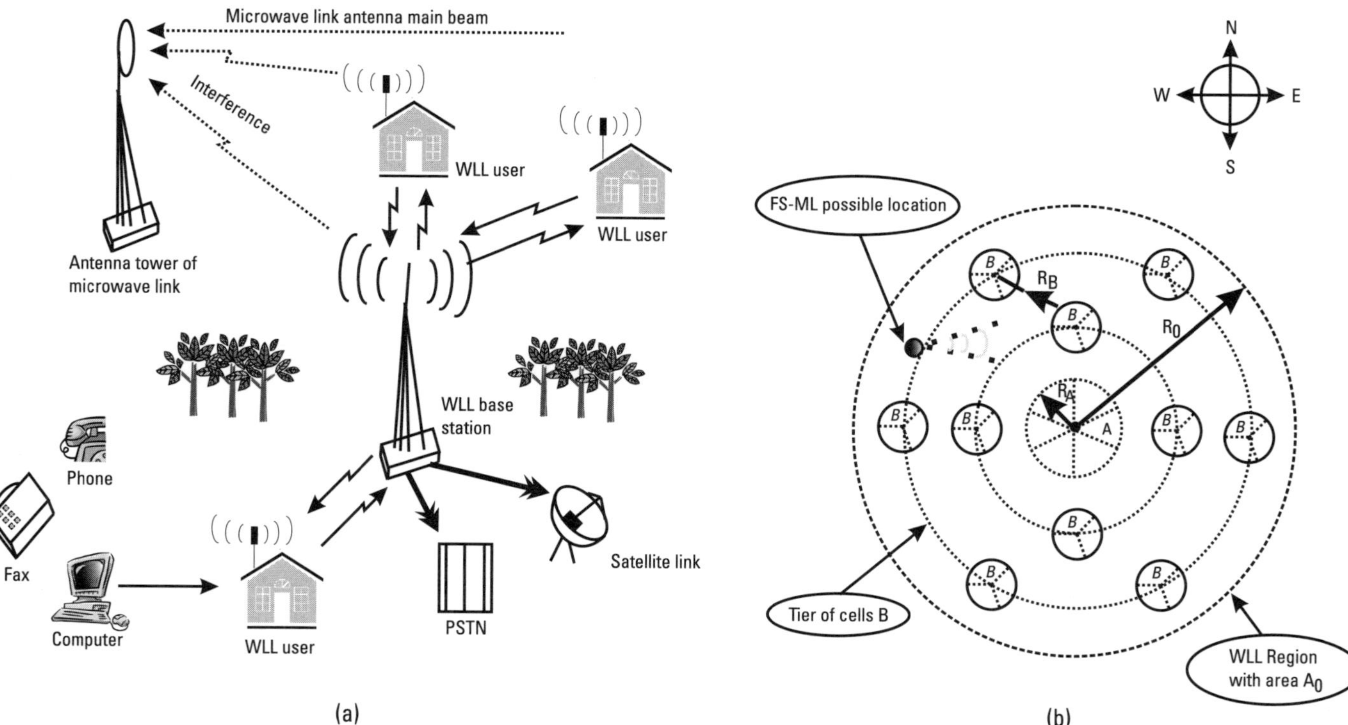

Figure 5.17 WLL application. (After: [19]. © 2001 John Wiley & Sons, Inc.)

frequency modulation (FS-FDM/FM) telephone service and total signal input at the fixed service microwave link (FS-ML) receiver is given by

$$s(t) = s_d(t) + s_I(t) + n(t)$$

where

$$s_d(t) = \sqrt{2P_0} \cos(\omega_0 + t\varphi(t))$$

where

$$f_c = \frac{\omega_0}{2\pi}$$

P_0 = carrier power;

ω_0 = carrier frequency;

$\varphi(t) = 2\pi \int x_{FDM}(\tau) \otimes h_p(\tau) \, d\tau.$

where

x_{FDM} is the modulating FDM signal;

$h_p(t)$ is the pre-emphasis impulse response.

and

$$s_I(t) = \sum_k \sqrt{2P_{I,k}} \, b^k(t - \tau_k) c^{(k)}(t - \tau_k) \cos(\omega_{It} + \theta_k)$$

This summation represents the WLL spread-spectrum DS-CDMA signals of the users, which are also interferers. The transmission used is a spread-spectrum DS-CDMA system via the base station, and the system affected is an FS-ML where $b^k(t)$ is the modulating signal, $c^k(t)$ is the pseudorandom sequence, and τ_k are time delays.

a) Interference Noise at the FDM/FM Receiver Output

It is shown in [19] that the combined signal to enter the demodulator (limiter/discriminator) of the FS-ML system, which is affected by the WLL users, is given by

$$s(t) = \text{Re } E\left\{\sqrt{2P_0}\, \exp\left[j\omega_0 t + j\varphi(t)\right]\left\{1 + \sum_k I^{(k)}(t)\, \exp\left[-j\varphi(t) + j\theta_n\right]\right\}\right\}$$

$$= \text{Re } E\left\{\sqrt{2P_0}\, A(t)\, \exp\left[j\omega_0 t + j\varphi(t) + j\lambda(t)\right]\right\} \tag{5.149}$$

where

$$I^{(k)}(t) = \sqrt{\frac{P_{I,k}}{P_0}}\, \left[b^{(n)}(t - \tau_n)\, c^{(n)}(t - \tau_n)\, \exp\left(j\omega_\Delta t\right)\right] \otimes h_{IF}(t) \tag{5.150}$$

and

$$\lambda(t) = \text{Im}\left\langle ln\left\{1 + \sum_{k=1}^{K} I^{(k)}(t)\, \exp\left[-j\varphi(t) + j\theta_n\right]\right\}\right\rangle \tag{5.151}$$

$$f_\Delta = \frac{\omega_1 - \omega_0}{2\pi} = \frac{\omega_\Delta}{2\pi}$$

At the demodulator output we get the signal

$$u_D(t) = \frac{1}{2\pi}\, \frac{d}{dt}\left[\varphi(t) + \lambda(t)\right] \tag{5.152}$$

where the first component represents the desired signal and the second one is the interference noise.

Under the assumption that in real working conditions, the following restriction is valid:

$$\left|\sum_{n=1}^{N} I^{(n)}(t)\right|_{\text{max}} < 1 \tag{5.153}$$

and the expression for the interference noise at the limiter-discriminator (L-D) output could be written in the following form:

$$\lambda(t) = \text{Im } E\left\{ \sum_{m=1}^{\infty} \frac{(-1)^{m-1}}{m} \left\{ \sum_{k=1}^{K} I^{(k)}(t) \exp\left[-j\varphi(t) + j\theta_n\right] \right\}^m \right\}$$

$$(5.154)$$

Similar expressions have been derived in Section 5.3.1. By applying the multinomial theorem for the autocorrelation function of the interference noise $R_\lambda(\tau) = E\lambda(t)\lambda^*(t + \tau)$ (see Appendix A), we find:

$$R_\lambda(\tau) = \sum_{m=1}^{\infty} \frac{1}{4m^2} \sum_{m_1+\ldots+m_n=N} \left(\frac{m!}{m_1! \ldots m_N!} \right)^2 \qquad (5.155)$$

$$\times \left[\prod_{n=1}^{m_n} R_1^{m_n}(\tau) R_0^{m_n}(\tau)^* + \prod_{n=1}^{m_n} R_I^{m_n}(\tau)^* R_0^{m_n}(\tau) \right]$$

where

$$R_I^{m_n}(\tau) = E\{\exp\left[jm_n\varphi(t) - jm_n\varphi(t + \tau)\right]\} \qquad (5.156)$$

and

$$R_I^{m_n}(\tau) = E\left\{ \left[I^{(n)}(t) I^{(n)}(t + \tau)^* \right]^{m_n} \right\} \qquad (5.157)$$

Taking into account (5.153), for the autocorrelation function of the interference noise, $R_\lambda(\tau) = E\langle \lambda(t)\lambda^*(t + \tau) \rangle$, we have [19]:

$$R_\lambda(\tau) \cong \frac{1}{8} \text{Re} \left\{ \sum_{n=1}^{N} E\langle R_I^{(m_n=1)}(\tau) R_0^{(m_n=1)}(\tau)^* \rangle \right\} \qquad (5.158)$$

By applying the Wiener-Khintchine theorem to the autocorrelation of $\lambda(t)$, and taking into account (5.153), the interference noise PSD at the FDM-FM receiver output is

$$S_{IN}(f) \cong \frac{f^2}{4|H_P(jf)|^2} \frac{P_{l,k}}{P_0} \left\{ \left[S_{CDMA}(f-f_\Delta)|H_{IF}(jf)|^2 \otimes S_{FM}(-f) \right] \right.$$

$$\left. + \left[S_{CDMA}(-f-f_\Delta)|H_{IF}(jf)|^2 \otimes S_{FM}(f) \right] \right\} \qquad (5.159)$$

where $S_{CDMA}(f)$ is the Fourier transform of the auto correlation of interfering CDMA signal.

The interference noise power in a telephony channel of the FS-ML system centered at f_{ch} may be written as

$$N_I = \frac{2bS_{IN}(f_{ch})}{(\Delta f_0)^2}\ (mWp) \tag{5.160}$$

where $b = 1.7$ kHz is the telephone channel psophometric band and the units of noise are in milliwatts in that band phorphometrically weighted. Δf_0 is the FM signal test tone deviation.

b) Probability of Error at the DML Receiver Output

In the case that the main telephone link was also digital (CDMA), we observe from the analysis so far that C/I, S/I, and probability of error or BER are powerful design tools. In addition to designing wireless systems of acceptable performance, lately they have been used as quality measures [16] and as measures for optimization of other aspects of wireless systems, such as resource allocations and more specifically channel assignment. In the next chapter, we shall encounter the metrics to be used as a reference level in our effort to develop algorithms, which will lead to interference suppression.

For evaluating the interference effects on FS-DML due to the WLL, we used the formulas for the probability of error in AWGN, because the CDMA interference was considered white noise. Generally, this is a questionable approximation, but the relatively flat PSD of CDMA signal in the bandwidth of interest gives us a certain degree of confidence in the interference analysis.

If we denote the PSD of total CDMA signal by

$$N_I = \frac{P_{I,k}}{B_d} \int_{f_\Delta - B_d/2}^{f_\Delta + B_d/2} S_{CDMA}(f)\ df \tag{5.161}$$

where B_d is the bandwidth of digital microwave link (DML) signal, BER for M-QAM, at the output of DML receiver are given by [19]:

$$P_{e,M-QAM} \cong \tag{5.162}$$

$$\frac{1}{ld(M)} \left\{ 1 - \left[1 - \left(1 - \frac{1}{\sqrt{M}} \right) \frac{1}{2} \, erfc \sqrt{\frac{3}{2(M-1)} \cdot \frac{E_{b,DML} \, ld(M)}{N_0 + N_I}} \right]^2 \right\}$$

where $E_{b,DML}$ is the mean energy per bit of DML signal.

References

[1] Stavroulakis, P., "Interference Analysis of Communications Systems," *IEEE Press,* New York, 1980.

[2] Karagiannidis, G. K., et al., "Cochannel Interference Analysis for a Rician Signal in L Nakagami Interference with Arbitrary Parameters,"*Proc. 1999 International Workshop on Mobile Communications Focused on MTS and IMT-2000: Chania,* Crete, Greece, June 24–26, 1999.

[3] Fuenzalida, J. C., O. Shimbo, and W. L. Cook, "Time-Domain Analyses of Intermodulation Effect Caused by Nonlinear Amplifiers," *Comsat Tech. Review,* Vol. 3, 1973, pp. 89–141.

[4] Pontano, B. A., J. C. Fuenzalida, and N. K. M. Chitre, "Interference into Angle Modulated System Carrying Multichannel Telephony Signals," *IEEE Trans. On Commun.,* Vol. com-21, June 1973, pp. 714–726.

[5] Rosenbaum, A. S., "PSK Error Performance with Gaussian Noise and Interference," *Bell System Tech. J.,* Vol. 48, February 1969, pp. 413–422.

[6] Proakis, J. G., *Digital Communications,* second edition, New York: McGraw-Hill Book Company, 1989.

[7] Yardim, A., et al., "Design of Efficient Receiver FIR Filters for Joint Minimization of Channel Noise, ISI and Adjacent Channel Interference," IEEE Global Telecomm. Conference, London, November 18–22, 1996.

[8] Feher, K., *Advanced Digital Communications,* Norcross, GA: Noble Publishing Corp., 1997.

[9] Rappaport, T. S., *Wireless Communications,* Upper Saddle River NJ: Prentice Hall, 1996.

[10] Yuhas, Ben, and Nirwan Ansari, *Neural Neworks in Telecommunications,* Boston, MA: Kluwer, 1994.

[11] Toskala, Antti, et al., "Cellular OFDM/CDMA Downlink Performance in the Link and System Levels," *IEEE VTC '97,* Phoenix, AZ, May 4–7, 1997.

[12] Vatalaro, F., et al., "CDMA Cellular Systems Performance with Imperfect Power Control and Shadowing," *IEEE VTC '96,* Atlanta, GA, April 28–May 1, 1996.

[13] Sathyendran, G. W. Tunnichoffe, and A. R. March, "Multi-Layered Underlay Overlay Frequency Planning Scheme for Cellular Networks," *IEEE VTC,* 1997.

[14] Khan, Farooq, and Djamal Zeghlach, "Multilevel Channel Assignment (MCA) for Wireless Personal Communications," *IEEE VTC '97*, Phoenix, AZ, May 4–7, 1997.

[15] Wu, Jung-Shyr, Jen-Kung Chung, and Yu-Chuan Yang, "Performance Improvement for a Hotspot Embedded in CDMA Systems," *IEEE VTC '97*, Phoenix, AZ, May 4–7, 1997.

[16] Nakano, Keisuke, et al., "Teletraffic Modelling in CDMA Cellular Systems," *IEEE VTC '97*, Phoenix, AZ, May 4–7, 1997.

[17] Bjelajac, Branko, "CIR Based Dynamic Channel Allocation Schemes and Handover Prioritisation for Mobile Satellite Systems," *IEEE VTC '96*, Atlanta, GA, April 28–May 1, 1996.

[18] Ariz, H. M., R. Tafarolli, and B. G. Evans, "Comparison of Total System Capacity for Band Sharing Between CDMA Based Non-Geostationay Satellite PCN's Under Imperfect Power Control Conditions," *IEEE VTC '97*, Phoenix, AZ, May 4–7, 1996.

[19] Stavroulakis, P., *Wireless Local Loops, Theory and Applications*, New York: John Wiley, 2001.

6

Interference Suppression Techniques

6.1 Introduction

In Chapters 4 and 5, we analyzed the interference concept from the mathematical point of view. We presented the mathematical tools that can be used in an appropriate and intelligent way in every situation of wireless system applications to achieve a unique goal. We shall review in a general context these tools and try to categorize them in such groups so that each group presents a discrete methodology. This way, we set the stage for using these methodologies in real life applications in Chapter 7. Multiuser problems are used in many cases as a model without excluding, however, the applicability to all types of wireless applications when a particular methodology offers an improvement.

A solution to the multiuser interference problem would be to design the user codes to have more stringent cross-correlation properties, because indeed if the signal was truly orthogonal this interference would not exist. Unfortunately, it is not theoretically possible that any set of codes will exhibit zero cross correlation in the asynchronous case. Thus, the multiuser interference case, which presents another interference situation, must be dealt with through a different viewpoint. In Chapter 4, we saw how we handle fading. In this chapter, we shall see that some of the techniques are the same and show similarities and differences for all types of the interference analyzed in Chapter 5. In the following chapter, we shall present various methods that simultaneously handle both in practical implementations, which can be used in real wireless systems. Moreover, in this chapter, a similar technique is presented. It will be pointed out as to whether it applies to one or both cases.

The most popular approach is to employ interference suppression (cancellation), that is, to attempt removal of the multiuser interference from each user's received signal before making data decisions [1, 2]. In principle, the interference cancellation (IC) schemes are considered in the literature fall into two categories: serial (successive) and parallel cancellation. One way to achieve this is coordinated processing of the received signal with a successive cancellation scheme in which the interference caused by the remaining users is removed from each user in succession. One disadvantage of this scheme is that a specific geometric power distribution must be assigned to the users in order that each see the same signal power to background noise plus interference. The first user to be processed sees all the interference from the remaining M-1 users, whereas each user downstream sees less and less interference as the cancellation progresses. Another disadvantage of this scheme has to do with the required delay necessary to fully accomplish the IC for all users in the system. Because the IC proceeds serially, a delay on the order of *Mbit* times is required to complete the job.

Parallel processing of multiuser interference simultaneously removes from each user the interference produced by the remaining users accessing the channel. In comparison with the serial processing scheme, the delay required to complete the operation is at most a few bit times because the IC is performed in parallel for all users,. In some schemes, the common point is that at each stage of the iteration, an attempt was made for each user to completely cancel the interference caused by all of the other users. This technique is referred to as brute force or total interference cancellation. This is not necessarily the best philosophy. Rather, when the interference estimate is poor (as in the early stages of interference cancellation), it is preferable not to cancel the entire amount of estimated multiuser interference. This technique is referred as weighted interference cancellation. The motivation behind this approach can also be derived from maximum likelihood estimate considerations. Various methods have appeared in the literature over the past few years [1–33], which will be explained in the following sections.

6.2 Interference Reduction/Mitigation

The problem of suppression of any type of interference in wireless communication systems can be encountered by many ways and methods [3, 4]. It is therefore important to classify the methods and explain which methods are applicable to which type of problem. More of this will be explained in Chapter 7.

One classification of the reduction interference methods is indirect and direct methods. When mitigation is required for the cases of additional fading, we utilize fading compensation techniques or distortion mitigation, as we showed in Chapter 4 and we shall see some detail in this chapter.

6.2.1 Indirect Reduction Methods

The indirect methods can reduce the possible interferer signals in a macroscopically, predetective point of view. The interference suppression in that case is achieved from the choice of the architecture design of the systems and the radio frequency interface (i.e., antenna pattern) [5–7]. In other words, the interference reduction is achieved automatically as the signal enters the detector.

All the effort of the engineer designing cellular mobile systems or any wireless system in general is based on achieving high user capacity with acceptable QoS. It is well known that cellular system capacity, for example, can be increased by reducing the cell cluster size N. That move, however, increases the cochannel interference. Several techniques for controlling cochannel interference have been proposed in literature, as we shall see in the following sections and in Chapter 7. In this section, we will study three of the techniques, which stand out for this category: narrow-beam adaptive antenna design, functional cell-loading factor, and power control. Narrow-beam adaptive antennas ("smart antennas") at the base stations significantly reduce cochannel interference by steering a high gain in the direction of the desired mobile station and very low gains in the direction of the undesired cochannel mobile stations. Another technique that has been proposed is based on a fractional cell-loading factor, which reduces the probability that a given channel is in use in the cochannel cells, which, consequently, reduces the total cochannel interference level for a particular channel. Power control has also been considered to control cochannel interference, allowing cluster size reduction and capacity improvement.

6.2.1.1 Narrow-Beam Antennas

When adaptive narrow-beam antennas are used at base stations in both the forward and reverse links, beams are steered toward the desired in-cell users. Consider the forward link of a cellular systems with cluster size N, T tiers of cochannel cells and cell radius R. Assuming hexagonal shapes for cells, the ith tier of cochannel cells has $6i$ cells. A mobile station located at the cell boundary, as shown in Figure 6.1, experiences worst-case cochannel interference [8].

Assuming that all base stations are equipped with omnidirectional antennas and transmit the same power $P_t = 1$, the total area mean cochannel interference I_T at a mobile located at the cell boundary is

$$I_T = \underbrace{\frac{1}{d_{1,1}^{\gamma}} + \ldots + \frac{1}{d_{1,6}^{\gamma}}}_{\text{from the 1st tier}} + \underbrace{\frac{1}{d_{2,1}^{\gamma}} + \ldots + \frac{1}{d_{2,12}^{\gamma}}}_{\text{from the 2nd tier}} + \ldots + \underbrace{\frac{1}{d_{T,1}^{\gamma}} + \ldots + \frac{1}{d_{T,6T}^{\gamma}}}_{\text{from the Tth tier}}$$

$$(6.1)$$

where γ is the path loss exponent and $d_{i,k}$ is the transmitter to receiver distance between the kth base station in the ith tier, where k assumes the

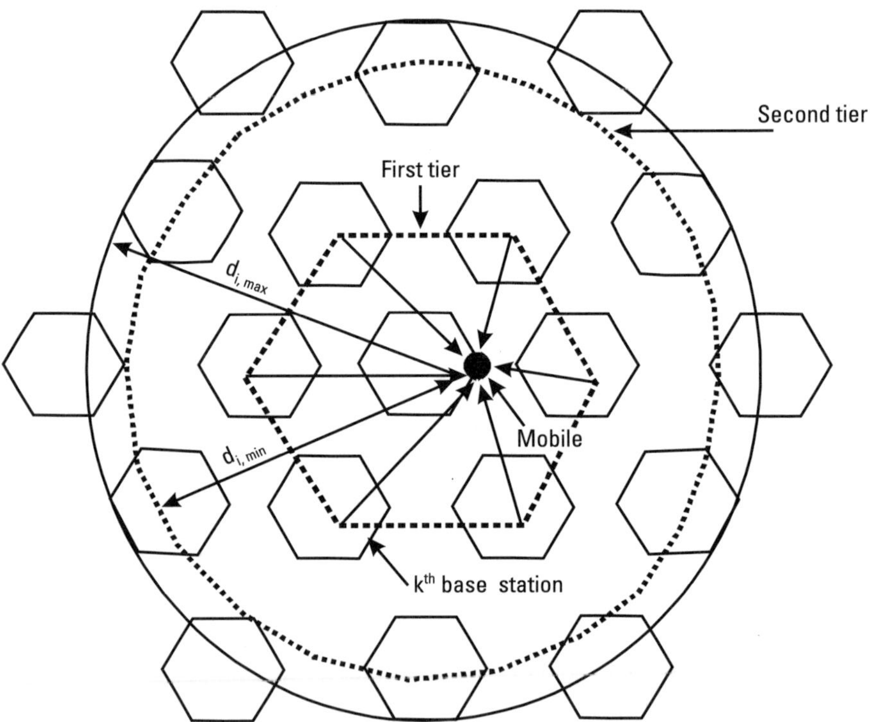

Figure 6.1 Cochannel cells in the forward link cellular system: $d_{i,k}$ is the transmitter to receiver distance between the kth cochannel base station ($k = 1, 2, \ldots, 6i$) in tier I and the mobile. (After: [8].)

value $k = 1, 2, \ldots, 6i$. Because the base stations in the first tier are closer to the mobile at the cell boundary than the other base stations, we use the exact distances $d_{1,k}$ in (6.1) for the base stations in the first tier. For more distant tiers, we approximate all distances between the base stations in a given tier i and the mobile as $d_{i,k} = \overline{d_i} = (d_{i,\max} + d_{i,\min})/2$ for all k, where $d_{i,\max} = i\sqrt{(3N)}\,R$ and $d_{\min} = i\,3\sqrt{N}R/2$ are the maximum and minimum distances, as shown in Figure 6.1.

Thus,

$$\overline{d_i} = i\left(\frac{\sqrt{3}+2}{4}\right)\sqrt{3N}R = i\overline{D}, \quad \overline{D} = \frac{2+\sqrt{3}}{4}\sqrt{3N}R \qquad (6.2)$$

Let I_T denote the total area mean cochannel interference received from the stations in the first tier.

$$I_1 = \frac{1}{d_{1,1}^{\gamma}} + \ldots + \frac{1}{d_{1,6}^{\gamma}} \qquad (6.3)$$

Also, let I_2 denote the total mean cochannel interference from tier 2, 3, $\ldots$, T, using the approximation in (6.2)

$$I_2 = \frac{12}{(2\overline{D})^{\gamma}} + \frac{18}{(3\overline{D})^{\gamma}} + \ldots + \frac{6T}{(T\overline{D})^{\gamma}} = \frac{6}{\overline{D}^{\gamma}}\left(\sum_{i=1}^{T}\frac{1}{i^{\gamma-1}} - 1\right) \qquad (6.4)$$

Thus,

$$I_T = I_1 + \frac{6}{\overline{D}^{\gamma}}\left(\sum_{i=1}^{T}\frac{1}{i^{\gamma-1}} - 1\right) \qquad (6.5)$$

The fraction of the total cochannel interference I_T that corresponds to the interference from the first tier is given by the ratio $\Gamma = I_1/I_T$. Table 6.1 presents the computed values of ratio Γ for cluster sizes $N = 1$, 3, 4, and 7 and path loss exponents $\gamma = 3$, 4, and 5, when T tends to infinity.

Note that the sum in (6.5) does not converge when T tends to infinity for a path loss exponent of two. This means that the fraction of total interference that corresponds to the interference from the first tier goes to

Table 6.1
Ratio of the Interference Γ from the Base Stations in the First Tier I_1
to the Total Interference I_T for Cluster Sizes $N = 1, 3, 4,$ and 7
and Path Exponents $\gamma = 3, 4,$ and 5

γ	$\Gamma = I_1/I_T$ (%)			
	$N = 1$	$N = 3$	$N = 4$	$N = 7$
3	72.0	62.3	60.4	58.4
4	92.4	85.8	84.0	82.0
5	98.0	94.7	93.5	92.1

zero ($\Gamma \rightarrow 0$) when free space propagation ($\gamma = 2$) is assumed. We see from Table 6.1 that for path exponent $\gamma = 4$, the area mean interference from the first tier accounts for at least 82% of total interference. Denote SIR_T as SIR computed using the total interference I_T, and denote SIR_1 as SIR computed using the interference from the first tier I_1. We have

$$SIR_1 = 10 \, \log\left(\frac{S}{I_1}\right) = SIR_T - 10 \, \log(\Gamma) \qquad (6.6)$$

where $S = 1/R^\gamma$ is the desired area mean signal received at the mobile. Therefore, the error caused by considering only the first tier when computing the area mean SIR is less than 1 dB ($10 \, \log 0.82 \approx -0.9$ dB) for path loss $\gamma = 4$ and cluster size $N = 1, 3, 4,$ and 7. This situation was analyzed in detail in Section 5.3.2.1.

It is shown that using only the first tier of cells induces a worst-case error of less than 1.1 dB in the estimation of SIR, regardless of the cluster size, when 40 dB/decade of path loss is assumed, and a worst-case error less than 2.3 dB for 30 dB/decade of path loss. It should be noted that the methodology presented here might be generalized for an arbitrary path loss value.

Assuming that all cochannel cells in the first tier are active, the total forward link interference power at the mobile at the center cell is given by

$$I^f = I_2^f + \ldots + I_7^f \qquad (6.7)$$

where I_i^f is the interference power received from the ith cochannel base station. Likewise, for the reverse link, the total reverse link interference power received at the base station at the center cell is given next, assuming that the interference signals are incoherent so that the power can be summed.

$$I^r = I_2^r + \ldots + I_7^r \tag{6.8}$$

where I_i^r is the interference power received from the ith cochannel mobile station. This is a realistic assumption for wireless signals, as the phase shift of the individual interference signals may be assumed to be independent and vary significantly due to scattering and travel distance. The cochannel interference received at the mobile at the center cell, caused by a given cochannel base station, is attenuated by the antenna gain when the mobile is not within the main lobe of the antenna of the cochannel base station transmission.

In Figure 6.2, we observe that the cochannel interference signals from base stations 2, 4, 6, and 7 are attenuated due to the use of narrow-beam antennas. However, there is no reduction in the interference caused by base stations 3 and 5. The same principle is valid for the reverse link.

It is obvious that the extent of cochannel interference reduction depends on the beamwidth (BW) and the sidelobe level (SLL) of the base station antennas. If the antenna is implemented using an array of antennas, the BW and SLL will depend on the number of elements in the array.

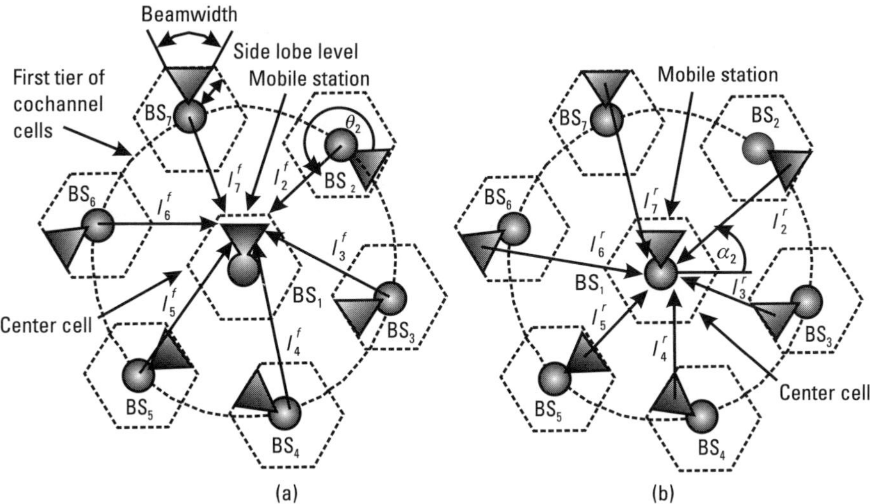

Figure 6.2 Narrow-beam antennas in cellular system: (a) forward, and (b) reverse links. (After: [8].)

6.2.1.2 Base Station Antenna Height Reduction

Reducing the base station antenna height is another method for reducing cochannel interference. The power gain (or loss) due to increasing (or lowering) the antenna height is given by the following formula:

$$\text{Antenna height gain (loss)} = 20 \log \frac{h'_{e1}}{h_{e1}} \tag{6.9}$$

where

h'_{e1} is the new effective antenna height;

h_{e1} is the old effective antenna height.

In some circumstances, such as on a fairly flat ground or in a valley, lowering the antenna height effectively reduces cochannel interference. When the antenna is located on the top of a high hill or on the top of a mountain, the reduction of cochannel interference due to lowering the antenna height is negligible.

When the base station antenna is located in a forested area, special care must be taken for the antenna not to be lower than the trees in the vicinity. Otherwise, excessive attenuation of the desired signal would occur.

6.2.1.3 Fractional Loading Factor

The total cochannel interference at a given mobile or base station depends on the k cochannel cells that are using the same pair of forward and reverse channels as the cell where the interference level is being measured [9]. This number k is related to the loading factor of each cell, which defines the probability that a given channel is in use within a cell. Considering the first tier of cochannel cells when a given channel is in use, k out of six cochannel cells interfere. The random variable k is binomially distributed, and the probability of having n $(0 \leq n \leq 6)$ interferers is, therefore,

$$P_n = \text{Prob}\,\{k = n\} = \binom{6}{n} p_{ch}^n (1 - p_{ch})^{6-n} \tag{6.10}$$

The loading factor p_{ch} is a function of the offered traffic A (in Erlangs), blocking probability P_B, and number of channels N_c assigned to each cell or sector.

$$p_{ch} = \frac{A(1 - P_B)}{N_c} \tag{6.11}$$

Assuming that blocked calls are cleared, the quantities A, P_B, and N_c are related to each other through the Erlang B formula [34–35].

$$P_B = \frac{\dfrac{A^{N_c}}{N_c!}}{\displaystyle\sum_{i=0}^{N_c} \frac{A^i}{i!}} \tag{6.12}$$

As the loading factor increases, the probability of having six cochannel cells active also increases, which corresponds to a higher total cochannel interference level. Therefore, the number of interferers and consequently, the total interference, depend upon the loading factor.

The loading factor increases as the cluster size decreases. This means that cluster size reduction has a twofold effect, as far as interference is concerned:

1. The interference increases because cochannel cells are closer to each other.

2. Due to the increase in the loading factor, the probability that cochannel cells are using the same channel increases, which implies that the total interference increases.

It should be clear, then, that the loading factor plays an important role in the total system cochannel interference, which could enable a small reuse factor to be used.

The fractional loading factor technique [9] aims at reducing the cochannel interference level by lowering the loading factor. The reduction of the loading factor is achieved by hard limiting the number of channels that may be used simultaneously in a cell. However, the hard limit imposed on the instantaneous channel usage reduces the maximum possible carried traffic and thus the maximum capacity of each cell.

While the use of a low loading factor reduces the total cochannel interference, it also reduces the system capacity, as only a fraction of the channels assigned to a cell are allowed to be used at the same time. This leads to another important trade-off, which can be explained as follows. The reduction in interference level, which is required for smaller reuse factor and thus capacity improvement, is related to the loading factor reduction and the corresponding capacity loss, and this relationship varies as a function of the cluster size.

When the fractional loading factor technique is used, an appropriate call admission control must be employed in order to keep the cell-loading factor at the desired level.

6.2.1.4 Transmitter Power Control

Controlling the transmitter power is a frequently used tool to combat cochannel interference. In most modern systems, both base stations and mobile units have the capability of real time (dynamic) adjustment of their transmitters' power. There are several reasons why this tool may be effective in order to enhance the performance of a cellular system [10, 11]:

1. Cochannel interference management: By proper power adjustment, the effects of cochannel interference can be reduced. This allows for a denser reuse of frequencies and thus higher capacities.

2. Enhanced adjacent channel protection: Transmitter power control can be used to combat "near-far" problems, as we saw in Section 5.2.5, where two signals on separate channels but with a large difference in signal level may interfere. These systems suffer from *adjacent channel interference.* The aim of the power control scheme in these systems is to maintain the received power levels from all mobile units within a cell at a constant level, thus combating near-far problems.

3. Reduced power consumption: In mobile units, battery power is a scarce commodity. By using a minimum of transmitter power to achieve the required transmission quality, the battery life may be prolonged.

It should be noted that the power level transmitted by the base station or the mobile station is mainly controlled by the mobile switching center (MSC). The base station or the mobile station can perform limited control. In either case, the power level directly affects the C/I. For this case, an increase of this ratio means reduction of cochannel interference.

The system is assumed to use perfectly orthogonal signals (channels). In Figure 6.3, the link gains of two different transmitter-receiver pairs belonging to two different cochannel cells is illustrated. G_{ij} denotes the power gain from the base station in cell j to the mobile station using this channel in

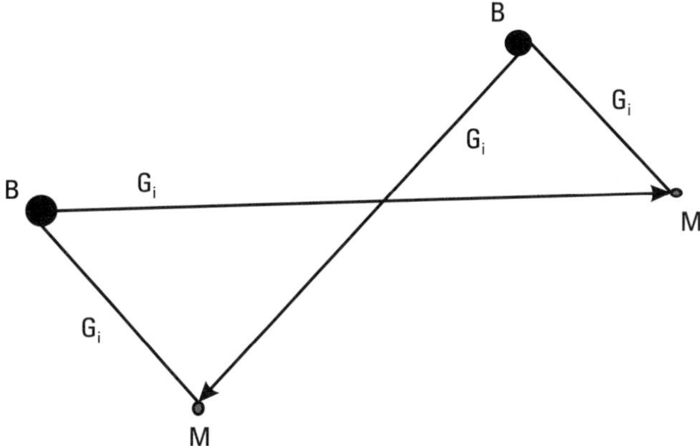

Figure 6.3 Link gains. (After: [9].)

cell i. It should be noted that the gains G_{ii} correspond to the desired communication links, whereas the G_{ij}, $i \neq j$, correspond to unwanted interference links. In general $G_{ij} \neq G_{ji}$.

The normalized downlink (base station to mobile station) gain, is given by

$$Z_{ij} = \frac{G_{ij}}{G_{ii}} \tag{6.13}$$

Furthermore, the thermal noise power at the receiver is denoted by N_i'. The normalized receiver noise is given by

$$N_i = \frac{N_i'}{G_{ii}} \tag{6.14}$$

The C/I at the mobile station i is given by:

$$\left(\frac{C}{I}\right)_i = \frac{G_{ii}P_i}{\displaystyle\sum_{\substack{j=1 \\ i \neq j}}^{K_0} G_{ij}P_j + N_i'} = \frac{P_i}{\displaystyle\sum_{\substack{j=1 \\ i \neq j}}^{K_0} P_j \frac{G_{ij}}{G_{ii}} + N_i} = \frac{P_i}{\displaystyle\sum_{j=1}^{K_0} P_j Z_{ij} - P_i + N_i} \tag{6.15}$$

where P_i, P_j are the power levels transmitter from the base stations BS_i, BS_j, respectively, and K_0 is the number of interferers.

Control of the Power Transmitted by the Base Station

When the signal received by the mobile station is very strong, the MSC reduces the transmitted power of both the base station and the mobile unit. In this way, the cochannel reuse distance decreases and so does the cochannel interference, as well as the adjacent-channel interference.

Control of the Power Transmitted by the Mobile Station

When a mobile station is approaching a base station, the power level of the mobile unit should be reduced for the following reasons:

1. Reducing the chance of generating intermodulation products from a saturated receiving amplifier;

2. Reducing the chance of interfering with other cochannel base stations;

3. Reducing the near-far interference ratio.

6.2.1.5 Diversity

Diversity is another powerful communication receiver technique that provides wireless link improvement at relatively low cost [12–15]. Unlike equalization, diversity requires no training overhead because the transmitter does not require a training sequence. Diversity exploits the random nature of radio propagation by finding independent signal paths for communication. In virtually all applications, diversity decisions are made by the receiver and are unknown to the transmitter, as we saw in Chapter 4.

Diversity receivers are used to reduce both multipath fading and interference. Diversity can take different forms, but the one usually used in mobile communications is *space diversity*. Two antennas are placed at a distance between each other. Both antennas receive the same signal and the results are compared in order to produce the correct output. This method is called the *selection method,* and is depicted in Figure 6.4. Diversity in mobile systems can be implemented with a very small antenna separation on the order of half a wavelength.

The selection method requires the same number of receivers as the number of diversity branches. Alternatively, a single receiver can be used, which is called a *switching* or *scanning* receiver. By using this type of receiver, different diversity techniques can be used in order to compare or combine the results and produce the correct output.

According to the first method, the received signal level is compared to a threshold. If the signal level falls below the threshold, the receiver

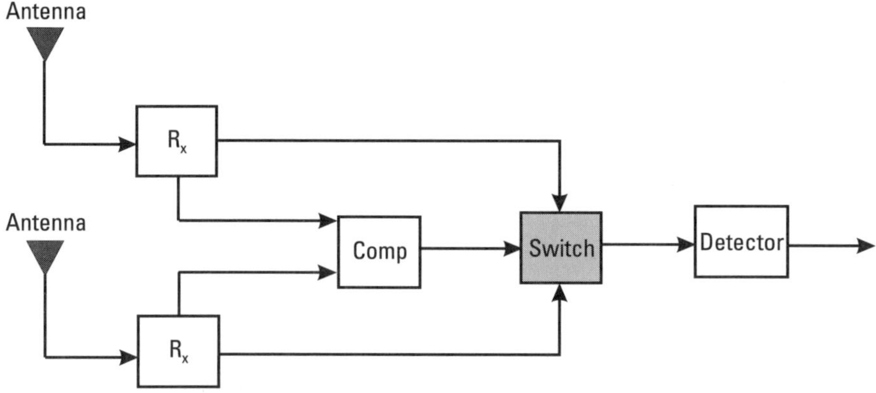

Figure 6.4 Selection method in space-diversity receivers. (After: [9].)

switches from one diversity branch to the other. This threshold may be fixed (Figure 6.5) or dynamically adjusted (Figure 6.6). A fixed threshold may be suitable for a small area, but is not necessarily suitable for the entire service area. Therefore, the threshold value should be adjusted dynamically as the vehicle moves.

6.2.1.6 Discontinuous Transmission

Discontinuous transmission was originally developed for satellite systems. The goal is to achieve mobile station power reduction and reduction of interference as well. During a normal conversation, the participants speak only 50% of the time. Each direction of transmission is occupied about 50% of the time. Discontinuous transmission is a mode of operation, whereby

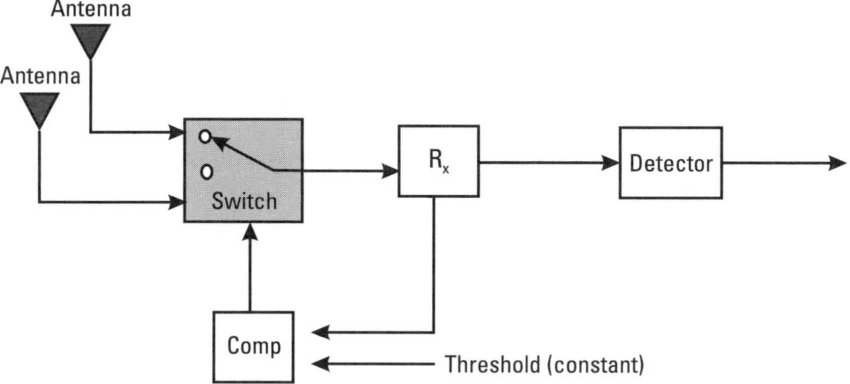

Figure 6.5 Switching receiver with a fixed threshold.

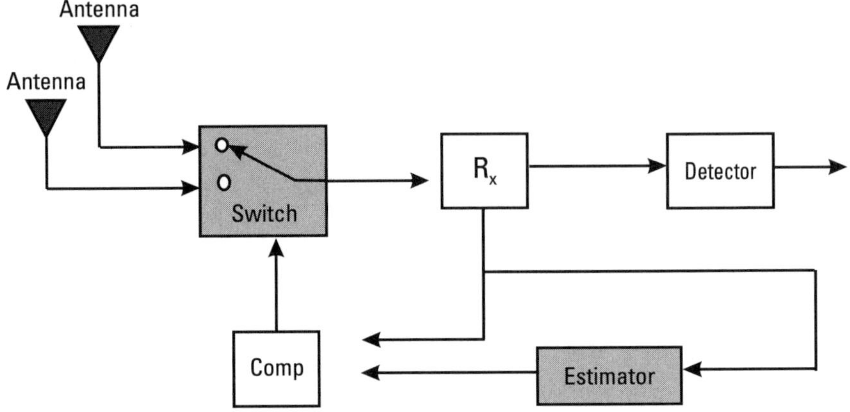

Figure 6.6 Switching receiver with a dynamically adjusted threshold.

the transmitters are switched on only when useful information is to be transmitted. The difficulty with the method is to find techniques to distinguish noisy speech from real noise, even in a noisy environment [16–17]. The background acoustic noise has to be evaluated in order to transmit characteristic parameters to the receiving side. The receiving side generates a similar noise called *comfort noise* during periods where the radio transmission is cut.

According to the second method, the two received signals are phase aligned and summed for maximum received level. Thus, this method is called the *phase-sweeping* method (Figures 6.7 and 6.8). The sweeping rate must be higher than twice the highest frequency of the modulation signal. This method becomes even more attractive when multiple diversity branches are used.

6.2.2   Direct Reduction Methods

The direct methods reduce the incoming interfering signals from a microscopic (postdetection) point of view. In this case, the interference reduction is achieved by including in the system hardware or software components specified to deal with the interference problem. In this class of methods, we shall point out the most representative ones, which have been analyzed in some detail in Chapters 4 and 5 and will be used in applications in Chapter 7.

Here, we will review the same methodology from the interference suppression angle. These methods deal with the design of appropriate receive

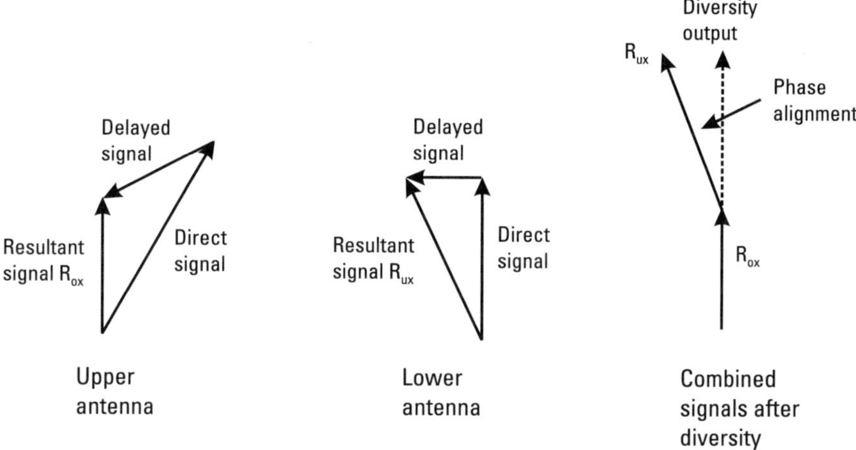

Figure 6.7 Phase-sweeping concept.

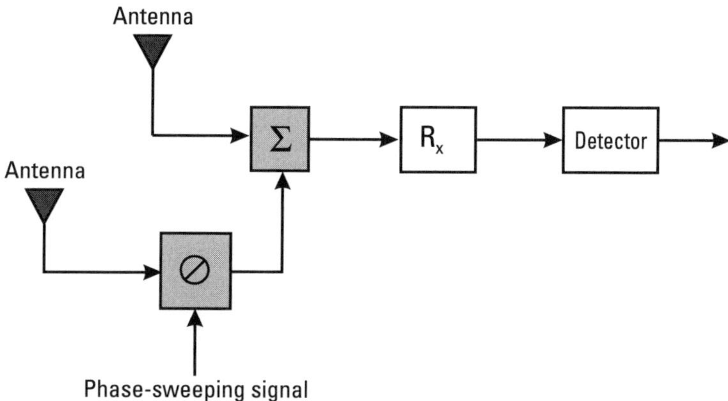

Figure 6.8 Phase-sweeping method in space-diversity receivers.

filters in an adaptive way, such as the finite impulse response filter (FIR), the reduction of MAI with some kind of blind algorithm, the creation of an immune transmission signal through the use of frequency hopping, with the distortion mitigation and combat, or with the direct SNR loss. All of these methods more or less somehow utilize advanced digital processing techniques, which have some ultimate goal to result in interference rejection, as do the error correction and precoding techniques.

6.2.2.1 Frequency Hopping

Frequency hopping (FH) involves a periodic change of transmission frequency. A frequency hopping signal may be regarded as a sequence of modulated data bursts with time-varying, pseudorandom carrier frequencies [18, 19]. The set of possible carrier frequencies is called the hopset. Hopping occurs over a frequency band that includes a number of channels. Each channel is defined as a spectral region with a central frequency in the hopset and a bandwidth large enough to include most of the power in a narrowband modulation burst (usually FSK) that has the corresponding carrier frequency. The bandwidth of a channel used in the hopset is called instantaneous bandwidth. The bandwidth of the spectrum over which the hopping occurs is called the total hopping bandwidth (spread spectrum). The transmitter carrier sends data to seemingly random channels, which are known only to the desired receiver. On each channel, small bursts of data are sent using conventional narrowband modulations before the transmitter hops again.

If only a single carrier frequency is used on each hop, digital data modulation is called single channel modulation. Figure 6.9 shows a single channel frequency hopping spread spectrum (FH-SS) system. The time duration between hops is called the hop duration or the hopping period and is denoted T_h. The total hopping bandwidth and the instantaneous bandwidth are denoted and W_{ss} and B, respectively.

Then the processing gain for FH systems will be given by

$$\text{Processing gain} = W_{ss}/B \qquad (6.16)$$

After frequency hopping has been removed from the received signal, the resulting signal is said to be dehopped. If the frequency pattern produced by the receiver synthesizer, as shown in Figure 6.9(b), is synchronized with the frequency pattern of the received signal, then the mixer output is a dehopped signal at a fixed difference frequency. Before demodulation, the dehopped signal occupies a particular hopping channel. The noise and interference in that channel are translated in frequency so that they enter the demodulator. Thus, it is possible to have collisions in a FH system where an undesired user transmits in the same channel at the same time at the desired user, as we discussed in Section 3.8.3.1.

Frequency hopping may be classified as fast or slow. Fast frequency hopping occurs if there is more than one frequency hop during each transmitted symbol. Thus, fast frequency hopping implies that the hopping rate equals or exceeds the information symbol rate. Slow frequency hopping occurs if one or more symbols are transmitted in the time interval between frequency hops.

The frequency channel occupied by a transmitted symbol is called the transmission channel. The channel that would be occupied if the alternative

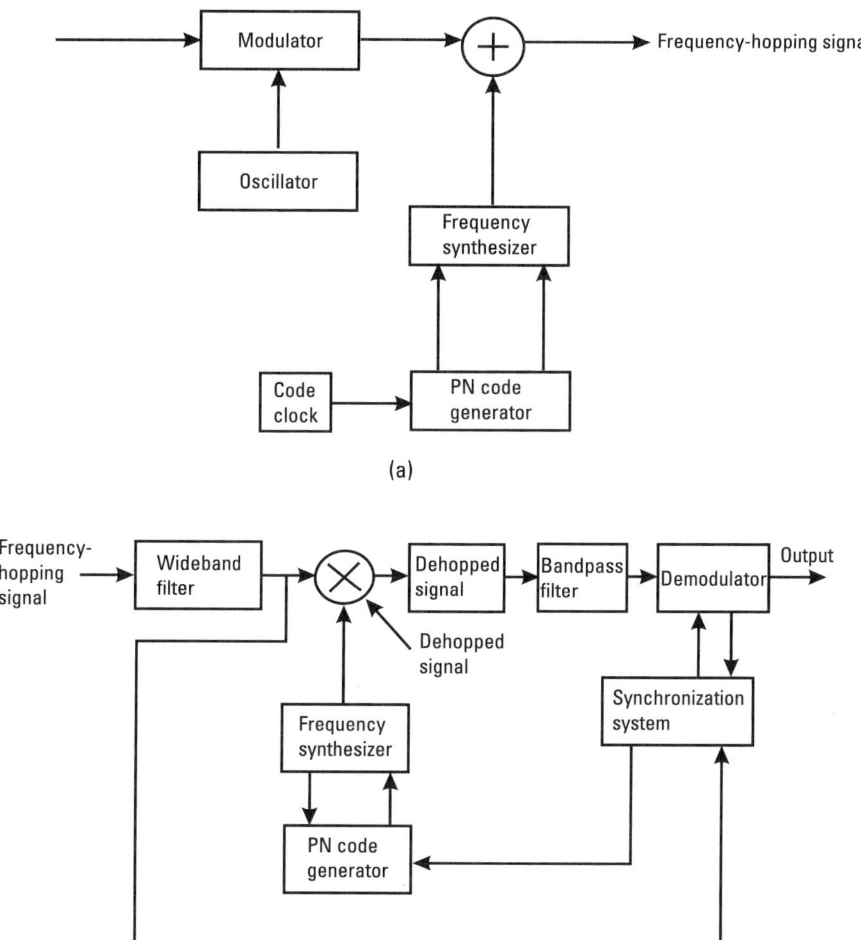

Figure 6.9 Block diagram of frequency hopping system with single channel modulation. (After: [8].)

symbol were transmitted is called the complementary channel. The frequency hop rate of an FH-SS system is determined by the frequency agility of receiver synthesizers, the type of information being transmitted, the amount of redundancy used to code against collisions, and the distance to the nearest potential interferer.

A source that causes interference on a particular frequency is unlikely to cause interference on the next frequency of the hop sequence as well.

Therefore, the power of the narrowband interfering signal is spread over the wider bandwidth of the spreading signal. Thus, interference takes the form of fading, which can be reduced by using error correction techniques. In this way, the system performance is enhanced [19].

Frequency hopping can be even more effective for combating cochannel interference if cochannel cells hop in an uncoordinated, rather than in a cyclic, way. The corresponding hopping mode is called (pseudo) random hopping. Uncorrelated hopping sequences are used in different cochannel cells. Hence, the probability of a collision (interference) between two cochannel cells decreases even further. By utilizing FH, the probability of cochannel interference becomes inversely proportional to the number of frequencies used in the hopping sequence. Thus, the gain of FH increases as the number of frequencies used in the hopping sequence increases. The two hopping modes are illustrated in Figures 6.10 and 6.11.

6.2.3 Distortion Mitigation

If the channel introduces signal distortion as a result of fading, the system performance can exhibit an irreducible error rate. When larger than the

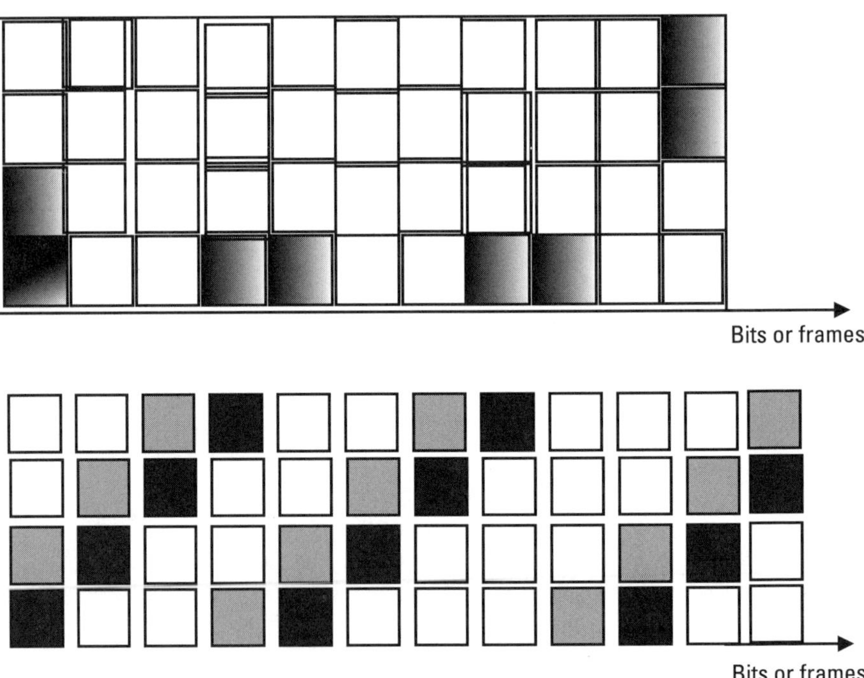

Figure 6.10 Cyclic frequency hopping.

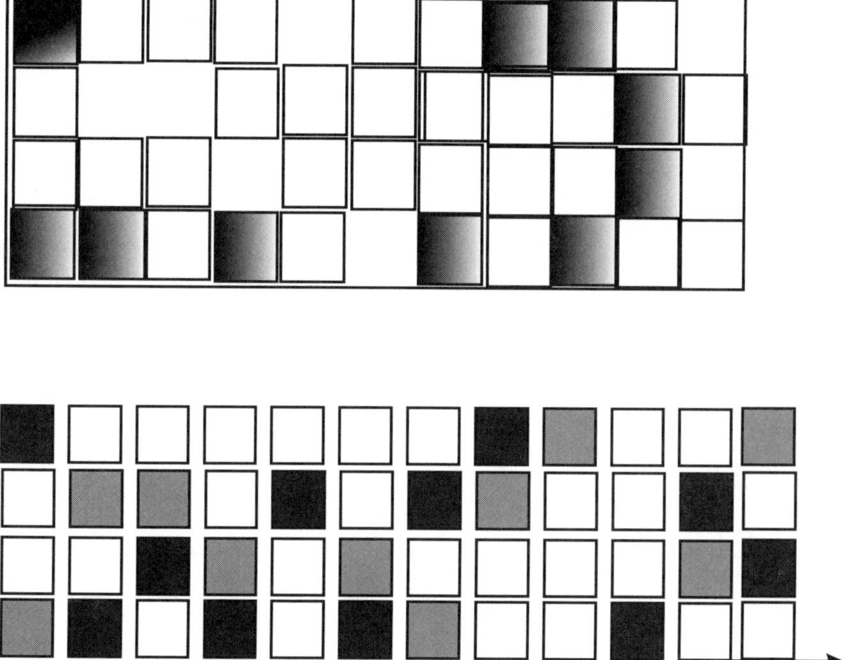

Figure 6.11 (Pseudo) random frequency hopping.

desired error rate, no amount of E_b/N_0 will help achieve the desired level of performance. In such cases, the general approach for improving performance is to use some form of mitigation to remove or reduce the distortion. The mitigation method depends on whether frequency selective fading or fast fading causes the distortion, as we discussed in detail in Chapters 4 and 5. In general, the mitigation approach to be used should follow two basic steps:

1. Provide distortion mitigation and combat distortion.
2. Provide diversity to combat loss of SNR.

6.2.3.1 Distortion Combat

In this category, we considered in Chapter 4 the following:

1. Frequency selective distortion, which includes adaptive equalization (e.g., decision feedback, Viterbi equalizer), spread spectrum (DS or FH), orthogonal FDM (OFDM), and pilot signals.

2. Fast-fading distortion, which includes robust modulation, signal redundancy to increase signaling rate, and coding and interleaving.

6.2.3.2 SNR Combat Loss

In this category, we considered in Chapter 4 the following:

1. Diversity, which includes time (e.g., interleaving), frequency (e.g., bandwidth (BW) expansion, spread-spectrum FH or DH with rate receiver), spatial (e.g., spaced receive antennas), and polarization.
2. Fast fading and slow-fading compensation, which includes some type of diversity to get additional uncorrelated estimates of signal and error-correction coding.

Another very important classification of the reduction interference methods is:

1. Linear methods;
2. Nonlinear methods.

6.2.3.3 Linear Methods

Optimal multiuser detection can make CDMA systems not be interference-limited, but it is too complex to be implemented [20–24]. To reduce the implementation complexity, therefore, most researchers have focused on finding suboptimal multiuser receivers, such as subtractive interference cancellation receivers and linear multiuser receivers. Linear receivers are multiple input-multiple output or single-input, single-output equalizers applied to multiuser systems. Both adaptive and nonadaptive techniques exist. Interference cancellation types are the most typical of nonlinear receivers. If, however, we separate the channel estimation or interference estimation procedure with some type of equalization or a procedure that entails a weighted finite summation, we can construct a combined technique, which provides a better performance.

The following discussion shows how a method (MMSE), which belongs to the linear methods, can be used to combine both [34–39].

Subtractive interference cancellation receivers estimate multiple access interference (MAI) and subtract it from the received signal. Thus, they improve system performance significantly over the conventional matched filter (MF). These receivers include the successive interference cancellation (SIC) receiver and parallel interference cancellation (PIC) receiver. In multicell environments, out-of-cell interference severely limits the benefits of subtractive interference cancellation [40]. That is because subtractive interference cancellation receivers require the information of spreading code, timing, amplitude, and phase of all of the stations and cannot suppress out-of-cell interference. To overcome the limitation of these receivers, an integrated scheme of a subtractive interference cancellation receiver and an adaptive minimum mean square error (MMSE) receiver is proposed [36].

Consider a pilot symbol-aided BPSK DS-CDMA system over AWGN and Rayleigh fading channels. L users share the channel. It is assumed that the ith user the desired one. The received signal sampled at $t = nT_c$ time-aligned to the ith user, where T_c is the chip duration, can be represented as [31]

$$r_i(n) = c_i(n)\, b_i(n/N)\, s_i(n) + \sum_{j \neq i} I_j(n) + z(n) \qquad (6.17)$$

where $c_i(n)$, $b_i(n/N)$, $s_i(n)$ are the complex channel coefficients, the transmitted data taking on values ± 1, the spreading code of the ith user, respectively

$\sum_{j \neq i} I_j(n)$ is the MAI contributed by the other users;

$z(n)$ is additive white Gaussian noise.

Here, N denotes the processing gain (i.e., $N = T_b/T_c$, where T_b is the bit duration and $[n/N]$ denotes the smallest integer greater than n/N). The signal samples over bit duration T_b is taken to be a signal vector $\vec{r}_i(m)$ at time $t = mT_b$. Each pilot symbol is periodically inserted into data symbol streams at every M data symbols.

In the multistage PIC receiver, let $I_j^{(k)}(n)$ and $\hat{I}_j^{(k)}(n)$ be the interference and its estimate at the kth stage, respectively, then the interference suppressed input data for the $(k + 1)$th stage is obtained as

$$r_i^{(k+1)}(n) = r_i(n) - \sum_{j \neq t} I_j^{(k)}(n) \text{ with } r_i^{(1)}(n) = r_i(n) \qquad (6.18)$$

where $i = 1, 2, \ldots L$.

The performance of the multistage PIC receiver heavily depends on the accuracy of the interference estimate. As the stage increases, bit decision and channel estimation are getting more reliable, so the accuracy of the regenerated signal in the PIC is improved. Although a conventional adaptive MMSE receiver provides performance improvement over the conventional matched filter (MF) receiver in a static channel, it has severe performance degradation in fading channels. To overcome this drawback, the adaptive constrained MMSE receiver is shown to give better results [36]. The criterion that the mth bit was received is given by the constrained MMSE.

$$E\left[|\bar{e}_i(m)|^2\right] = E\left[|\tilde{d}_i(m) - w_i^H(m)\,r_i(m)|^2\right] \tag{6.19}$$

$$\text{subject to } w_i^H(m)\,s_i = 1$$

Here $(-)^H$ denotes Hermitian operation (complex conjugate and transpose operation) where $\tilde{d}_i(m)$ is the reference signal multiplied by the channel estimate $e_i(m)$, $\underline{w}_i(m)$ is the tap weight vector, and $\underline{s}_i$ is the spreading code vector. The channel estimate obtained at the adaptive filter output is more accurate than that obtained at the adaptive filter input. The constraint prevents the convergence problem when channel estimation is accomplished at the adaptive filter output.

The constrained MMSE criterion in (6.19) can be adaptively implemented using the orthogonal decomposition-based LMS algorithm.

$$\text{Let } \underline{w}_i(m) = \underline{s}_i + \underline{x}_i(m) \tag{6.20}$$

where $w_i(m)$ is the adaptive component of the tap weights vector and orthogonal to $\underline{s}_i$, then the orthogonal decomposition-based LMS algorithm is given by:

$$\underline{x}_i(m + 1) = \underline{x}_i(m) + \mu \cdot \bar{e}_i^*(m) \cdot \underline{r}_{x_i}(m) \tag{6.21}$$

where $\underline{r}_{x_i}(m) = \underline{r}_i(m) - \dfrac{r_i^H(m)\,\underline{s}_i}{\underline{s}_i^H\,\underline{s}_i}\,\underline{s}_i$ is the projection of the received signal vector $\underline{r}_i(m)$, on $\underline{x}_i(m)$ μ is the step size, and $(.)^*$ denotes complex conjugate.

Simulation results show that this adaptive constrained MMSE receiver provides significant performance improvements on the basis of BER over the conventional MF receiver and adaptive MMSE receivers, even at low signal to interference plus noise ratio (SINR) [36].

6.2.3.4   Equalization

ISI caused by multipath in bandlimited time-dispersive channels distorts the transmitted signal, causing bit errors at the receiver. That type of interference has been recognized as the major obstacle to high data transmission over mobile radio channels. Equalization is a technique used to combat ISI, as discussed in detail in Section 4.5.1. In this section, we shall present the general structure of the equalizers and form the basis for discussing another method of equalization based on hidden Markov models (HMMs) [25–31]. This concept was presented in Chapter 4 in the context of fading compensation. In this section we shall present the general structure of a channel equalizer and show how the same result can be achieved using a new technique called hidden Markov modeling [2]. In radio channels, a variety of adaptive equalizers can be used to cancel interference while providing diversity. Because the mobile fading channel is random and time varying, equalizers must track the time-varying characteristics of the mobile channel; thus, they are called adaptive equalizers.

The operating modes of an adaptive equalizer include training and tracking. The transmitter sends a fixed-length training sequence so that the receiver's equalizer may average to a proper setting. The training sequence is typically a pseudorandom binary signal or a fixed prescribed bit pattern. Immediately following this training sequence, the user data is sent, and the adaptive equalizer at the receiver utilizes a recursive algorithm to evaluate the channel and estimate filter coefficient to compensate for the channel. The training sequence is designed to permit an equalizer at the receiver to acquire the proper filter coefficients in the worst possible channel conditions so that when the training sequence is finished, the filter coefficients are near the optimal values for reception of user data. As user data is received, the adaptive algorithm of the equalizer tracks the changing channel. As a consequence, the adaptive equalizer is continually changing its filter characteristics over time.

The timespan over which an equalizer converges is a function of the equalizer algorithm, the equalizer structure, and the rate of time change of the multipath radio channel. Equalizers require periodic retraining in order to maintain effective ISI cancellation and are commonly used in digital communication systems where user data is segmented into short time blocks. An equalizer is usually implemented at baseband or at IF in a receiver. Because the baseband complex envelope expression can be used to represent bandpass waveforms, the channel response, demodulated signal, and adaptive equalizer algorithms are usually simulated and implemented at baseband.

Figure 6.12 shows a block diagram of a communication system with an adaptive equalizer in the receiver.

If $s(t)$ is the original information signal, and $h(t)$ is the combined complex baseband impulse response of the transmitter, channel, and the

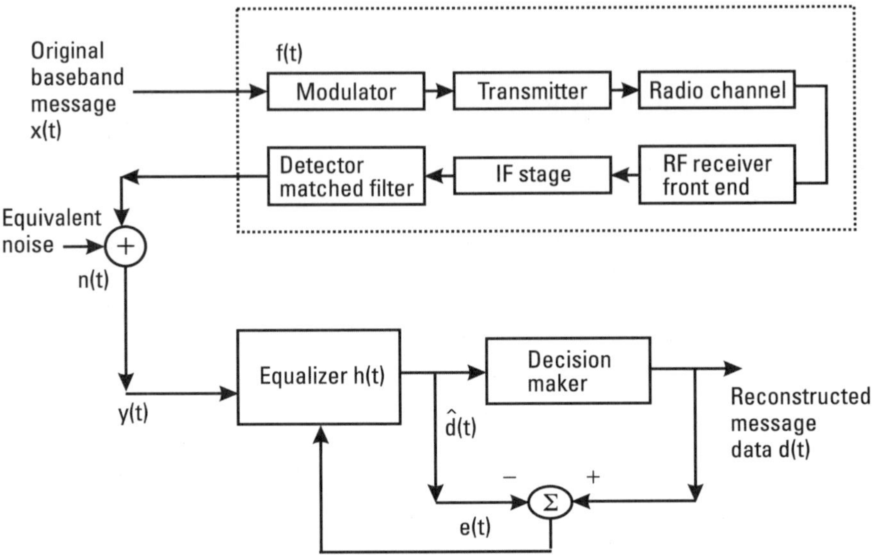

Figure 6.12 Block diagram of a simplified communications system using an adaptive equalizer at the receiver.

RF/IF sections of the receiver, the signal received by the equalizer may be expressed as

$$y(t) = s(t) \otimes h_c^*(t) + n_b(t) \tag{6.22}$$

where $h_c^*(t)$ is the complex conjugate of $h_c(t)$, $n_b(t)$ is the baseband noise at the input of the equalizer and $\otimes$ denotes the convolution operation. If the impulse response of the equalizer is $h_{eq}(t)$, then the output of the equalizer is

$$\bar{y}(t) = s(t) \otimes h_c^*(t) \otimes h_{eq}(t) + n_b(t) \otimes h_{eq}(t) \tag{6.23}$$

$$= s(t) \otimes g(t) + n_b(t) \otimes h_{eq}(t)$$

where $g(t)$ is the combined impulse response of the transmitter, channel, RF/IF sections of the receiver, and the equalizer. The complex baseband impulse response of a transversal filter equalizer is given by

$$h_{eq}(t) = \sum_n c_n \delta(t - nT) \tag{6.24}$$

where

c_n = the complex filter coefficients of the equalizer.

The desired output of the equalizer is $s(t)$, the original source date. Assume that $n_b(t) = 0$. Then, in order to force $\bar{y}(t) = s(t)$ in (6.23), $g(t)$ must be equal to

$$g(t) = h_c^*(t) \otimes h_{eq}(t) = \delta(t) \tag{6.25}$$

The goal of equalization is to satisfy (6.25). In the frequency domain, (6.25) can be expressed as

$$H_{eq}(f)/H_c^*(f) = 1 \tag{6.26}$$

where $H_{eq}(f)$ and $H_c(f)$ are Fourier transforms of $h_{eq}(t)$ and $h_e(t)$, respectively.

Equation (6.26) indicates that an equalizer is actually an inverse filter of the channel. If the channel is frequency selective, the equalizer enhances the frequency components with small amplitudes and attenuates the strong frequencies in the received frequency spectrum. This provides a flat, composite, received frequency response and linear phase response. For a time-varying channel, an adaptive equalizer is designed to track the channel variations so that (6.25) is approximately satisfied.

Channel Equalization Based on HMMs

In this subsection, we shall consider some form of blind equalization, where the channel inputs are modeled by a set of HMMs. An HMM is a finite-state Bayesian model with a Markovian state prior and a Gaussian observation likelihood (see Appendix B). An N-state HMM can be used to model a nonstationary process, such as speech, as a chain of N stationary states connected by a set of Markovian state transitions, as shown in Figure 6.13.

The HMM-based channel equalization problem can be stated as follows: Given a sequence of NP-dimensional channel output vectors $Y = [y(0), \ldots, y(N-1)]$, and separating them, the prior knowledge that the channel input sequence is drawn from a set of U HMMs $M = \{M_i, i = 1, \ldots, U\}$, estimate the channel response and the channel input.

The joint posterior PDF of an input word M_i and the channel vector h can be expressed as

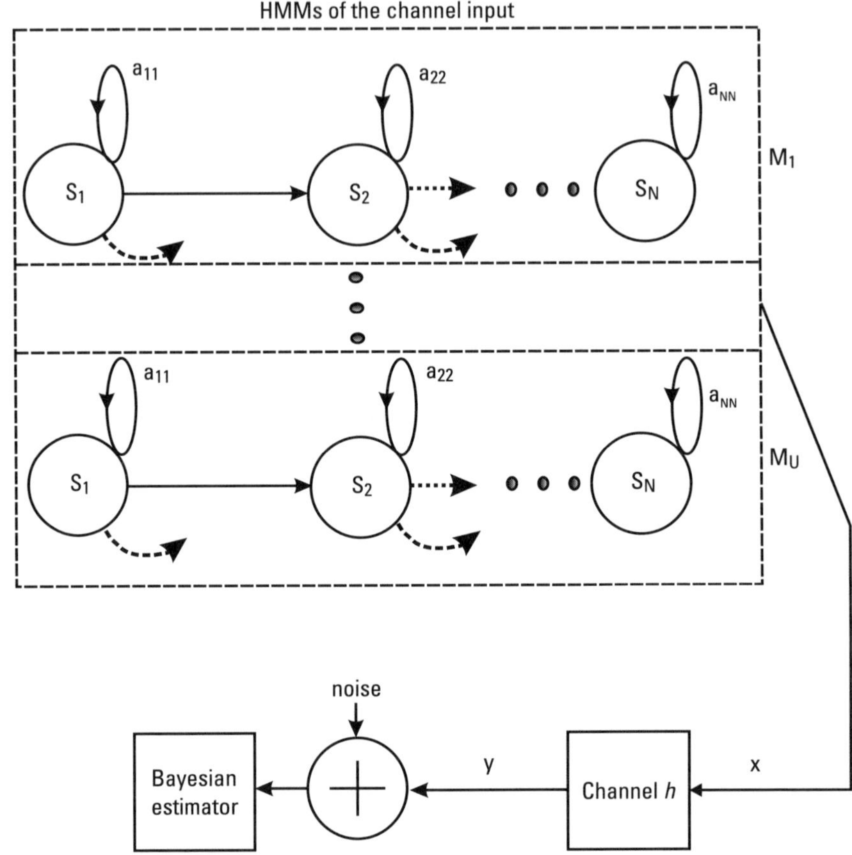

Figure 6.13　Channel input modeled by a set of HMMs. (After: [2]. © 2000 John Wiley & Sons, Inc.)

$$f_{M,H|Y}(M_i, h | Y) = P_{M|H,Y}(M_i | h, Y) f_{H|Y}(h | Y) \qquad (6.27)$$

Simultaneous joint estimation of the channel vector h and classification of the unknown input word M_i is a nontrivial exercise. The problem is usually approached iteratively by making an estimate of the channel response, and then using this estimate to obtain the channel input as follows. From Bayes' rule, the posterior PDF of the channel h conditioned on the assumption that the input model is M_i, and given the observation sequence Y, can be expressed as

$$f_{H|M,Y}(h | M_i, Y) = \frac{1}{f_{Y|M}(Y | M_i)} f_{Y|M,H}(Y | M_i h) f_{H|M}(h | M_i)$$

$$(6.28)$$

The likelihood of the observation sequence, given the channel and the input word model, can be expressed as

$$f_{Y|M,H}(Y|M_i,h) = f_{X|M}(Y - h|M_i) \tag{6.29}$$

where it is assumed that the channel output is transformed into capital variables so that the channel distortion is additive. For a given input model M_i and state sequence $s = [s(0), s(1), \ldots, s(N-1)]$, the maximum likelihood channel estimate is given by [2]:

$$\hat{h}^{ML}(Y, s) = \sum_{m=0}^{N-1} \left(\sum_{k=0}^{N-1} \Sigma_{xx,s(k)}^{-1} \right)^{-1} \Sigma_{xx,s(m)}^{-1} (y(m) - \mu_{x,s(m)}) \tag{6.30}$$

where $\mu_{x,s}$ and $\Sigma_{xx,s(k)}$ is the mean and covariance matrix of the Gaussian observation PDF of the HMM state s of model M_i.

Note that when all of the state observation covariance matrices are identical, the channel estimate becomes

$$\hat{h}^{ML}(Y, s) = \frac{1}{N} \sum_{m=0}^{N-1} (y(m) - \mu_{x,s(m)}) \tag{6.31}$$

The maximum likelihood (ML) estimate of (6.31) is based on the ML state sequence s of M_i. In the following section, we consider the conditional mean estimate over all state sequences of a model.

In the following, we will consider three implementation methods for HMM-based channel equalization.

Method I: Use of the Statistical Averages Taken over All HMMs

A simple approach to blind equalization, similar to that proposed by [27], is to use as the channel input statistics the average of the mean vectors and covariance matrices, taken over all of the states of all of the HMMs.

$$\mu_x = \frac{1}{UN_S} \sum_{i=1}^{U} \sum_{j=1}^{N_S} \mu_{M_i, j}, \quad \Sigma_{xx} = \frac{1}{UN_S} \sum_{i=1}^{U} \sum_{j=1}^{N_S} \Sigma_{M_i, j} \tag{6.32}$$

where $\mu_{M_i, j}$ and $\Sigma_{M_i, j}$ are the mean and the covariance of the jth state of the ith HMM, and U and N_S denote the number of models and number of states per model, respectively. The maximum likelihood estimate of the channel, $\hat{h}^{ML}$, is defined as

$$\hat{h}^{ML} = (\bar{y} - \mu_x) \tag{6.33}$$

where $\bar{y}$ is the time-averaged channel output. The estimate of the channel input is

$$\hat{x}(m) = y(m) - \hat{h}^{ML} \tag{6.34}$$

Using the averages over all states and models, the maximum a posteriori (MAP) channel estimate becomes

$$\hat{h}^{MAP}(Y) = \sum_{m=0}^{N-1} (\Sigma_{xx} + \Sigma_{hh})^{-1}\Sigma_{hh}(y(m) - \mu_x) + (\Sigma_{xx} + \Sigma_{hh})^{-1}\Sigma_{xx}\mu_h \tag{6.35}$$

Method II: Hypothesized Input HMM Equalization

In this method, for each candidate HMM in the input set, a channel estimate is obtained and then used to equalize the channel output, prior to the computation of a likelihood score for the HMM. Thus, a channel estimate h_w is based on the hypothesis that the input word is w. It is expected that a better channel estimate is obtained from the correctly hypothesized HMM, and a poorer estimate is obtained from an incorrectly hypothesized HMM. The hypothesized-input HMM algorithm is as follows in Figure 6.14.

For $i = 1$ to number of words U.

- Step 1: Using each HMM, M_i, make an estimate of the channel, $\hat{h}_i$, and estimate the channel input;

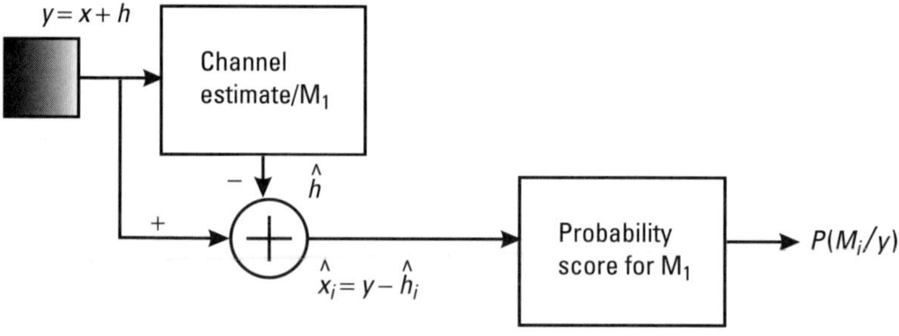

Figure 6.14　Input modeled by a set of HMMs. (After: [2]. © 2000 John Wiley & Sons, Inc.)

- Step 2: Using the channel estimate $\hat{h}_i$, estimate the channel input $\hat{x}(m) = y(m) - \hat{h}_i$;

- Step 3: Compute a probability score for model M_i, given the estimate $[\hat{x}(m)]$.

Select the channel estimate associated with the most probable word.

Method III: Decision-Directed Equalization

Blind adaptive equalizers are often composed of two distinct sections: an adaptive linear equalizer followed by a nonlinear estimator to improve the equalizer output. The output of the nonlinear estimator is the final estimate of the channel input, and is used as the desired signal to direct the equalizer adaptation. The use of the output of the nonlinear estimator as the desired signal assumes that the linear equalization filter removes a large part of the channel distortion, thereby enabling the nonlinear estimator to produce an accurate estimate of the channel input. A method of ensuring that the equalizer locks into and cancels a large part of the channel distortion is to use a startup equalizer training period during which a known signal is transmitted. Figure 6.15 illustrates a blind equalizer incorporating an adaptive linear filter followed by an HMM model classifier/estimator. The HMM classifies the output of the filter as one of a number of likely signals and provides an enhanced output, which is also used for adaptation of the linear filter. The output of the equalizer $z(m)$ is expressed as the sum of the input to the channel $x(m)$ and a so-called convolutional noise term $n(m)$ as

$$z(m) = x(m) + n(m) \tag{6.36}$$

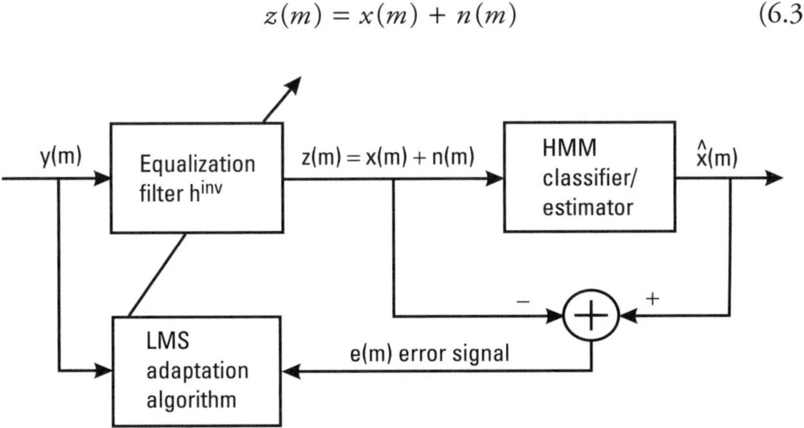

Figure 6.15 Decision directed equalizer. (After: [2]. © 2000 John Wiley & Sons, Inc.)

The HMM may incorporate state-based Wiener filters for suppression of the convolutional noise $n(m)$ [3]. Assuming that the LMS adaptation method is employed, the adaptation of the equalizer coefficient vector is governed by the following recursive equation:

$$\hat{h}^{-1}(m) = \hat{h}^{-1}(m-1) + \mu e(m) y(m) \qquad (6.37)$$

where $\hat{h}^{-1}(m)$ is an estimate of the optimal inverse channel filter, μ is an adaptation step size, and the error signal $e(m)$ is defined as

$$e(m) = \hat{x}^{HMM}(m) - z(m) \qquad (6.38)$$

where $\hat{x}^{HMM}(m)$ is the output of the HMM-based estimator and is used as the correct estimate of the desired signal to direct the adaptation process.

6.2.4 Nonlinear Methods

Besides linear methods (e.g., the transversal equalizer in the case of ISI), nonlinear methods based on interference estimation and subtraction are known. In general, the nonlinear approaches offer better performance compared to linear methods with comparable complexity—for example, the decision feedback equalizer (DFE) for interference elimination and the approach of successive interference cancellation for MAI elimination. The basic idea of these nonlinear approaches is to estimate the interference part of the signal and to subtract it from the signal as shown in Figure 6.16. In the section to follow, we shall present a method of interference elimination by first estimating the interfering signal [27–40].

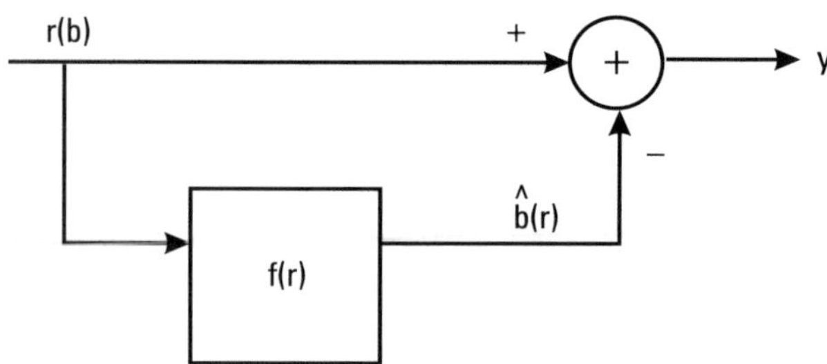

Figure 6.16 General block diagram for an interference elimination scheme. (After: [27].)

6.2.4.1 Interference Estimation/Elimination

The crucial part of the interference elimination problem is an accurate estimation of the interfering signal [27–44]. The optimization criterion is the minimization of the mean square error of the estimation.

The input signal of the estimation device, which represents one portion of the interfering signal part, can be written in signal space description as

$$r = b + n \tag{6.39}$$

where

$b =$ is the transmitted symbol (of an interferer or an interfering signal part);

$n =$ additive Gaussian noise (consisting of channel noise and may be additional interference).

This device derives the estimate $\hat{b}(r)$ for the transmitted symbol b in order to minimize the residual estimation error $b - \hat{b}$ in the output signal.

$$y = r - \hat{b} = b - \hat{b} + n \tag{6.40}$$

In the mean square sense, we seek to minimize J

$$J = E\{(b - \hat{b})^2\} \to \min \tag{6.41}$$

The absolute minimum of (6.41)—the minimal residual interference power after elimination—is achieved by $\hat{b}$ being a real number rather than restricting it to the transmit symbol alphabet. Therefore, this should be considered as a parameter estimation problem rather than a detection problem, as is usually the case. The general solution of this problem is given next.

$$\hat{b}(r) = E\{b/r\} = \int_{-\infty}^{\infty} b \cdot p(b/r)\, db \tag{6.42}$$

which is the conditional expectation value of b given r, and $p(b/r)$ is the conditional probability density of b conditioned on the knowledge of r etc.

Using Bayes' theorem, we obtain

$$p(b/r) = \frac{p(r/b) \cdot p(b)}{p(r)} = c(r) \cdot p(r/b) \cdot p(b) \tag{6.43}$$

To evaluate (6.42) and (6.43), the corresponding PDF $p(b)$ is required. In this case, assuming that we have equally distributed antipodal transmit symbols $bE\{-1, 1\}$, the PDF of b is given by

$$p_b(b) = 0.5 \cdot [\delta(b + 1) + \delta(b - 1)] \tag{6.44}$$

If it is assumed as Gaussian noise, the conditional PDF of the signal r given b is

$$p(r/b) = p(n) = \frac{1}{\sqrt{2\pi}\sigma} e^{-\frac{(r-b)^2}{2\sigma^2}} \tag{6.45}$$

where σ^2 is the variance of the additive noise n.

The unknown $c(r)$ in (6.43) can be determined by exploiting the property

$$\int p(b/r)\, db = 1$$

Inserting the distribution into (6.42), the optimum estimation results in the nonlinear characteristic

$$\hat{b}(r) = \tanh\left(\frac{r}{\sigma^2}\right) \tag{6.46}$$

This optimum characteristic depends on the SNR at the estimation device, having a smooth characteristic for small SNR and approaching the "hard" sign-function (two-point characteristic) for large SNR. The resulting minimum mean square error is given by

$$J_{\min} = E\{(\hat{b}(r) - b)^2\} = \frac{1}{\sqrt{2\pi}\sigma} \int_{-\infty}^{\infty} \left[\tanh\left(\frac{r}{\sigma^2}\right) - 1\right]^2 e^{-\frac{(r-1)^2}{2\sigma^2}}\, dr$$

$$\tag{6.47}$$

The minimization of J has resulted in (6.47), which is nothing but the interference estimate. This can now be subtracted from the original signal that contains the interference. An application of this method to CDMA signals is shown in Figure 6.17. The result will be the information signal without interference.

The input signal $r(k)$ is first demodulated (despread) for all but the wanted user 1 (r_i, $i = 2 \ldots, L$), the data $\hat{b}_i$ is estimated, respread, and subtracted from the original signal. Note that in this case the interference cancellation is done in one step rather than in the usual successive manner, where the signal of the third user is demodulated after cancellation of the second user and so forth for all other users. This one-slot approach may perform slightly worse than the successive one, but it offers the possibility of a parallel implementation or the use of an efficient common dispreading algorithm (e.g., the fast Hadamard transform) when using Walsh functions as spreading codes [27].

The received signal in time-discrete representation is given by

$$r(k) = \sum_{i=1}^{L} b_i \cdot c_i(k) + n(k) \tag{6.48}$$

as the sum of the data-modulated signals of all users and AWGN from the channel. Assuming our proposed receiver structure, the signal for the wanted first user is given by

$$y(k) = r(k) - \sum_{i=2}^{L} \hat{b}_i \cdot c_i(k) = b_1 \cdot c_1(k) + \sum_{i=2}^{L} (b_i - \hat{b}_i) \cdot c_i(k) + n(k) \tag{6.49}$$

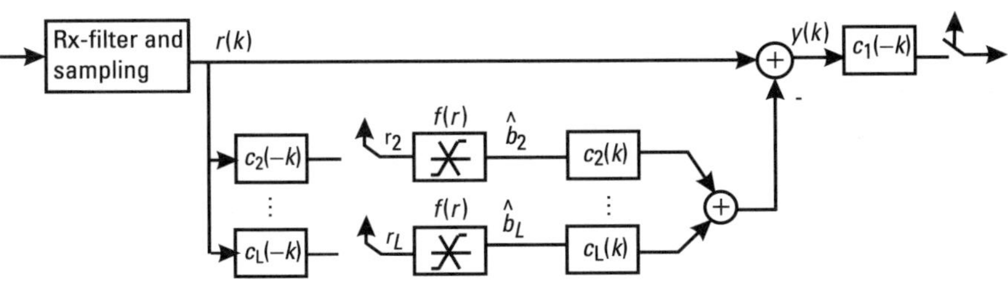

Figure 6.17 The receiver structure for the proposed interference cancellation scheme. (After: [27].)

Minimizing the contribution of every user $b_i - \hat{b}_i$, $i \neq 1$, which can be solved in an optimal manner using the optimum signal estimation, minimizes the interference term.

The signal r_j in front of the decision device after despreading and sampling for any user to be canceled is given by

$$r_j = b_j + \sum_{i \neq j} w_{ij} b_j + n \qquad j \neq 1 \tag{6.50}$$

where w_{ij} = cross correlation coefficient between the sequences $c_i(k)$ and $c_j(k)$ of user i and j, respectively.

Because of the randomness of the spreading sequences, the quite large spreading length, and the sufficient number of users, the interference in (6.50) can be approximated as zero mean Gaussian distributed noise by invoking the central limit theorem. The variance is given by the total power of interference divided by the spreading gain N.

Therefore, (6.50) is identical to (6.39), with the total noise consisting of both the interference and the thermal noise, thus having the variance

$$\sigma^2 = \sigma_1^2 + \sigma^2 = \frac{L-1}{N} + \frac{1}{E_b/N_0} \tag{6.51}$$

The second term is due to the thermal noise n having assumed two-sided spectral density N_o. Note that the signal energy per user is normalized to 1, whereas the total interference power equals $L - 1$.

6.2.4.2 Recursive Narrowband Interference Estimation Using Kalman Filtering

Another way to estimate interference is the well known Kalman filtering mechanism, which is a combination of HMMs and the recursive estimate-maximize (EM) method, forming a powerful estimation technique [45]. This algorithm cross couples two optimal filters—an HMM and a Kalman filter (HMM-KF) (see Appendix B) as shown in Figure 6.18. The HMM estimator (which is described in Appendix B) yields filtered state estimates $s_{k|k}$ of the spread-spectrum signal. Given the spread-spectrum signal $s_{k|k}$ and the associated error variance $p_{s_{k|k}}$ of $w_k \equiv s_k - s_{k|k}$, our objective now is to compute state and parameter estimates of the narrowband interference. The signal model is given by [45]:

$$y_k - s_{k|k} = i_k + w_k + n_k \tag{6.52}$$

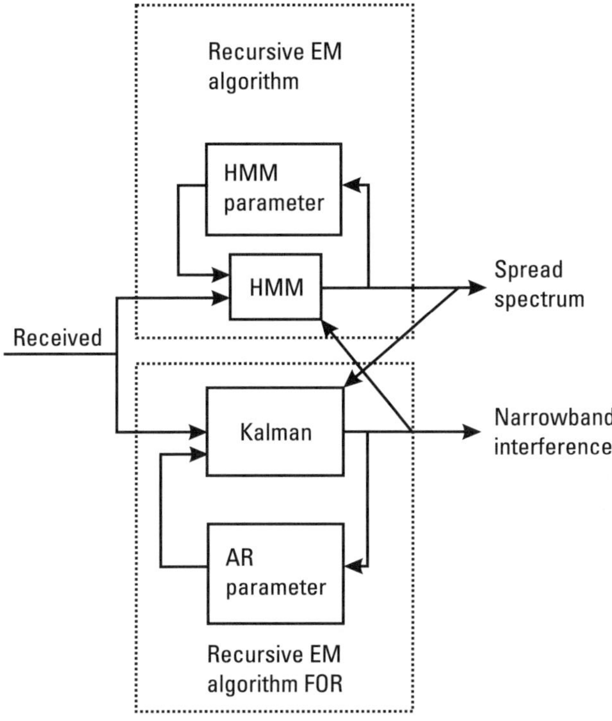

Figure 6.18 HMM-KF interference estimator. (After: [45].)

where y_k is the observation and i_k is the narrowband interference signal. $w_k \sim N(0, p_{s_{k|k}})$ is modeled as a zero mean white Gaussian process with variance $p_{s_{k|k}}$ and assumed independent of the observation noise $n_k \sim N(0, \sigma_n^2)$ and process noise $e_k \sim N(0, \sigma_e^2)$. Equation (6.52) can be represented as the following state space model.

State Space Model

$$x_k = Fx_{k-1} + Ge_k \tag{6.53}$$

$$y_k - s_{k|k} = Hx_k + w_k + n_k$$

where the state vector $x_k = (i_k, i_{k-1}, \ldots, i_{k-p})'$

$$F = \begin{pmatrix} -D' & 0 \\ I_{p \times p} & 0_{p \times 1} \end{pmatrix}, \quad D = (d_1, \ldots, d_p)' \tag{6.54}$$

$$G = (1 \quad 0_{1 \times p})', \quad H = (1 \quad 0_{1 \times p})$$

The recursive EM estimator recursively updates the narrowband interference autoregressive coefficients, the narrowband interference process noise, and observation noise. The recursive EM parameter estimate at k is denoted as

$$\phi_{KF}^{(k)} \equiv \left(D^{(k)},\ \sigma_e^{2(k)},\ \sigma_n^{2(k)}\right) \tag{6.55}$$

Remark: As shown in the Appendix B, the recursive HMM estimator also provides an update formula for the observation noise. Thus, we can average the two estimates or use either one of the two updates at each time instant.

Given the signal model of (6.53), the state and parameter estimation procedure for i_k is as follows [45]

1. *State estimation.* Conditional mean estimates of x_k can be obtained by a Kalman Filter [45] (see Appendix B): The estimate of the narrowband interference $i_{k|k}$ can be given by the first element of the vector $x_{k|k}$, while the error covariance $p_{i_{k|k}}$ is given by the element $(1\ 1)$ of $P_{k|k}$. The parameter estimation procedure given in the following subsection requires the evaluation of quantities such as

$$\overline{i_{k-m}\,i_{k-n}}^{(k-1)} \equiv E\left\{i_{k-m}\,i_{k-n}\,|\,Y_k,\ S_{k|k},\ \phi_{KF}^{(k-1)}\right\} \tag{6.56}$$

   which are computed from elements of $x_{k|k}$ and $P_{k|k}$ as follows

$$\overline{i_{k-m}\,i_{k-n}}^{(k-1)} = x_{k|k}[m]\,x_{k|k}[n] + P_{k|k}[m,\,n],$$

   where

$$m,\,n \in \{0,\,\ldots,\,p\} \tag{6.57}$$

   where $W[m,\,n]$ denotes the element $(m+1,\,n+1)$ of the matrix W and $w[m]$ denotes the $m+1$th element of vector w.

2. *Parameter estimation.* In the recursive EM algorithm, $Y_k = Z_{k,obs}$ and $I_k = Z_{k,mis}$ where $Z_{k,obs}$ is the observed data and $Z_{k,mis}$ is the missed data.

   Thus, given $S_{k|k}$, which are obtained from the recursive HMM (see Appendix B), we can calculate

$$L_k^{KF}(\phi) = E\left\{ \ln f\left(y_k, i_k \mid Y_{k-1}, I_{k-1}, \phi \mid Y_k, \phi_{KF}^{(k-1)}\right)\right\}$$

$$= -\frac{1}{2} \ln\left(2\pi\left(p_{s_{k|k}} + \sigma_n^2\right)\right) - \frac{1}{2\left(p_{s_{k|k}} + \sigma_n^2\right)} \overline{\left(y_k - s_{k|k} - i_k\right)^2}^{(k-1)}$$

$$- \frac{1}{2} \ln\left(2\pi\sigma_e^2\right)$$

$$- \frac{1}{2\sigma_e^2} \overline{\left(\sum_{m=0}^{p} d_m i_{k-m}\right)^2}^{(k-1)} \tag{6.58}$$

where $\overline{(\cdot)}^{(k-1)} \equiv E\left\{\cdot \mid Y_k, S_{k|k}, \phi_{KF}^{(k-1)}\right\}$ is the conditional expectation operator given the data Y_k, the spread-spectrum estimates $S_{k|k}$ and using the current model estimate $\phi_{KF}^{(k-1)}$.

Ignoring the terms $\partial^2 L_k^{KF}(\phi)/\partial\sigma_e^2 \partial d_m$ for all $m = 1, \ldots, p$, we have

$$I_{com}\left(\phi_{KF}^{(k-1)}\right) = \text{blockdiag}\left(I_{D^{(k-1)}}, I_{\sigma_e^{2(k-1)}}, I_{\sigma_n^{2(k-1)}}\right) \tag{6.59}$$

$$S\left(\phi_{KF}^{(k-1)}\right) = \left(S'_{D^{(k-1)}}, S'_{\sigma_e^{2(k-1)}}, S'_{\sigma_n^{2(k-1)}}\right) \tag{6.60}$$

Thus, $\phi_{KF}^{(k)}$ can be recursively updated as follows.

Autoregressive Coefficients

The update equation for the vector $D^{(k)}$ is given by [45]

$$D^{(k)} = D^{(k-1)} + I_{D^{(k-1)}}^{-1} S_{D^{(k-1)}} \tag{6.61}$$

where

$$S_{D^{(k-1)}} = -\frac{1}{\sigma_e^{2(k-1)}} \begin{pmatrix} \sum_{m=0}^{p} d_m^{(k-1)} \overline{i_{k-m} i_{k-1}}^{(k-1)} \\ \vdots \\ \sum_{m=0}^{p} d_m^{(k-1)} \overline{i_{k-m} i_{k-p}}^{(k-1)} \end{pmatrix} \tag{6.62}$$

and

$$I_D{}^{(k-1)} = \rho I_D{}^{(k-2)} + \frac{1}{\sigma_e^{2(k-1)}}$$

(6.63)

$$\times \begin{pmatrix} \overline{i_{k-1}\,i_{k-1}}^{(k-1)} & \overline{i_{k-1}\,i_{k-2}}^{(k-1)} & \cdots & \overline{i_{k-1}\,i_{k-p}}^{(k-1)} \\ & \overline{i_{k-2}\,i_{k-2}}^{(k-1)} & \cdots & \overline{i_{k-2}\,i_{k-p}}^{(k-1)} \\ & & & \vdots \\ \text{symetric} & & & \overline{i_{k-p}\,i_{k-p}}^{(k-1)} \end{pmatrix}$$

Process Noise

The update equation for $\sigma_e^{2(k)}$ is given by

$$\sigma_e^{2(k)} = \sigma_e^{2(k-1)} + I_{\sigma_e^{2(k-1)}}^{-1} S_{\sigma_e^{2(k-1)}}$$

(6.64)

where

$$S_{\sigma_e^{2(k-1)}} = \frac{\overline{\left(\sum_{m=0}^{p} d_m^{(k-1)}\, i_{k-m}\right)^2}^{(k-1)}}{2\left(\sigma_e^{2(k-1)}\right)^2} - \frac{1}{2\sigma_e^{2(k-1)}}$$

(6.65)

$$I_{\sigma_e^{2(k-1)}} = \rho I_{\sigma_e^{2(k-2)}} + \frac{\overline{\left(\sum_{m=0}^{p} d_m^{(k-1)}\, i_{k-m}\right)^2}^{(k-1)}}{\left(\sigma_n^{2(k-1)}\right)^3} - \frac{1}{2(\sigma_n^{2(k-1)})^2}$$

(6.66)

With no forgetting factor ($\rho = 1$), then the update equation for the process noise is given by

$$\sigma_e^{2(k)} = \sigma_e^{2(k-1)} + \frac{1}{k}\left(\overline{\left(\sum_{m=0}^{p} d_m^{(k-1)}\, i_{k-m}\right)^2}^{(k-1)} - \sigma_e^{2(k-1)}\right)$$

(6.67)

It is straightforward to show that [26]:

$$\overline{\left(\sum_{m=0}^{p} d_m^{(k-1)} i_{k-m}\right)^2}^{(k-1)} = 2 \sum_{m=0}^{p} \sum_{n=0}^{p} d_m^{(k-1)} d_n^{(k-1)} \overline{i_{k-m} i_{k-n}}^{(k-1)}$$

$$- \sum_{m=0}^{p} d_m^{2(k-1)} \overline{i_{k-m}^2}^{(k-1)} \tag{6.68}$$

Observation Noise

The update equation for $\sigma_n^{2(k)}$ is given by

$$\sigma_n^{2(k)} = \sigma_n^{2(k-1)} + I_{\sigma_n^{2(k-1)}}^{-1} S_{\sigma_n^{2(k-1)}} \tag{6.69}$$

where

$$S_{\sigma_n^{2(k-1)}} = \frac{\overline{\left(\sum_{m=0}^{p} d_m^{(k-1)} i_{k-m}\right)^2}^{(k-1)}}{2\left(\sigma_n^{2(k-1)} + p_{s_{k|k}}\right)^2} - \frac{1}{2(\sigma_n^{2(k-1)} + p_{s_{k|k}})} \tag{6.70}$$

$$I_{\sigma_n^{2(k-1)}} = \rho I_{\sigma_n^{2(k-2)}} + \frac{\overline{\left(\sum_{m=0}^{p} d_m^{(k-1)} i_{k-m}\right)^2}^{(k-1)}}{\left(\sigma_n^{2(k-1)} + p_{s_{k|k}}\right)^3} - \frac{1}{2(\sigma_n^{2(k-1)} + p_{s_{k|k}})^2} \tag{6.71}$$

With no forgetting factor ($\rho = 1$), and if we ignore the error in $s_{k|k}$, (i.e., $p_{s_{k|k}} = 0$ for all k), then the update equation for the observation noise is given by [26]:

$$\sigma_n^{2(k)} = \sigma_n^{2(k-1)} + \frac{1}{k}\left(\overline{\left(\sum_{m=0}^{p} d_m^{(k-1)} i_{k-m}\right)^2}^{(k-1)} - \sigma_n^{2(k-1)}\right) \tag{6.72}$$

The HMM-KF Algorithm

1. Initialization: At $k = 0$, initialize the HMM and the KF estimator as follows:
   - Step 1.1 Initialize KF parameter estimates: ϕ_{KF}^0 defined in (6.55).
   - Step 1.2 Initialize KF state estimate: $x_{0|0}$ and the associated error covariance $P_{0|0}$.
   - Step 1.3 Initialize HMM parameter estimates: ϕ_{HMM}^0, defined in Appendix B.
   - Step 1.4 Initialize HMM state estimate: $s_{0|0}^{MAP}$, $s_{0|0}^{CM}$ and the associated error variance $p_{s_{0|0}}$ defined in Appendix B, respectively.

2. State and parameter update. At each time instant $k = 1, \ldots ,$ update the state and parameter estimates for the HMM and the KF estimator as follows:
   - Step 2.1 Narrowband interference state prediction: Compute $x_{k|k-1}$ and $P_{k|k-1}$ using a Kalman filter.
   - Step 2.2 Spread-spectrum state update: Compute $s_{k|k}^{MAP}$, $s_{k|k}^{CM}$ and $p_{s_{k|k}}$ using expressions from Appendix B.
   - Step 2.3 Spread-spectrum parameter update: If A, q and σ_n^2 are unknown, then compute estimates $A^{(k)}$, $q^{(k)}$, and $\sigma_n^{2(k)}$, using Appendix B.
   - Step 2.4 Narrowband interference state update: Compute $x_{k|k}$ and $P_{k|k}$ by using the procedure used in step 2.1 (Kalman filter recursively), respectively.
   - Step 2.5 Narrowband interference parameter update: If D, σ_e^2, and σ_n^2 are unknown, then compute estimates $D^{(k)}$, $\sigma_e^{2(k)}$, and $\sigma_n^{2(k)}$, using (6.61), (6.64), and (6.69), respectively.
   - Step 2.6 Set $k \to k + 1$ goes back to step 2.1.

Remark: It is quite obvious that step 2.3 is ignored when the spread-spectrum parameters are known. The same applies for step 2.5. This algorithm is shown in Figure 6.19.

Computational Complexity

The main cost of the HMM-KF algorithm is based on computing the following variables:

- The HMM state filter recursion α_k in (B.13) in Appendix B. This requires $O(M^2)$ computations at each time instant.

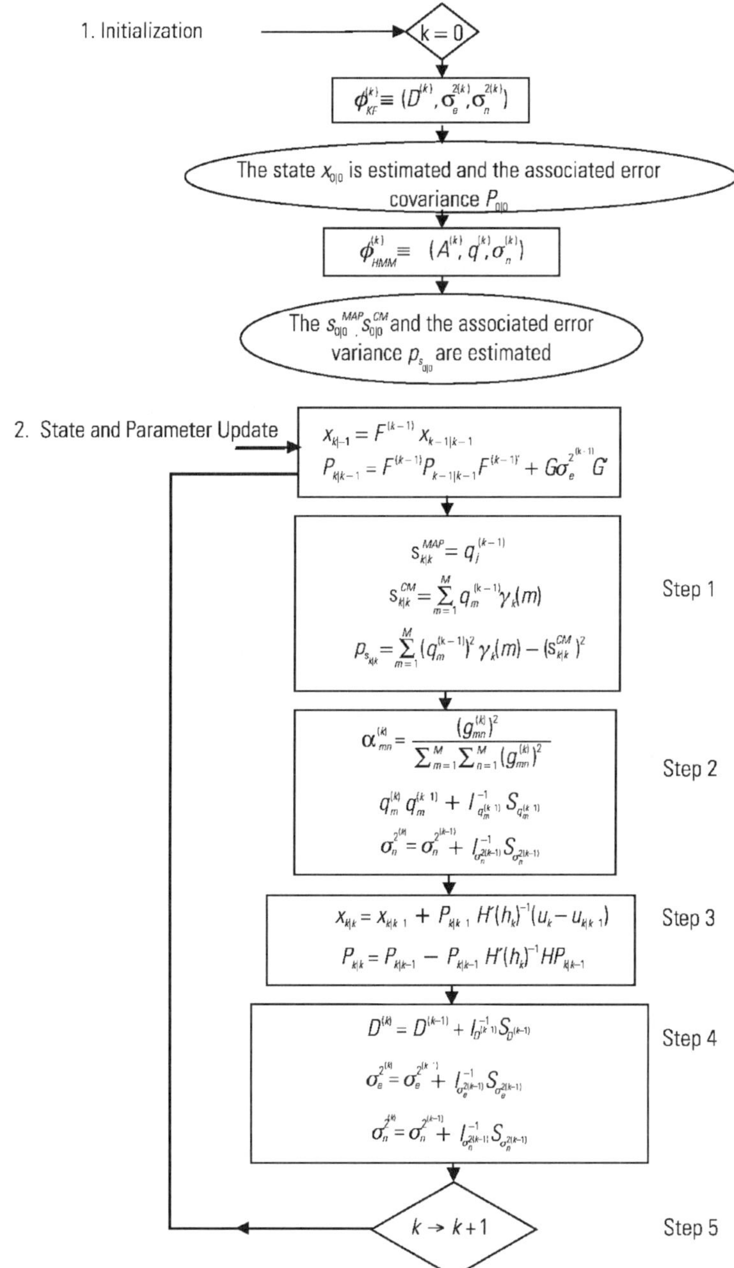

Figure 6.19 HMM-KF algorithm.

- The Kalman filter state and covariance recursions in the equations shown in steps 2.1 and 2.4 of the algorithm shown for the state and parameter updates calculations. This requires $O(p^2)$ computations at each time instant.

Simulation results have shown [45] that the HMM-KF algorithm presents a powerful tool for estimating interferences and outperforms the appropriate conditional mean filter for medium to high observation noise. It can be used effectively in the design of adaptive filters for narrowband suppression in spread spectrum CDMA systems.

6.3 Interference Avoidance

A class of receivers, which can adapt their modulation and demodulation in the presence of interference with the corresponding transmitter to achieve better performance on the basis of signal-to-interference ratio (SIR), has emerged the last few years [46–48]. This technique is called interference avoidance. Even though this procedure has been applied to system transmitters and receivers that are waveform agile, which is not the case of real life wireless systems—especially mobile, its conceptual simplicity will make it worthy of consideration in areas where it has not yet been tried. A brief conceptual description will be given next as to how avoidance improves SIR.

6.3.1 SIR Optimization Via Interference Avoidance

Consider the classical continuous-time digital communications model, in which during an interval $[0;\ T]$, a signal $b\sqrt{P}s(t)$ is transmitted where $b = \pm 1$ equiprobably, P is the received power, and $s(t)$ is the a signal waveform with some energy. A receiver recovery is given by

$$r(t) = b\sqrt{P}s(t) + z(t) \tag{6.73}$$

where $z(t)$ is an independent interference stochastic waveform that may be composed of both thermal noise and interfering signals of other transmitters.

For a single bit, the fundamental problem is to build a receiver, which guesses b with minimum probability of error. Alternatively, when b is one bit in a stream of coded bits, we would like to produce a soft estimate of b with high SIR. When $z(t)$ is composed of known waveforms in addition to independent Gaussian noise, that is,

$$z(t) = \sum_i b_i \sqrt{p_i} \, s_i(t) + N(t) \tag{6.74}$$

Multiuser receivers have been designed for a variety of objectives (e.g., minimum probability of error, maximum SIR, or zero interference from other users). These multiuser systems share the property that the receiver does as best it can given the set of transmitter signals $s_i(t)$.

We define the covariance of the noise process $z(t)$ as

$$R_Z(t, \tau) = E[z(t)z(\tau)] \tag{6.75}$$

and then seek a set of orthonormal functions $\Phi_i(t)$ for which $E[\langle \Phi_i(t), z(t) \rangle \langle \Phi_i(\tau), z(\tau) \rangle]$ yields uncorrelated projections. Precisely, we require

$$\int_0^T \Phi_j(t) \left(\int_0^T R_Z(t, \tau) \Phi_i(\tau) \, d\tau \right) dt = \lambda_j \delta_{ij} \tag{6.76}$$

The solution to this integral equation requires

$$\lambda_i \Phi_i(t) = \int_0^T R_Z(t, \tau) \Phi_i(\tau) \, d\tau \tag{6.77}$$

Because integral equations are in general difficult to solve, it is useful to derive an equivalent discrete representation of (6.77). This will allow us to use simple methods from linear algebra. So let us assume that $z(t)$, and therefore the function set $\{\Phi_i(t)\}$, can be well approximated by a finite set of orthonormal basis functions $\{\Phi_n(t)\}$ on the interval $[0, T]$. That is, we assume that the process $z(t)$ has no significant energy outside some finite signal space. As an example, a process "almost" limited to bandwidth $\pm W$ has a basis function set with about $2WT$ orthonormal functions. Likewise, for a synchronous CDMA system with N chips per bit, the appropriate orthonormal set consists of the N time-shifted chip pulses. One could also use a space-time orthogonalization for reception/transmission antenna diversity and/or a frequency-time orthogonalization for a frequency-hopped system.

Regardless of the specifics, once we assume a convenient finite basis function set, or signal space for the interval, the $\Phi_i(t)$ can then be represented by the finite sum

$$\Phi_i(t) = \sum_{n=1}^{N} \phi_{in} \Psi_n(t) \tag{6.78}$$

and likewise

$$i(t) = \sum_{n=1}^{N} i_n \Psi_n(t) \tag{6.79}$$

with $\phi_{in} = \langle \Phi_i(t), \Psi_n(t) \rangle$ and $i_n = \langle i(t), \Psi_n(t) \rangle$.

This allows the reduction of (6.77) to a standard matrix eigenvalue/eigenvector equation of the form

$$E\left[\mathbf{zz}^T \right] \phi_i = \mathbf{R}\phi_i = \lambda_i \phi_i \tag{6.80}$$

where $\underline{\phi}_i = [\phi_{i1} \ldots \phi_{iN}]^T$ and $\underline{z} = [z_1 \ldots z_N]^T$.

R is the matrix with elements

$$r_{kn} = \int_0^T \int_0^T R_i(t,\ \tau) \Psi_k(t) \Psi_n(\tau)\ dt d\tau.$$

Each eigenvector corresponds to an eigenfunction of (6.80), and it is easily verified that *each eigenvalue is the amount of interference signal energy carried by that eigenfunction*. It is also easy to verify that because $R_Z(t,\ \tau)$ is an autocorrelation function, $\mathbf{R}$ is symmetric and positive semidefinite. This implies that $\mathbf{R}$ has nonnegative eigenvalues and an associated full set of orthonormal eigenvectors which span $\Re^N$. The receiver observes the signal $r(t)$ as input on the interval $[0,\ T]$. Projecting the received signal onto the interference eigenfunctions $\Phi_1(t), \ldots, \Phi_n(t)$, we obtain the vector output

$$\mathbf{r} = b\underline{s} + \underline{z} \tag{6.81}$$

where $\underline{s}$ and $\underline{z}$ have n components, $s_n = \langle s(t), \Phi_n(t) \rangle$, $z_n = \langle z(t), \Phi_n(t) \rangle$, and the z_n are mutually uncorrelated.

At this point, it is instructive to consider the detection of b when $z(t)$ is a Gaussian interference process. Because we chose the interference eigenfunctions $\{\Phi_1(t), \ldots, \Phi_N(t)\}$ to yield uncorrelated (and therefore independent) interference components z_n, optimal decision rule becomes

$$\sum_{n=1}^{N} \frac{s_n r_n}{\lambda_n} = \sum_{n=1}^{N} \frac{r_n}{\sqrt{\lambda_n}} \frac{s_n}{\sqrt{\lambda_n}} \overset{\text{say }1}{\underset{\text{say }0}{\gtrless}} 0 \tag{6.82}$$

The implied receiver is called a whitening filter because it can be viewed as an initial rescaling of the input to make interference components ($\{z_n\}$), already uncorrelated, have equal energy $\left(\{z_n/\sqrt{\lambda_n}\}\right)$—just as would be the case for a white noise process without rescaling. A matched filter on the rescaled signal vector components $z_n/\sqrt{\lambda_n}$ is then performed to complete the detection process. We now note that in a CDMA system where $z(t)$ consists of known signature waveforms of the other users and AWGN, the vector $\underline{c}$ with components $c_n = s_n/\lambda_n$ is a scaled version of the well-known MMSE linear filter, and the decision rule (6.82) is the MMSE multiuser detector [46]. We see that the filter output (and decision statistic) is

$$X = \underline{c}^T \underline{r} = \sum_n c_n r_n = \left(\sum_{n=1}^{N} \frac{s_n^2}{\lambda_n}\right) b + \sum_{n=1}^{N} \frac{s_n z_n}{\lambda_n} \tag{6.83}$$

and that the output signal to interference ratio (SIR) is given by

$$SIR_x = \sum_{n=1}^{N} \frac{s_n^2}{\lambda_n} \tag{6.84}$$

The MMSE filter maximizes the output SIR over all linear filters [45]. Thus, the classic whitening approach and the newer MMSE approach lead to the same result. However, (6.84) also demonstrates that it is possible to obtain a higher output SIR by altering the components s_n of the desired signal $s(t)$. That is, when $s(t)$ is subject to an energy constraint $\sum_n s_n^2 = 1$, we can maximize SIR_X by choosing $s_n = 1$ for any $\lambda_n = \lambda^*$ $= \min_k \lambda_k$. In this case, we have

$$s(t) = \Phi_n(t) \tag{6.85}$$

Equivalently, we could distribute the signal energy in some arbitrary way over all such $\Phi_n(t)$. Regardless, this result has a simple intuitively pleasing physical interpretation: To obtain maximum SIR, place all the signal energy where there is least interference, or maximize the signal power where the interference is minimal.

We call the process of altering the signal waveform of a user interference avoidance, and for a single user with a given interference process, the method is straightforward. We now examine the implications of this simple Karhunen-Loeve–inspired rule for an ensemble of users.

6.3.2 Interference Avoidance for Multiple Users

We now consider a multiuser system in which the received signal $r(t)$ explicitly includes M users and white Gaussian noise. Given the existence of a finite set of N orthonormal basis functions $\Psi_i(t)$ for the signal space, we can express the received signal as the vector

$$\mathbf{r} = \sum_{i=1}^{M} \sqrt{P_i}\, b_i \mathbf{s}_i + \mathbf{n} \tag{6.86}$$

where n is the projection of the AWGN onto the basis.

The classic communications scenario presumes that each user signature $s_i(t)$ is fixed. Assuming software radio transceivers, we now allow the use of tailored signature waveforms $s_i(t)$. Without loss of generality, we assume each $s_i(t)$ has unit energy. It has been shown that for a set of users' rates $R_1, \ldots, R_M$ belonging to the information theoretic achievable rate region C, the *sum* capacity is given by the formula [47]:

$$C_s = \max_{(R_1,\ldots,R_M)\in C} \sum_{i=1}^{M} R_i = \frac{1}{2} \log\left[\det\left(I_n + \sigma^{-2} SPS^T\right)\right] \tag{6.87}$$

In (6.72), I_N is the $N \times N$ identity matrix, P is the diagonal matrix of users powers P_k. $\mathbf{S} = [\underline{s}_1, \ldots, \underline{s}_M]$ is the $N \times M$ matrix with columns $\underline{s}_i$, and σ^2 is the Gaussian white noise variance.

6.3.3 Capacity and Total Square Correlation

It is shown [47] that the sum capacity for equal received powers is maximized if the signature sequences are chosen such that if $M \le N$, that

$$\mathbf{S}^T\mathbf{S} = \mathbf{I}_M \tag{6.88}$$

and if $M \geq N$,

$$\mathbf{SS}^T = (M/N)\mathbf{I}_N \tag{6.89}$$

The *user* capacity of a CDMA system is defined in terms of the maximum number of admissible users. Given the signal space dimensionality N and a common SIR target β, M users are said to be admissible if there are positive powers p_i and signature sequences s_i such that each user has an SIR at least as large as β. The user capacity was found for two kinds of linear receiver structures [11, 47]: matched filters and MMSE filters. The user capacity with MMSE receivers is maximized if the signature sequence set is chosen to satisfy (6.88) or (6.89) if $M \geq N$ and that the MMSE filter is the matched filter in these cases. Thus, the user capacity of a system with matched filter receivers is the same as that using MMSE filters.

We define total square correlation (TSC) as

$$\text{TSC} = \text{Trace}\,[(\mathbf{SS}^T)^2] = \sum_{i=1}^{M} \sum_{j=1}^{M} \left(\mathbf{s}_i^T \mathbf{s}_j\right)^2 \geq \frac{M^2}{N} \tag{6.90}$$

and it is related to sum capacity [11, 47]. First we define the eigenvalues of $\sigma^2 \mathbf{I}_N + \mathbf{SS}^T$, as λ_i, $i = 1, \ldots, N$ and rewrite sum capacity as

$$C_s = -N \log \sigma + \frac{1}{2} \sum_{i=1}^{N} \log \lambda_i \tag{6.91}$$

Now note that if $\{\lambda_i\}$ are the eigenvalues of $\sigma^2 \mathbf{I}_N + \mathbf{SS}^T$ then

$$\text{Trace}\,[(\sigma^2 \mathbf{I}_N + \mathbf{SS}^T)^2] = \sum_{i=1}^{N} \lambda_i^2 \tag{6.92}$$

because the eigenvalues of $(\sigma^2 \mathbf{I}_N + \mathbf{SS}^T)^2$ are $\{\lambda_i^2\}$.

In summary, we note that the function described in (6.91) is Schur concave, while that of (6.92) is Schur convex. Because any constraints on the eigenvalues must be identical, and in fact form a convex set, we can conclude that any set $\{\lambda_i\}$, which maximizes (6.91), must also minimize (6.92), and vice versa. Therefore, minimization of TSC is completely equivalent to maximization of C_s assuming a convex constraint on the $\{\lambda_i\}$. For those unfamiliar with maximization and Schur convexity, an alternate development based on Lagrange methods is provided in [46].

6.3.4 Iterative Methods of TSC Reduction

A number of methods might be used to determine codeword sets, which minimize TSC. Here we explore simple iterative methods that can be applied by each transmitter/receiver pair asynchronously and independently. For a single user k, we observe that $\mathbf{SS}^T = \mathbf{R}_k + \mathbf{s}_k \mathbf{s}_k^T$ where $\mathbf{R}_k = \Sigma_{i \neq k} \mathbf{s}_i \mathbf{s}_i^T$, the correlation matrix of the interference faced by user k, is analogous to the matrix $\mathbf{R}$ introduced in Section 6.3.1. When user k replaces its signature vector $\mathbf{s}_k$ with a vector $\mathbf{x}$, the resulting difference in TSC is

$$\Delta = \mathrm{Trace}\left[\left(\mathbf{R}_K + \mathbf{s}_k \mathbf{s}_k^T\right)^2\right] - \mathrm{Trace}\left[\left(\mathbf{R}_k + \mathbf{x}\mathbf{x}^T\right)^2\right] \qquad (6.93)$$

After some linear algebraic manipulations, we find that $\Delta \geq 0$ if

$$2\mathbf{s}_k^T \mathbf{R}_k \mathbf{s}_k + |\mathbf{s}_k|^2 \geq 2\mathbf{x}^T \mathbf{R}_k \mathbf{x} + |\mathbf{x}|^2 \qquad (6.94)$$

which reduces to

$$\mathbf{s}_k^T \mathbf{R}_k \mathbf{s}_k \geq \mathbf{x}^T \mathbf{R}_k \mathbf{x} \qquad (6.95)$$

if $|\mathbf{x}| = |\mathbf{s}_k|$, as we will hereafter assume.

When the interference faced by user k includes AWGN with power spectral density σ^2, we may replace $\mathbf{R}_k$ by $\mathbf{Z}_k = \mathbf{R}_k + \sigma^2 \mathbf{I}$ if desired, as the terms depending on σ^2 cancel each other in (6.95) because $|\mathbf{x}| = |\mathbf{s}_k|$. In terms of TSC minimization, operations on $\mathbf{R}_k$ or $\mathbf{Z}_k$ are equivalent.

Note that (6.95) defines a class of replacement algorithms whereby a given user can reduce (or at least not increase) the total squared correlation, assuming other users' codewords remain fixed during the replacement. Each user may use such an algorithm sequentially until all users have updated their codewords. At that point the cycle may begin anew. Cycles (iterations) would then be repeated until there was no further change in the TSC by individual codeword updates. Consideration of this process raises at least two questions. First, what is an example of such an algorithm? Second, do such algorithms eventually minimize TSC?

In answer to the first question we present two algorithms. We call the first algorithm the MMSE algorithm because we replace s_k by the normalized MMSE receiver filter.

$$\mathbf{c}_k = \left(\mathbf{s}_k^T \mathbf{Z}_k^{-2} \mathbf{s}_k\right)^{-1/2} \mathbf{Z}_k^{-1} \mathbf{s}_k \qquad (6.96)$$

Details of the algorithm and its convergence properties, as well as the fact that the MMSE algorithm is an interference-avoidance algorithm, are given in [24]. We call the second algorithm, which we introduce here as the eigen-algorithm because we replace s_k with $x = \phi_k^*$ where ϕ_k^* is a minimum eigenvalue eigenvector of $\mathbf{R}_k$. Using (6.95), we see that both algorithms guarantee $\Delta \geq 0$: For the MMSE algorithm for a proof that $\Delta \geq 0$, see [24], and for the eigen-algorithm, $\Delta \geq 0$ follows from the Rayleigh quotient because both the right- and left-hand sides of the condition are underbounded by $(\phi_k^*)^T \mathbf{R}_k \phi_k^*$. Note also from (6.84) that one step of the eigen-algorithm maximizes the SIR of user k by allowing nonzero signal energy only along those basis functions with absolute minimum λ_n.

Because both algorithms decrease the TSC monotonically, and because TSC is bound below by the Welch bound, both must converge. At fixed points of both algorithms, each s_k is an eigenvector of Z_k. The resulting codeword set is unique for neither algorithm. For example, any rotation of the codeword set will have the same cross-correlation properties. When $M \leq N$, the signatures converge to an orthonormal set. When $M \geq N$, the algorithms may converge to a Welch bound equality (WBE) signature set S satisfying (6.89).

Alternatively, the algorithms may converge to a local minimum for TSC. In [44, 46–48], mild conditions are derived under which the MMSE algorithm converges. For the eigen-algorithm, a modification of the procedure guarantees convergence to a global optimum. In numerical experiments, both algorithms have always converged to the optimal signature set when starting from randomly chosen initial waveforms.

The intuition behind all interference-avoidance algorithms that obey (6.95) is embodied by the simple requirement $s_k^T \mathbf{R}_k s_k \geq \mathbf{x}^T \mathbf{R}_k \mathbf{x}$, (i.e., the replacement vector x attempts to reduce the interference from the ensemble of other user vectors and noise). From the standpoint of implementation, in the MMSE algorithm, user k must identify $\mathbf{Z}_k^{-1} s_k$. In the eigen-algorithm, user k seeks a minimum eigenvalue eigenvector ϕ_k^* of $\mathbf{R}_k$. These points, taken together, suggest that the class of algorithms governed by (6.95) could be implemented by blind techniques at the receiver, along with a feedback channel to the transmitter. Specifically, in the MMSE algorithm, the receiver for user k could be a blind adaptive MMSE filter, based on the observable $\mathbf{Z}_k$. Likewise, for the eigen-algorithm, $\mathbf{x}$ can be found by minimizing $\mathbf{x}^T \mathbf{Z}_k \mathbf{x}$, which can also be implemented using blind techniques. Thus, interference-avoidance algorithms are based on a measurable quantity—the interference/noise signal correlation $\mathbf{Z}_k$.

In the MMSE algorithm, a codeword replacement by user k requires first that the receiver filter for user k converge. Further, the MMSE filter coefficients c_k must be communicated to the transmitter via a feedback channel. Consequently, at each iterative step, the speed of the algorithm is limited because the convergence to the MMSE filter may require several hundred bits and several hundred bits may be needed for the feedback transmission of the new signature. These same conclusions will also hold for the eigen-algorithm. Therefore, these signature adaptation algorithms operate on a slower time scale than the algorithms for multiuser interference suppression. Thus, if the channel is not stable for a sufficient number of bit intervals, it is not clear whether interference avoidance will offer an advantage. However, for channels that are stable over a sufficient number of bit intervals, signature adaptation may offer potentially large capacity increases. This analysis shows interference avoidance is a study tool to stay and will be used more and more as the radio hardware sophistication supporting advanced signal processing increases [46–48].

References

[1] Stavroulakis, P., *Interference Analysis of Communication Systems,* New York: IEEE Press, 1980.

[2] Vaseghi, S. V., *Advanced Digital Signal Processing and Noise Reduction,* New York: John Wiley, 2000.

[3] Lau, H. K., and S. N. Cheung, "Performance of a Pilot Symbol-Aided Technique in Frequency-Selective Rayleigh Fading Channels Corrupted by Cochannel Interference and Gaussian Noise," *IEEE VTC,* Atlanta, GA, 1996.

[4] Sasaki, M., et al., "Cochannel Interference Reduction Techniques for Land Mobile Communications," *IEICE RCS 90-91,* Japan, 1994, pp. 41–48.

[5] Tutschku, K., "Interference Minimization Using Automatic Design of Cellular Communication Networks," *IEEE, VTC,* Ottawa, Canada, 1998.

[6] Tsoulos, G. V., M. A. Beach, and Simon C. Swales, "Performance Enhancement of DS-CDMA Microcellular Networks with Adaptive Antennas," *IEEE, VTC,* Atlanta, GA, 1996.

[7] Fuhl, J., A. Kuchar, and E. Bonek, "Capacity Increase in Cellular PCS by Smart Antennas," *IEEE, VTC,* Phoenix, AZ, 1997.

[8] Rappaport, T. S., *Wireless Communication, Principles and Practice,* Upper Saddle River, NJ: Prentice-Hall, 1996.

[9] Prasad, R., *Universal Wireless Personal Communications,* Norwood, MA: Artech House, 1998.

[10] Hamred, K., and G. Labedz, "AMPS All Transmitter Interference to CDMA Mobile Receiver," *IEEE, VTC,* Atlanta, GA, 1996.

[11] Viswanath, P., V. Anantharam, and D. Tse, "Optimal Sequences, Power Control and Capacity of Spread Spectrum Systems with Multiuser Receivers," *IEEE Trans. on Information Theory,* 1998.

[12] Grant, S. J., and J. K. Cavers, "Performance Enhancement Through Joint Detection of Cochannel Signals Using Diversity Arrays," *IEEE, VTC,* Atlanta, GA, 1996.

[13] Wong, P. Bill, and D. C. Cox, "Low Complexity Cochannel Interference Cancellation and Macroscopic Diversity for High Capacity Personal Communication Systems," *IEEE Trans. On Vehicular Technology,* Vol. 47, No.1, February 1998.

[14] Sheen, W., and T. Chien-Hsiang, "A Non-Coherent Tracking Loop with Diversity and Multipath Interference Cancellations for Direct-Sequence Spread-Spectrum Systems," *IEEE, VTC,* Atlanta, GA, 1996.

[15] Yoon, Y. C., R. Kohno, and H. Imai, "Cascaded Cochannel Interference Canceling and Diversity Combining for Spread Spectrum Multi-Access Over Multipath Fading Channels," *Symp Inf. Theory and its Applications,* September 1992.

[16] Sato, T., et al., "Sequential Interference Cancellation Systems Applying to Wideband CDMA Systems," *IEEE, VTC,* Atlanta, GA, 1996.

[17] Johansson, A. L., and A. Svensson, "Multistage Interference Cancellation in Multirate DS/CDMA on a Mobile Radio Channel," *IEEE, VTC,* Atlanta, GA, 1996.

[18] Berangi, R., P. Leung, and M. Faulkner, "Cochannel Interference Canceling of Constant Envelope Modulation Schemes in Cellular Radio Systems," *IEEE VTC,* Atlanta, GA, 1996.

[19] Santucci, F., and M. Pratesi, "Outage Analysis in Slow Frequency-Hopping Mobile Radio Networks," *IEEE 49th VTC,* Vol. 2, 1999, pp. 909–913.

[20] Bravo, A. M., "Limited Linear Cancellation of Multiuser Interference on DS/CDMA Asynchronous Systems," *IEEE Trans. on Comm.,* Vol. 45, No. 11, November 1997.

[21] Schramm, P., and R. R. Muller, "Spectral Efficiency of CDMA Systems with Linear MMSE Interference Suppression," *IEEE Transactions on Communications,* Vol. 47, No. 5, May 1999.

[22] Patel, P., and J. Holtzman, "Analysis of Simple Successive Interference Cancellations Scheme in a DS/CDMA Systems," *IEEE Trans. on Selected Areas in Comm.,* Vol. 12, No. 10, October 1994.

[23] Honig, M., and V. Veerkachen, "Performance Variability of Linear Multiuse Detection for DS-CDMA," *Proc. of the IEEE, VTC,* Atlanta, GA, 1996.

[24] Rapajic, P. B., and B. S. Vucetic, "Linear Adaptive Transmitter–Receiver Structures for Asynchronous CD-MA," *European Trans. on Telecomm.* Vol. 6, No. 1, January 1995, pp. 21–28.

[25] Ariyavisitakul, S., J. H. Winters, and N. R. Sollenberger, "Joint Equalization and Interference Suppression for High Data Rate Wireless Systems," *IEEE VTC,* Houston TX, 1999.

[26] Krishnamurthy, V., and J. B. Moore, "On Line Estimation of Hidden Markov Parameters Based on the Kullback-Leibler Information Measure," *IEEE Trans. Signal Processing,* Vol. 41, August 1993.

[27] Frey, T., and M. Reinhardt, "Signal Estimation for Interference Cancellation and Decision Feedback Equalization," *IEEE VTC,* Phoenix, AZ, May 4–7, 1997.

[28] Buehrer, M. R., and B. D. Woerner,, "Analysis of Adaptive Multistage Interference Cancellation for CDMA Using an Improved Gaussian Approximation," *IEEE Trans. on Comm.,* Vol. 44, No. 10, October 1996.

[29] Jukka, R., "An Equalization Method Using Preliminary Decision for Orthogonal Frequency Division Multiplexing Systems in Channels with Frequency Selective Fading," *IEEE, VTC,* Atlanta, GA, 1996.

[30] Uesugi, M., S. Futagi, and K. Homma, "Interference Cancellation Method Using Decision Feedback Equalizer," *IEEE VTC,* Atlanta, GA, 1996.

[31] Kim, S. R., et al., "Incorporation of Adaptive Interference Cancellation into Parallel Interference Cancellation," *IEEE VTC,* Houston TX, 1999.

[32] Hui, A. L. C., and K. B. Letaief, "Successive Interference Cancellation for Multiuser Asynchronous DS/CDMA Detectors in Multipath Fading Links," *IEEE Transactions on Communications,* Vol. 46, No. 3, March 1998.

[33] Poor, V. H., and Xiaodong Wang, "Code-Aided Interference Suppression for DS/CDMA Communications–Part II, Parallel Blind Adaptive Implementations," *IEEE Transactions on Comm.,* Vol. 45, No. 9, September 1997.

[34] Cruickshank, D. G. M., "Suppression of Multiple Access Interference in a DS-CDMA System Using Wiener Filtering and Parallel Cancellations," *IEEE Proceedings on Comm.,* Vol. 143, No. 4, August 1996.

[35] Yukitoshi, Sanada, and Wang Qiang, "A Cochannel Interference Cancellation Technique Using Orthogonal Convolutional Codes," *IEEE Trans. On Comm.,* Vol. 44, No. 5, May 1996.

[36] Cameron, R., and B. Woerner, "Synchronization of CDMA Systems Employing Interference Cancellation," *IEEE, VTC,* Atlanta, GA, 1996.

[37] Stranch, P., and B. Mulgrew, "Nonlinear Interference Cancellation Using a Radial Basis Function Network," *Globecom,* '98, Sydney, Australia, November 8–12, 1998.

[38] Glisic, S. G., et al., "Multilayer LMS Interference Suppression Algorithms for CDMA Wireless Networks," *IEEE Transactions on Communications,* Vol. 48, No. 8, Aug. 2000.

[39] Jamal, K., and E. Dahlman, "Multi-Stage Serial Interference Cancellation for DS-CDMA, *IEEE VTC,* Atlanta, GA, 1996.

[40] Latva-aho, M., and J. Lilleberg, "Parallel Interference Cancellations in Multiuser CDMA Channel Estimation," *Wireless Personal Communications,* Kluwer Academic Publishers, Vol. 7, No. 213, August 1998.

[41] Madhow, U., and M. L. Honig, "MMSE Interference Suppression for Direct Sequence Spread Spectrum CDMA," *IEEE Trans. On Comm.,* 42 (12) December 1994.

[42] Muller, R. R., and J. B. Huber, "Capacity of Cellular CDMA System: Applying Interference Cancellation and Channel Coding," *IEEE, VTC,* Phoenix, AZ, 1997.

[43] Honig, M., U. Madhow, and S. Verdu, "Blind Adaptive Multiuse Detection for DS-CDMA," *Proc. of the IEEE, VTC,* Atlanta, GA, 1996.

[44] Madhow, V., and M. L. Honig, "MMSE Interference Suppression for Direct Sequence Spread Spectrum CDMA," *IEEE Trans. On Comm.*, December 1994.

[45] Krishnamurthy, V., and A. Logothetis, "Adaptive Nonlinear Fillers for Narrowband Interference Suppression in Spread Spectrum CDMA Systems," *IEEE Trans. On Comm.*, Vol. 47, No. 5, May 1999.

[46] Rose, C., "Sum Capacity and Interference Avoidance: Convergence via Class Warfare" in *CISS 2000,* Princeton, NJ, March 2000.

[47] Rupf, M., and J. L. Massey, "Optimum Sequence Multisets for Synchronous Code-Division Multiple Access Channels," *IEEE Trans. on Info. Theory IT,* Vol. 40, No. 4, July 1994.

[48] Rose, C., Sennur Ulukus, and Roy Yates, "Interference Avoidance in Wireless Systems," Technical Report, Winlab, Rutgers University, August 11, 1999.

7

Applications

7.1 Introduction

This section will deal with applications that have been chosen to present typical cancelers, which employ some of the suppression techniques presented in Chapters 4, 5, and 6, in one way or another. The main objectives were the following:

1. Find cancelers that can be implemented in practical wireless systems and can use the analytical results developed in Chapter 6.

2. Present all of the analytical results presented so far to applied practical wireless systems.

3. Provide the means of convincing the reader of the importance of practical and universally available suppression techniques to wireless systems designers.

Emphasis is given to multiuser systems without excluding the possibility of applying these techniques to all types of wireless systems, as we pointed out in the previous chapters.

Generally speaking, the spread-spectrum communication systems have been studied to realize high-capacity mobile communication systems. These systems allow several users to simultaneously use the same frequency band, on one hand, but imperfect orthogonality between the spreading codes results in transmission performance degradation due to interference, on the other hand. It is well known that the predetection combining techniques, such as maximal radio combining, are effective for improving transmission performance under the nonfrequency-selective fading condition. Equalization with combining diversity is a promising approach if the impairment due to ISI is included. There are two kinds of joint processing equalization and diversity. One is the decision feedback equalizer with combining diversity. Transversal filters in the diversity branches play the part of feedforward filters in decision feedback equalizer. The transversal filters optimally combine the received signals, and this diversity is referred to as transversal combining. The other kind is the maximum likelihood sequence estimation with diversity combining. Maximum likelihood sequence estimation combines the absolute squared a priori estimation errors derived in the respective diversity branches as the metric, and this diversity is referred to as metric combining. The metric combining scheme requires much higher complexity than the transversal combining scheme because the complexity of the Viterbi algorithm in the metric combining scheme grows exponentially with channel memory length.

Moreover, to satisfy the high capacity the third generation mobile communication systems should provide, it is necessary to adapt very efficient frequency-reuse techniques. The reuse efficiency is limited according to the amount of cochannel interference. It was shown in Chapter 5 that transversal combining cancels not only ISI but also cochannel interference through optimal combining control with the MMSE criterion. It can be considered that decision feedback equalizer with transversal combining diversity is an extended equalizer that cancels cochannel interference. On the contrary, conventional metric combining schemes are unable to cancel cochannel interference. This is because a priori estimation error of each diversity branch containing cochannel interference is squared. The combined errors (i.e., metrics of the branches) are only the results of accumulation, and the cochannel interference components cannot be removed. Therefore, transversal combining is superior to metric combining under cochannel conditions [1]. In the case where k cochannel signals exist, however, transversal combining requires $(k + 1)$ diversity branches to cancel all cochannel signals. The same argument is theoretically valid for adaptive array antenna techniques [2, 3].

The maximum likelihood sequence estimation type canceler was principally developed as a multiuser detector for parameter invariant transmission [4]. In the literature, proposed novel adaptive interference canceling equalizers can be found that utilize recursive least-squares maximum-likelihood sequence estimation (RLS-MLSE) schemes suitable for mobile communications. Mobile radio channels can be characterized by fast frequency selective fading with white Gaussian noise. The RLS-MLSE scheme, which has been proven to be effective for adaptive equalization, was derived from the maximum likelihood estimation theory [5]. The theoretical analysis results in a combination of the Kalman filter for the parameter estimation and MLSE for the symbol sequence estimation. The recursive least square algorithm is derived by approximating the unknown process noise covariance by a Kalman filter.

A distinctive advantage of the MLSE-type canceler over the transversal combining-type canceler is that the former cancels the cochannel signal even when the desired and cochannel signals come from the same direction, while the latter does not in the same degree because of its linear combining property.

Several variations of the MLSE-type canceler have been reported, as we shall subsequently see. These are an extension utilizing a blind algorithm, a combination of trellis coded modulation, and simplifications of the MLSE and adaptive algorithms.

7.2 Interference-Canceling Equalizer for Mobile Radio Communication

In order to verify the basic characteristics of cancelers, the following two conditions are important: the total user number is less than four, and the delay component number is the same for every user and less than two. The first condition is referred as the one-path model or the Rayleigh fading channel and the second is referred to as the two-path model or the frequency selective fading channel, as we saw in previous chapters.

7.2.1 Configuration of Interference-Canceling Equalizer

The configurations of an MLSE equalizer and a MLSE interference-canceling equalizer (ICE) are shown in Figure 7.1. Both are based on the RLS-MLSE scheme [6].

In the RLS-MLSE equalizer, the MLSE part outputs candidates of the desired signal code sequence. A candidate is denoted by m. The candidate code sequence is transformed to a replica via a modulated signal. The a priori estimation error $\alpha_{1m}(i)$ between received signal $y_s(i)$ sampled at $t = iT$, and the replica given by the output of the transversal filter is

$$\alpha_{1m}(i) = y_s(i) - C_{1m}^{H}(i)X_{1m}(i-1) \tag{7.1}$$

where

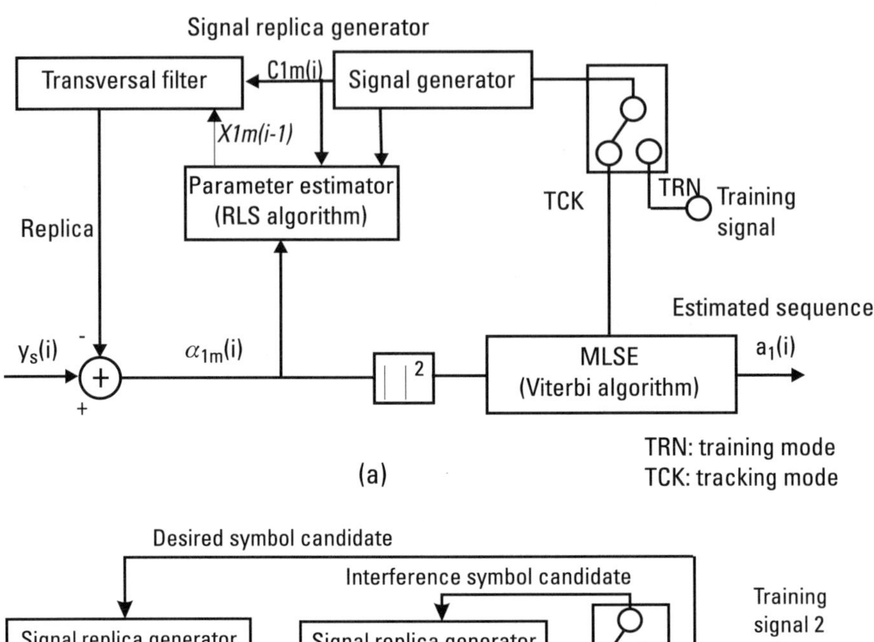

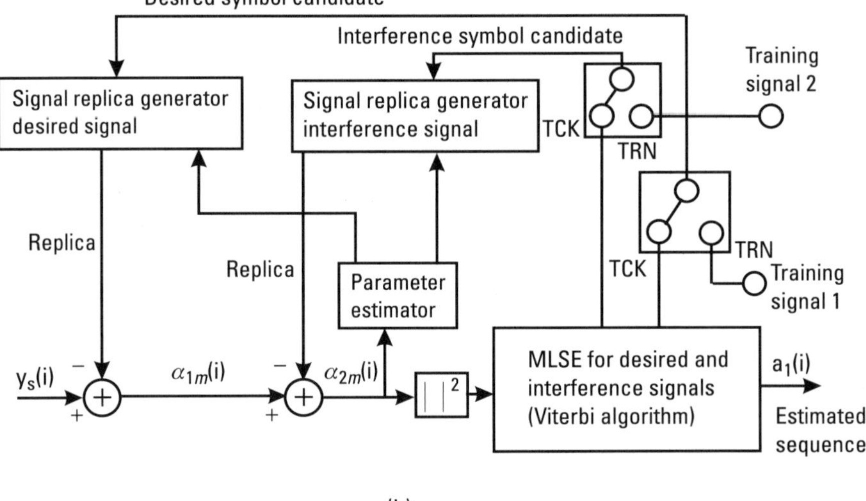

Figure 7.1 Adaptive equalizer and adaptive ICE using RLS-MLSE: (a) equalizer of RLS-MLSE, and (b) ICE of RLS-MLSE. (After: [6].)

$$H = \text{complex conjugation and transposition;}$$

$$X_{1m}(i-1) = \text{impulse response vector of the channel estimated at } i-1 \text{ for the } m\text{th candidate;}$$

$$C_{1m}(i) = m\text{th candidate of the modulation vector of the received symbol sequence.}$$

The squared value $|\alpha_{1m}(i)|^2$ is used for branch metric computation in MLSE. The Viterbi algorithm is employed for MLSE so the mth Viterbi algorithm state accompanies a corresponding estimate $X_{1m}(i-1)$, which is updated by the RLS algorithm using $C_{1m}(i)$. In the training mode, the impulse response vector $X_{1m}(i-1)$ converges through the use of the known modulation vector $C_{1m}(i) = C_1(i)$, where $C_1(i)$ is the training signal. In the tracking mode, $X_{1m}(i-1)$ is updated by a conventional adaptive algorithm with the candidate of modulation vector $C_{1m}(i)$ provided from MLSE.

$X_{1m}(i-1)$ is realized with a symbol-spaced tap transversal filter. It has been reported that there is practical difficulty with the symbol-spaced tap transversal filter (i.e., the timing phase jitter problem). Employing a fractionally spaced tap transversal filter against this problem is effective, but more taps are needed in the transversal filter and this degrades BER performance.

The equalizer generates the replica of the received signal, which contains the desired signal and its ISI components. The error is computed by subtracting this replica from the actual received signal. The resultant metric of the correct candidate includes no ISI. Hence, RLS-MLSE shows excellent performance in ISI-dominant environments. In environments dominated by cochannel interference, however, the cochannel components are not removed from a priori error $\alpha_{1m}(i)$. These cochannel interference components are equivalently treated as noise in MLSE, so performance is severely degraded [1].

When a single cochannel interference signal exists, the interference canceling equalizer shown in Figure 7.1(b) additionally generates a replica of cochannel components. This replica is subtracted from a priori error $\alpha_{1m}(i)$, which still contains the cochannel interference component. The resultant a priori error $\alpha_{2m}(i)$ is

$$\alpha_{2m}(i) = \alpha_{1m}(i) - C_{2m}^H(i)X_{2m}(i-1) \tag{7.2}$$

where

$$C_{2m}^H(i)X_{2m}(i-1) = \text{is the replica of the cochannel signal;}$$

$$X_{2m}(i-1) = \text{the impulse response vector of the cochannel interference signal;}$$

$$C_{2m}^H(i) = \text{a candidate of the modulation vector of the cochannel interference signal.}$$

The index m is uniquely assigned to the mth candidate pair, the combination of $C_{1m}(i)$ and $C_{2m}(i)$. Interference-canceler equalizer uses the value $|\alpha_{2m}(i)|^2$ as the branch metric of Viterbi algorithm.

The adaptive algorithm updates the impulse response vector pair, which is the set of $X_{1m}(i-1)$ for the desired signal and $X_{2m}(i-1)$ for the cochannel interference signal. As the adaptive algorithm, either RLS, a simplified version of RLS such as ensemble-averaged inverse-matrix least squares (EILS), or LMS, is applicable [6]. In order to reduce the complexity of the RLS algorithm, the EILS algorithm is employed here. The conventional RLS algorithm requires a large amount of processing to update the Kalman gain vector and the inverse of the covariance matrix of the modulation vectors $C_{1m}(i)$ and $C_{2m}(i)$. The EILS algorithm significantly reduces the total amount of processing required for these operations by replacing the covariance matrix with a fixed matrix, which can be theoretically calculated beforehand. In the training mode, $X_{1m}(i-1)$ and $X_{2m}(i-1)$ are converged by using the known training signals $C_{1m}(i)$ and $C_{2m}(i)$. In the tracking mode, $C_{1m}(i)$ and $C_{2m}(i)$ are generated by MLSE as the candidates of the transmitted symbol sequences corresponding to Viterbi algorithm states and transitions. The adaptive algorithm uses $C_{1m}(i)$ and $C_{2m}(i)$ to update $X_{1m}(i-1)$ and $X_{2m}(i-1)$ simultaneously. The combined impulse response vector for both the desired and the cochannel interference signal is given by

$$X_m^H(i-1) = \left[X_{1m}^H(i-1)\, X_{2m}^H(i-1) \right] \tag{7.3}$$

The modulation vector candidate for both the desired and the cochannel interference signal is given by

$$C_m^H(i) = \left[C_{1m}^H(i)\, C_{2m}^H(i) \right] \tag{7.4}$$

By using those variables and a priori error $\alpha_{2m}(i)$ in (7.2), the EILS algorithm updates the impulse response vector $X_m(i-1)$ as follows [5]:

$$X_m^H(i) = X_m^H(i-1) + P_0 C_m^H(i)\, \alpha_{2m}(i) \tag{7.5}$$

where P_0 is the inverse of the covariance matrix, which can be calculated beforehand because the modulation vector $C_m(i)$ is not affected by fluctuations due to fading and noise. Thus, P_0 is given by

$$P_0 = \lim_{k \to \infty} \left\{ \sum_{i=1}^{k} \lambda^{k-1} C_m(i) C_m^H(i) \right\}^{-1} \tag{7.6}$$

where λ is the forgetting factor.

P_0 is a diagonal matrix if no coding scheme is applied to the transmitted data sequence.

For the correct candidate m in ICE, the cochannel interference component is removed from a priori error $\alpha_{1m}(i)$. Thus, the major component of a priori error $\alpha_{2m}(i)$ is just the noise and channel estimation error. As a result, interference canceler/equalizer shows excellent interference-suppression performance. Under multiple cochannel interference conditions, ICE has to generate an additional replica for each multiple cochannel interference, but the generalization of (7.3).

7.3 A Linear Interference Canceler with a Blind Algorithm for CDMA Systems

Orthogonalizing matched filter (OMF) [7] is one of the effective linear interference cancelers. OMF employs the constrained minimum mean square (CMMS) criteria for controlling the combining coefficients. It operates as the decorrelator or the orthogonality filter, under the interference-dominant condition and as the matched filter in the noise-dominant condition. CMMS minimizes the average OMF output power under a constraint.

The conventional OMF adaptive algorithm is blind if the received signal has a delay profile in which discrete and respective components are spaced by integer multiples of the chip duration. This condition is, however, very tight for practical applications, especially in mobile communication applications where fractional spacing between respective components always occurs.

7.3.1 Configuration and Operation of a Linear Interference Canceler

The configuration of a receiver structure utilizing OMF is shown in Figure 7.2 [7].

They operate in a DS-CDMA system whose conditions are: (i) the process gain is L, (ii) the root full-raised-cosine baseband pulse is applied to the spreading and receiver filters, and (iii) there are K users. The normalized spreading code pulse for the kth user is denoted by $C_k(t)$.

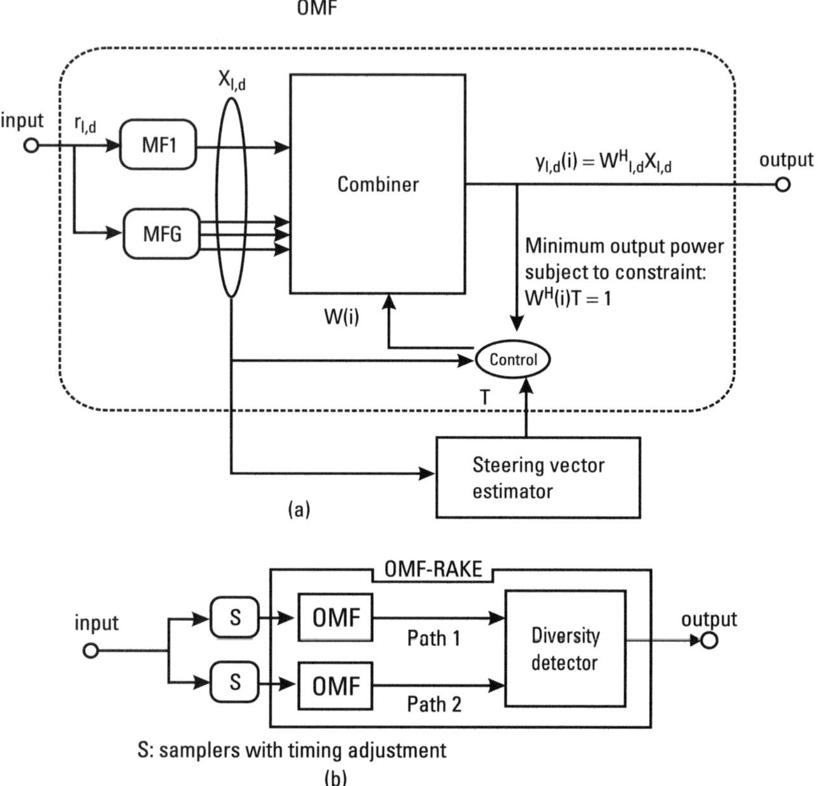

Figure 7.2 Receiver structure utilizing OMF: (a) basic configuration of OMF, and (b) RAKE receiver configuration (OMF-RAKE). (After: [7].)

In order to clearly show the relationship between the operation and the multipath model, we consider a simplified two-path propagation model that consists of a direct component. The delay time of each user is unique and denoted by $\tau_d = dT_c$ where the delay parameter d varies from 0 to 1, and T_c is the chip duration. For the kth user, the direct and the delay components are proportional to transmission coefficients $h_{k,0}$ and $h_{k,d}$, respectively.

In Figure 7.2(a), the received signal $r(t)$ is a combination of the signal from the K users. Double sampling per chip (i.e., $2L$ samples per symbol duration) is necessary for fractional matched filter (MF) into which the asynchronous timing received signals are fed. There are $2L$ despreading circuits whose despreading codes $\tilde{C}_1(t) = C_1(t)$. Code $\tilde{C}_1(t)$ is assigned to MF, and $\tilde{C}_l$, $2 \leq l \leq 2L$ for MFG in Figure 7.2. This orthogonalization generates basic vectors for a $2L$-dimensional code vector space.

The output of the lth despreading circuit $r_l(t)$ consists of the kth user, dth delay signal components denoted by $h_{l,k,d} = \rho_{l,k} h_{k,d}$, where $\rho_{l,k,d} = \langle \bar{C}_l(t), C_k(\tau_d) \rangle$ is the correlation between the lth despreading code and the spreading code of the kth user, dth delay component. The combiner with CMMS criteria combines the filter outputs and the combined signal is the OMF output.

The OMF-RAKE receiver shown in Figure 7.2(b) consists of two OMFs, one for the direct component extraction and the other for the one-chip delay component extraction. An optimal combiner combines the two OMF outputs. Timing of the despread codes for the second OMF is delayed by T_c.

7.3.1.1 Vector Representation

Using $2L$ dimensional despread received signal vector $\mathbf{X}(i)$ and noise vector $\mathbf{n}(i)$

$$\mathbf{X}^H(i) = \left[r_1^*(iT), r_2^*(iT), \ldots r_{2L}^*(iT) \right] \tag{7.7}$$

$$\mathbf{n}^H(i) = \left[n_1^*(iT), n_2^*(iT), \ldots n_{2L}^*(iT) \right] \tag{7.8}$$

and $K \times K$ impulse response matrix $\mathbf{H}(i)$ and K dimensional transmitting symbol vector $\mathbf{S}(i)$

$$[\mathbf{H}(i)]_{l,m} = \begin{cases} h_{m,0}(iT) & l = 2m - 1 \\ h_{m,1}(iT) & l = 2m \\ 0 & \text{elsewhere} \end{cases} \tag{7.9}$$

$$\mathbf{S}^H(i) = \left[s_1^*(iT), s_2^*(iT), \ldots s_K^*(iT) \right] \tag{7.10}$$

$\mathbf{X}(i)$ is given by

$$\mathbf{X}(i) = \begin{bmatrix} \rho_{1,1,0} & \rho_{1,1,1} & \cdots & \rho_{1,K,0} & \rho_{1,K,1} \\ \vdots & \vdots & & \vdots & \vdots \\ \rho_{2L,1,0} & \rho_{2L,1,1} & \cdots & \rho_{2L,K,0} & \rho_{2L,K,1} \end{bmatrix} \begin{bmatrix} h_{1,0}(iT)s_1(iT) \\ h_{1,1}(iT)s_1(iT) \\ \vdots \\ h_{K,0}(iT)s_K(iT) \\ h_{K,1}(iT)s_K(iT) \end{bmatrix}$$

$$= \mathbf{GH}(i)\mathbf{S}(i) + \mathbf{n}(i) \tag{7.11}$$

$$\mathbf{G} = [\mathbf{g}_{1,0} \mathbf{g}_{1,1} \cdots \mathbf{g}_{K,0} \mathbf{g}_{K,1}] \tag{7.12}$$

$$\mathbf{g}_{K,d}^H = \left[\rho_{1,K,d}^* \rho_{2,K,d}^* \cdots \rho_{2L,K,d}^* \right] \tag{7.13}$$

Statistical property of $s_k(iT)$ is expressed by

$$\langle s_{k_1}^*(iT) s_{k_2}(iT) \rangle = \delta_{k_1 k_2} \qquad (7.14)$$

where

$\langle \cdot \rangle$ = Expected value operator

Autocorrelation matrix $\mathbf{R}_N$ of $\mathbf{n}(i)$ is diagonal

$$\mathbf{R}_N = \langle \mathbf{n}(i)\mathbf{n}^H(i) \rangle = \sigma_n^2 \mathbf{I}_{2L} \qquad (7.15)$$

The canceler for user 1 extracts the desired signal by combining $X(i)$ by using $2L$ dimensional coefficient vector $\mathbf{W}$:

$$\mathbf{W}^H = \begin{bmatrix} w_1^* & \dots & w_{2L}^* \end{bmatrix} \qquad (7.16)$$

The combined signal $y(i)$ is:

$$y(i) = \mathbf{W}^H \mathbf{X}(i) \qquad (7.17)$$

7.3.1.2 CMMS Criterion

The CMMS criterion is used to minimize average output power $|y(i)|^2$ with a constraint for controlling the coefficient vector. The constraint is

$$\mathbf{T}^H\mathbf{W} = a, \; \|T\| > 0 \qquad (7.18)$$

Using steering vector $\mathbf{T}$, discussed in a later section, is necessary to avoid the primitive solution $\mathbf{W} = 0$ in minimization. In this case, the cost function J becomes

$$J = \langle |\mathbf{W}^H\mathbf{X}(i)|^2 \rangle + \lambda_L(\mathbf{W}^H\mathbf{T} - \alpha) \qquad (7.19)$$

$$= \mathbf{W}^H\mathbf{R}\mathbf{W} + \lambda_L(\mathbf{W}^H\mathbf{T} - \alpha)$$

where $\mathbf{R} = \langle \mathbf{X}(i)\mathbf{X}^H(i) \rangle$ and λ_L is the Lagrange multiple.

Finding the minimum of J, we obtain

$$J_0 = \mathbf{W}_0^H \mathbf{R} \mathbf{W}_0 \tag{7.20}$$

where the optimal w_0 is given by

$$\mathbf{W}_0 = -\lambda_L R^{-1} T \tag{7.21}$$

$$\lambda_L = \frac{-\alpha}{T^H R^{-1} T} \tag{7.22}$$

It is shown [7] that the signal to interference and noise ratio is given by

$$SINR = \frac{|V^H R^{-1} T|^2}{\sigma_d^2 T^H R^{-1} T - |V^H R^{-1} T|^2} \tag{7.23}$$

where

$V =$ correlation matrix between $d(i)$ and X (i.e., $V = E[X(i)\, d^*(i)]$

$$\sigma_d^2 = E\left\{|d(i)|\right\}^2 \tag{7.24}$$

when $T = V$ and $\alpha = V^H R^{-1} V$, then $\mathbf{W}_0$ becomes optimal

$$\mathbf{W}_0 = R^{-1} V \tag{7.25}$$

$$SINR_0 = \frac{V^H R_{NI}^{-1} V}{\sigma_d^2} \tag{7.26}$$

$$R_{NI} = R - \frac{VV^H}{\sigma_d^2} \tag{7.27}$$

where

R_{NI} is a correlation matrix of noise and interference components of $X(i)$.

However, V is unknown. It can be shown that if we express T as a finite set of the eigenvectors of R, β_i, in the form $T = \sum_i c_i \underline{\beta_i}$, and use (7.17), we obtain

$$\langle |y(i)| \rangle^2 = \mathbf{W}^H R \mathbf{W} = (T^H R^{-1} T)^{-1} \tag{7.28}$$

And using the expansion for T, we obtain

$$\langle |y(i)| \rangle^2 = \left[\sum_i \frac{|c_i|^2}{\lambda_i} \right]^{-1} \tag{7.29}$$

where λ_i are the eigenvalues corresponding to the eigenvector β_i.

A similar technique was shown in Chapter 6 using the interference-avoidance concept. The power of desired signal $y(i)$ is increased, as seen by (7.29), if we decrease $|c_i|^2$ where λ_i is small. Simulation results [7] show that this type of blind algorithm defines a canceler through a steering vector and avoids the drawbacks of a simple OMF, which requires integer multipliers of the chip duration of discrete signal components.

7.4 Indirect Cochannel Interference Canceler

The indirect cochannel interference canceler (ICIC) is a novel approach to the design of cochannel interference cancelers, independent from interference channel and timing [8]. The receiver is suitable for constant envelope modulation schemes with a dominant cochannel interferer. The ICIC receiver shows a good BER performance in the presence of cochannel interference; however, its performance in an AWGN channel is not likewise satisfactory.

7.4.1 Configuration of the Receiver

A mobile communication channel with a dominant cochannel interferer is shown in Figure 7.3.

Desired signal is transmitted over a Rayleigh fading channel. A cochannel interference signal, which is assumed to be dominant among all cochannel interference signals, passes through an independent fading channel and interferes with the desired signal. AWGN and other interference sources are

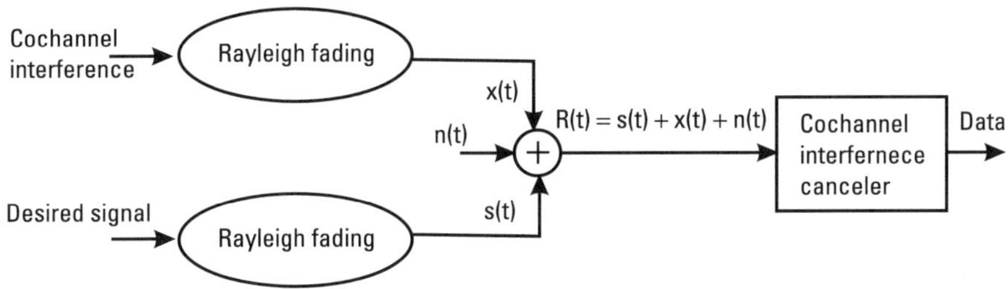

Figure 7.3 Model of a mobile communication channel with dominant cochannel interference and cochannel interference canceling. (After: [8].)

shown by $n(t)$. Furthermore, the desired signal and its cochannel interferer are assumed to have the same bit rate and modulation specifications with a constant envelope. The structure of a proposed ICIC receiver is shown in Figure 7.4.

The received baseband I/Q signals are individually sampled in both I and Q channels with a sampling rate of m samples per data symbol, then the possible desired signal pulse shapes $\{w_i(k)\}$, $i = 1, 2, \ldots, N$, produced by a local waveform generator, are canceled from received signal samples, $r(k)$, and from the noisy estimates of the cochannel interference. It is assumed that the information about desired signal channel and timing are known by the receiver.

Considering that at the receiver we have the desired signal and cochannel interference, the pulse shapes of the desired signal are generated having exact

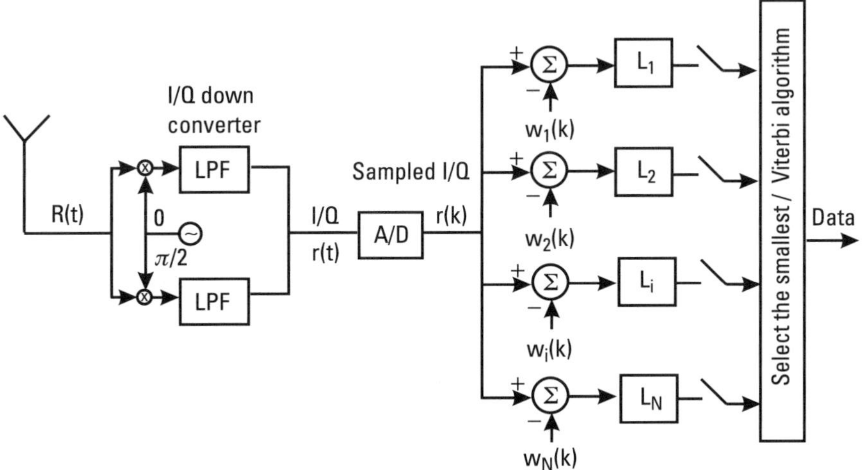

Figure 7.4 Block diagram of a cochannel interference canceler (After: [8].)

knowledge about channel and timing. In the case of constant envelope modulation schemes, we obtain a constant envelope cochannel interference (zero variance envelope), which can be used to identify the correct waveform.

The metric, which shows this constant envelope within a bit-timing interval, has been defined as:

$$L_i = \frac{T}{m+1} \sum_{k=0}^{m} \left| M_i^2(k) - \overline{M_i^2} \right| \tag{7.30}$$

where $T =$ bit timing interval

$$M_i(k) = \left| r(k) - w_i(k) \right|$$

$$\overline{M_i^2} = \frac{1}{m+1} \sum_{k=0}^{m} M_i^2(k)$$

Because the coefficient $\dfrac{T}{m+1}$ is identical for all the waveforms, it can be disregarded. The N metrics L_i, $(i = 1, 2, \ldots, N)$ as defined in (7.30) are computed in a bit time, and then the receiver can perform either bit-by-bit detection or sequence estimation. Using this envelope distance metric, the ICIC receiver is shown in [8] to be suitable for environments with dominant cochannel interference.

7.5 Adaptive Interference Canceler

One of the approaches for blind interference cancellation is to minimize the average output power subject to a constraint coefficient condition. This approach, known as OMF, assumes MF and filter bank orthogonal to the MF and then controls the taps of the filter bank to minimize the average combined output power. The conventional algorithm of this approach has the shortcoming of canceling not only the interference signal but also the desired signal once the desired signal starts to appear at the filter bank output due to nonorthogonality caused by multipath propagation. Thus, it is necessary to reject the desired signal contained in the OMF bank output, which necessarily increases the computational complexity.

A scheme used to overcome this drawback is incorporated in an OMF that uses a high pass filter (HPF) to remove the desired signal component from the filter bank output.

OMF is also used as the basis of a blind algorithm in Section 7.2 in order to cancel interference by using some kind of decomposition of the received signal through a steering vector.

7.5.1 Configuration of the Canceler

In order to reduce computational complexity, the structure of the canceler should follow the one shown in Figure 7.5 [9].

This consists of an MF and an adaptive filter (AF) generating the interference signal component. The interference is canceled by subtracting the AF output from the MF output. The tap coefficients of the AF are properly controlled to generate the interference component at the MF output. For this purpose, the signal waveform contained in the MF output is used as a reference. The interference components from other users are extracted by applying the received signal to the filter bank orthogonal to the desired signal, or the blocking matrix (BM).

The operation of the canceler can be summarized as follows. The tap coefficients vector $w_\alpha(i)$ of the AF at time iT_s is defined as

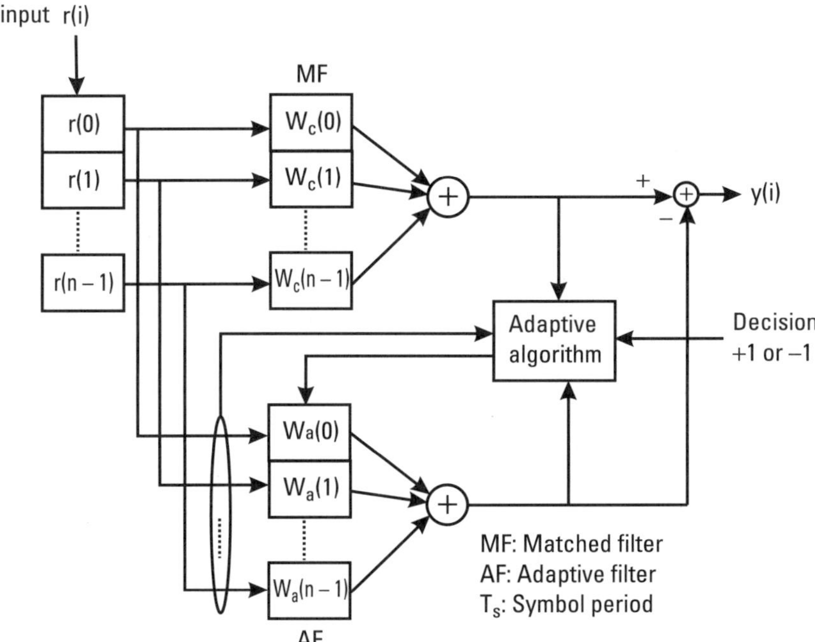

Figure 7.5 Structure of the interference canceler. (After: [9].)

$$w_\alpha(i) = (w_\alpha(0),\, w_\alpha(1),\, \ldots,\, w_\alpha(n-1))^T \tag{7.31}$$

Then, the output of the canceler $y(i)$ is given by

$$y(i) = \left(\underline{w}_c^{II} - \underline{w}_\alpha^{II}(i)\right)\underline{r}(i) \tag{7.32}$$

If the error signal is $e(i)$, the HPF output (reference) is $d(i)$, and the step size of LMS algorithm is μ, then the tap updating is given as follows

$$e(i) = d(i) - \underline{w}_\alpha^H(i)\,\underline{r}(i) \tag{7.33}$$

$$\underline{w}_\alpha(i+1) = \underline{w}_\alpha(i) + \mu\underline{r}(i)e^*(i) \tag{7.34}$$

As can be seen in Figure 7.5, the BM is no longer necessary and the processing to remove the desired signal is applied only to the MF output of this canceler. Thereby computational complexity is significantly reduced. In (7.34), $\underline{r}(i)$ is the sampled received vector.

7.6 Intersymbol Interference and Cochannel Interference Canceler Combining Adaptive Array Antennas and the Viterbi Equalizer in a Digital Mobile Radio

Several systems have been proposed to overcome the ISI and cochannel interference using MLSE. Because these systems estimate the impulse responses and transmitted symbols of both ISI and cochannel interference, they can suppress the cochannel interference, equalize ISI, and achieve path diversity gain. They have good BER performance at low C/I. However, their performance degrades when there are more than two interference stations, and the amount of their signal processing increases exponentially with the number of interference stations and/or maximum time delay of ISI and/or cochannel interference.

Another way to overcome the ISI and cochannel interference is to use an adaptive array. The adaptive array can reduce ISI and cochannel interference simultaneously. However, a conventional adaptive array cannot obtain path diversity gain because the ISI is regarded as interference and suppressed. To make up for this weak point in the conventional adaptive array, some systems that can select several paths have been proposed. However, the adaptive array that employs the simple combination of the path selection cannot obtain sufficient path diversity gain at low C/I [10–12].

The adaptive array of a system that combines an adaptive array and a MLSE employs a new training signal and new sampling timing in the training period.

7.6.1 System's Configuration

The configuration of the system is shown in Figure 7.6.

It is seen that the adaptive array part synthesizes the desired signal and ISI simultaneously by employing a new training signal and data sampling timing in the training period. A Viterbi equalizer (VE) performs MLSE to obtain path diversity gain without being affected by long-delayed ISI and CCI signals.

The system contains adaptive array antennas that have N antenna elements. The received signal contains the desired signal, ISI, cochannel interference, and thermal noise. The number of the stations that contain a desired station and interference stations is M. The number of arriving signals from the jth station is L_j. The received signal is then described as

$$\mathbf{X}(t) = \sum_{j=1}^{M} \sum_{k=1}^{L_j} \mathbf{H}_{j,k}\, s_j(t - \tau_{j,k}) + \mathbf{N}(t) \tag{7.35}$$

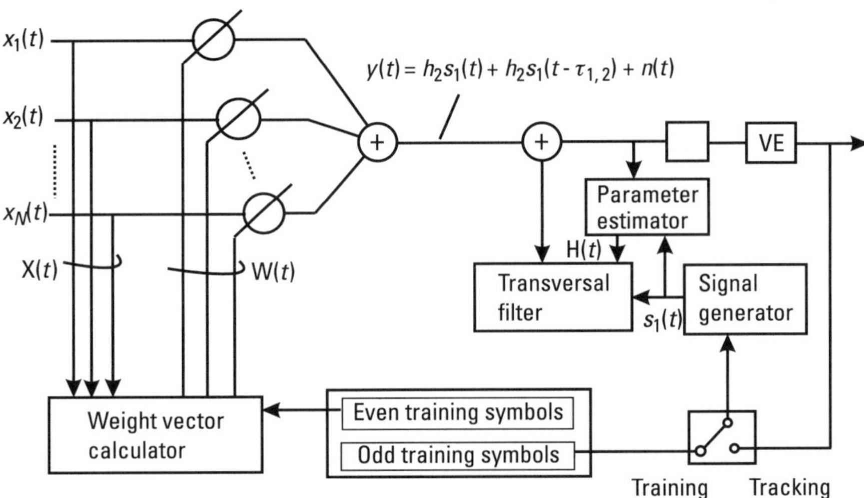

Figure 7.6 Block diagram of the interference canceler combining an adaptive array and an MLSE equalizer. (After: [10].)

where

$$\mathbf{X}(t) = [x_1(t), x_2(t), \ldots, x_N(t)]^T$$

$$\mathbf{H}_{j,k} = [h_{1,j,k}, h_{2,j,k}, \ldots, h_{N,j,k}]^T$$

$$\mathbf{N}(t) = [n_1(t), n_2(t), \ldots, n_N(t)]^T$$

where T denotes transpose, $s_j(t)$ is a signal transmitted from the desired station when $j = 1$, and a signal transmitted from interference stations when $j \neq 1$. After this, index j denotes the number of the station, $j = 1$ means the desired station and $j \neq 1$ means the interfering station. $x_i(t)$ is a received signal of the ith antenna element. $\mathbf{H}_{j,k}$ is an impulse response vector of the kth arrived path transmitted from the jth station. $h_{i,j,k}$ is an impulse response of the kth arrived path transmitted from the jth station received by the ith antenna element. $\tau_{j,k}$ is a time delay of the kth arrived path transmitted from the jth station. $n_i(t)$ is white Gaussian noise of the ith antenna element.

The MMSE criterion is employed to calculate the weight vector of the adaptive array. The MMSE adaptive array synthesizes a signal that correlates with the training signal held in the receiver. The synthesized signal $y(t)$ is described as

$$y(t) = \mathbf{W}^T \mathbf{X}(t) \tag{7.36}$$

$$\mathbf{W} = [w_1, w_2, \ldots, w_N]^T$$

where $\mathbf{W}$ is a weight vector and w_i is the weight of the ith antenna element.

The training sequence $d(t)$ is described as

$$d((2i - 1)T) = d(2iT), \quad i = 1, 2, N_d/2 \tag{7.37}$$

where T is one symbol length and N_d is the number of training symbols.

The first arrived signal and the one-symbol delayed ISI have the same sampled signal in the training period, by setting the sampling timing in the training period. Therefore, the system treats the desired and the one-symbol delayed ISI equivalently. Thus, the adaptive array can select both paths simultaneously. If the multipath has a desired path, the one-symbol delayed ISI, the more-than-one symbol delayed ISI, and cochannel interference, the synthesized signal is described as

$$y(t) = h_1 s_1(t) + h_2 s_1(t - T) + n(t) \qquad (7.38)$$

where

$$h_1 = \mathbf{H}_{1,1}^{T} \mathbf{W}_{opt}$$

$$h_2 = \mathbf{H}_{1,2}^{T} \mathbf{W}_{opt}$$

$$n(t) = \mathbf{N}^{T}(t) \mathbf{W}_{opt}$$

W_{opt} is an optimum-weight vector, $H_{1,1}$ and $H_{1,2}$ are impulse-response vectors of the desired signal and the one-symbol delayed ISI signal, respectively. Subsequently, the system estimates the transmitted signal using MLSE and achieves the path diversity gain. Simulations results in [10] show that BER performance is greatly improved over conventional adaptive arrays. The reason is that this canceler reduces ISI and CCI to a point that it achieves path diversity gain even if the CIR is low.

7.7 Hybrid Interference Canceler with Zero-Delay Channel Estimation for CDMA

For interference cancellation, the two main algorithms suggested are SIC [13–14] and PIC [15]. With reference to their first-generation designs, PIC, while having the virtues of parallel computation, suffers from a "ping-pong" effect—the BER in the even stages converges to one value and the BER for odd stages converges to a different value. More details are given in Section 7.9, where an attempt is made to alleviate this problem. The SIC has no such problem but suffers from considerable processing delay due to the sequential approach. For most practical scenarios, this delay is unacceptable; thus, a logical direction is to consider a combination of the two, which we generically call the hybrid interference canceler (HIC) [16].

7.7.1 HIC

The HIC incorporates the functional elements of serial, parallel and multistage cancellation into the architecture [17]. Figure 7.7 shows the block diagram of the HIC. K active users are split into G groups, and an iterative parallel-serial cancellation is performed.

It is shown that the joint process of cancellation and channel estimation in the gth group-interference cancellation unit (G-ICU) at the mth stage ($m > 1$) with p users can be expressed as follows [16]:

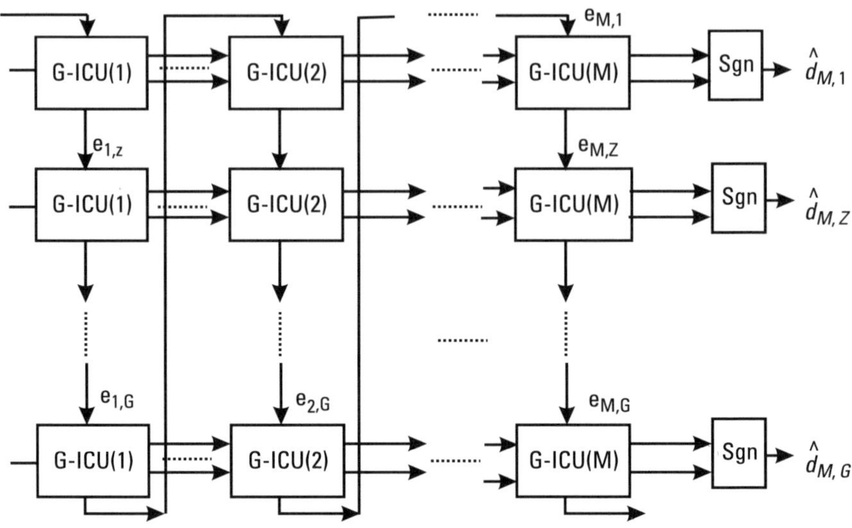

Figure 7.7 Block diagram of K-user, G-group, M-stage HIC detector. (After: [16].)

$$y_{m,g,k}(i) = \sum_{l=1}^{L} \hat{c}^{*}_{m,k,l}(i) \tag{7.39}$$

$$\left[\sum_{n=\tau_{k,l}+iN}^{\tau_{k,l}+(i+1)N-1} s^{*}_{k}(n-\tau_{k,l})e_{m,g}(n) + f_{x}(y_{m-1,g,k}(i))\hat{c}_{m-1,k,l}(i) \right]$$

$$\hat{d}_{m,g,k}(i) = \mathrm{sgn}\,(y_{m,g,k}(i)) \tag{7.40}$$

$$\Delta I_{m,g,k}(n) = \sum_{i=0}^{P} \sum_{l=1}^{L} [f_{x}(y_{m,g,k}(i))\bar{c}_{m,k,l}(i) - f_{x}(y_{m-1,g,k}(i))\bar{c}_{m-1,k,l}(i)]$$

$$\cdot\, u(n-iN-\tau_{k,l})s_{k}(n-\tau_{k,l}) \tag{7.41}$$

$$\Delta I_{m,g}(n) = \sum_{k=1}^{P} \Delta I_{m,g,k}(n) \tag{7.42}$$

$$e_{m,g+1} = e_{m,g}(n) - \Delta I_{m,g}(n) \tag{7.43}$$

where $y_{m,g,k}(i)$ is the soft decision, $\hat{d}_{m,g,k}$ is the corresponding estimated symbol, $f_{x}(\cdot)$ denotes the mapping function, x^{*} is complex conjugate of x, $e_{m,g}(n)$ is the residual signal after the mth stage, $(g-1)$th group cancellation

$(e_{1,1}(n)$ is thus the received signal $r(n))$. $\hat{c}_{m,k,l}(i)$ is the channel estimate of the kth user, lth path, at the mth stage, and $r_{k,l}$ are the corresponding transmission delays. The channel estimation method will be further clarified in the following. $\Delta I_{m,g,k}$ and $\Delta I_{m,g}$ denote the estimated MAI difference between the m and $(m-1)$th stage, for the particular user and group, respectively. For $m=1$, the computation is identical except $f_x(y_{0,g,k}(i)) = 0$. Also, for enhanced performance, the clipped-soft-decision (CSD) mapping function in the ICU [18] is implemented (e.g., for QPSK modulation and separate clipping of the I and Q channels), and the mapping for the I-channel is

$$\Re\{f_x(y_{m,g,k}(i))\} = \begin{cases} M & \Re\{y_{m,g,k}(i)\} > M \\ \Re\{y_{m,g,k}(i)\} & -M \le \Re\{y_{m,g,k}(i)\} \le M \\ -M & \Re\{y_{m,g,k}(i)\} < -M \end{cases}$$

$$(7.44)$$

where M is the clipping threshold magnitude, and $\Re\{\cdot\}$ denotes the real part. For unit received power, M for QPSK modulation is $1/\sqrt{2}$.

From (7.39)–(7.43), it can be seen that the cancellation process necessarily requires one to estimate the amplitude and phase of all KL paths. Multiuser detection capacity gains can be seriously degraded or even reversed if the channel estimates are incorrect [19]. Further, the inherently complex cancellation process immediately limits the channel estimation method to relatively simple algorithms to avoid causing excessive delays to delay-sensitive traffic. One approach to escape from that dilemma is to combine channel estimator and RAKE receiver to operate concurrently with the cancellation process. Figure 7.8 illustrates the implemented channel estimator.

Functionally, the wireless channel estimator first eliminates the data phase from the raw correlated signals via decision feedback. The phase-

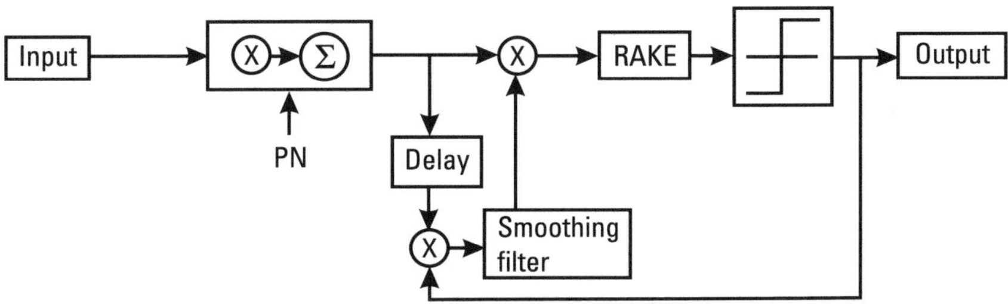

Figure 7.8 Structure of zero-delay channel estimator. (After: [16].)

eliminated signal is then passed through a smoothing filter block, which minimizes the correlation error. As a smoothing filter, a simple moving average is performed on the last W symbol estimates, with the window increasing in size from the beginning of each slot to maximum window size W_s, then sliding to the end of the slot with W remaining at W_s. In the next slot, the smoothing filter window will be cleared.

Thus, for the ith symbol

$$\hat{c}_{m,k,l}(i) = \begin{cases} \dfrac{1}{i - i_0} \displaystyle\sum_{w=i_0}^{i-1} \hat{c}_{m,k,l}(w) & \mod(i, S_s) \le W_s \\[2em] \dfrac{1}{W_s} \displaystyle\sum_{w=i-W_s}^{i-1} \hat{c}_{m,k,l}(w) & \text{otherwise} \end{cases} \tag{7.45}$$

where $i_0 \equiv \lfloor i/S_s \rfloor S_s$ and $\lfloor x \rfloor$ denotes the largest integer smaller than or equal to x, and S_s is the number of symbols within a slot. It is assumed that the data is formatted in slots consisting of a block of 40 symbols, the first four being pilot and the remaining 36 data. Pilot symbols are treated as symbols with known values, and they serve to increase the reliability of the channel estimation. The correlation process essentially obtains the raw estimates, that is

$$\hat{c}_{m,k,l}(i) = \sum_{l=1}^{L} \bar{d}_{m,g,k}^{*}(i) \tag{7.46}$$

$$\left[\sum_{n=\tau_{k,l}+iN}^{\tau_{k,l}+(i+1)N-1} s_k^{*}(n - \tau_{k,l}) e_{m,g}(n) + f_x(y_{m-1,g,k}(i)) \hat{c}_{m-1,k,l}(i) \right]$$

and

$$\hat{d}_{m,g,k}(i) = \begin{cases} d_k(i) & \mod(i, S_s) \le 4 \\ \hat{d}_{m,g,k}(i) & \text{otherwise} \end{cases} \tag{7.47}$$

Simulation results [16] have shown that this type of interference canceler consisting of serial, parallel, and multistage detection in a Rayleigh fading environment outperforms a conventional correlator detector and doesn't suffer from the considerable processing of the sequential approach of the successive interference cancelers [13, 14].

7.8 Cancellation of Adjacent Channel Signals in FDMA/TDMA Digital Mobile Radio Systems

ACI impairs the performance of digital wireless communications systems. ACI mitigation improves performance and capacity for cellular, mobile-satellite, and land mobile radio systems. Cellular capacity can be improved by decreasing reuse spacing or by employing more flexible channel allocation schemes, both of which would require better ACI mitigation at the receiver. Mobile-satellite systems are both power and bandwidth limited [20]. The power limitation implies restricted link margins, so that margins for interference are limited. ACI mitigation allows better link margin. ACI mitigation also allows higher capacity via closer channel reuse. In land mobile radio systems, each site serves a large geographical area, and different operators may use adjacent channels in the same area. As a result, the relative power levels of adjacent channel signals can be very large, requiring the receiver to provide adjacent channel protection (ACP), typically on the order of 55 to 65 dB [21]. By contrast, cellular systems are specified at 18 to 26 dB. To achieve large ACP values, power control, linear power amplification, and spectrally efficient modulation [22] have been proposed. ACI mitigation would ease system design and improve signal quality.

Prior solutions to ACI mitigation include narrowband filtering, equalization methods, and subtractive demodulation. Receive filtering methods are effective, but the ISI introduced limits performance gains. Equalization methods exploit cyclostationarity, adapting demodulation parameters to minimize the effects of noise and ACI. Subtractive approaches employ demodulation of the adjacent signal, subsequent regeneration, and subtraction from the received signal prior to demodulation of the desired signal [23].

In the present section, the idea of subtractive demodulation is introduced [23], and it can be shown that successive cancellation of baseband signals can be achieved on the basis of signal strength. The problem is formulated in the context of FDMA/TDMA mobile radio systems employing MLSE receivers. The receiver is evaluated in conjunction with GMSK modulation and a mobile radio environment. Practical considerations such as front-end IF filtering and channel impulse response estimation are taken into account in the performance evaluation.

7.8.1 Receiver's Configuration

By successive cancellation of adjacent channel signals, we mean that we can detect a user's signal in its band using conventional demodulation, and then remove or cancel its effect from the adjacent bands. Detection is limited by uncanceled ACI. Performing cancellation in order of decreasing signal strength minimizes this problem. Signal strength order is determined by channel estimation.

Two approaches to successive cancellation are considered. Unlike conventional single user demodulation, in which each user's signal is demodulated as if it were the only one present, these receivers process not only the channel of interest but also other adjacent frequency channels by using a bank of standard practical receiver filters [21]. The first approach, illustrated in Figure 7.9, uses a wideband receiver to receive a group of adjacent signals before frequency band channelization.

The wideband signal must be highly oversampled. The received signal is then passed through a bank of matched filters appropriate to receive signals on different carrier frequencies. It is desirable to partition the matched filters into two parts, one matched to the known pulse shape followed by one matched to the unknown medium [24]. This leads to traditional receiver designs, employing fixed analog filters followed by sampling and baseband signal processing.

For the GSM signal model, one sample per bit is sufficient. Signal strength order is determined via channel estimation. The information bits that belong to the strongest signal are detected using coherent MLSE based on the approach in [25]. However, this is suboptimal, as it assumes that the

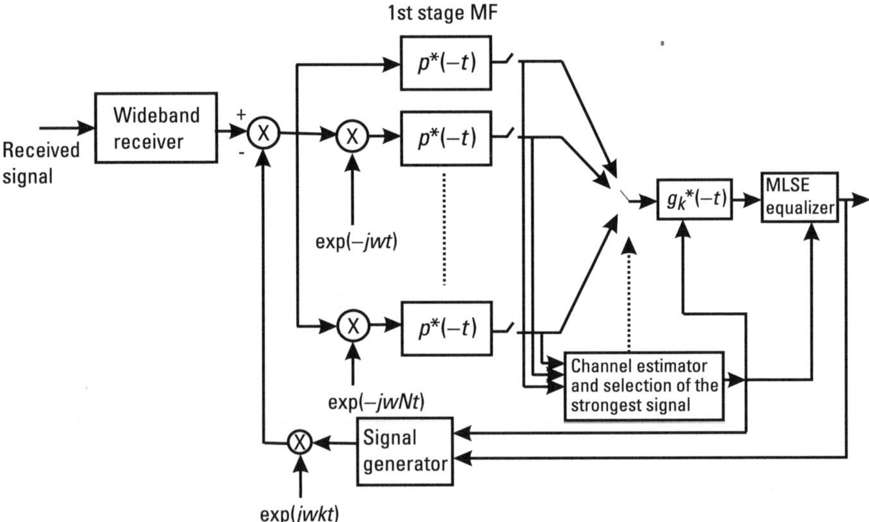

Figure 7.9 Successive cancellation (method 1). (After: [21].)

sum of the interfering adjacent signals can be modeled as white noise. The strongest signal is regenerated in highly oversampled form using the medium response estimates, detected bit sequence, and knowledge of the pulse shaping and carrier spacing. The regenerated signal is then subtracted from the total wideband received signal to obtain a reduced ACI received signal for the remaining user's signals. This process is repeated until the weakest user signal is detected. Though not explored here, multistage interference cancellation could then be applied to improve performance further.

The disadvantage of this approach is that subtraction occurs using a highly oversampled signal, and channelization filtering must be performed repeatedly. An equivalent, more efficient method can be obtained by employing successive cancellation at the sampled outputs of the matched filters.

Figure 7.10 illustrates the second approach, in which successive cancellation is applied after frequency band channelization.

The sampled outputs of each filter contain the desired and interfering signal terms plus the noise. Usually the strongest interfering signals arise from the immediate or secondary adjacent channels depending on the carrier spacing, and the effect of further away signals on the channel of interest can be ignored. Therefore, the strongest signal is detected and canceled from the baseband signals corresponding to the immediately adjacent signals. The same procedure is repeated until the weakest signal is detected. With this

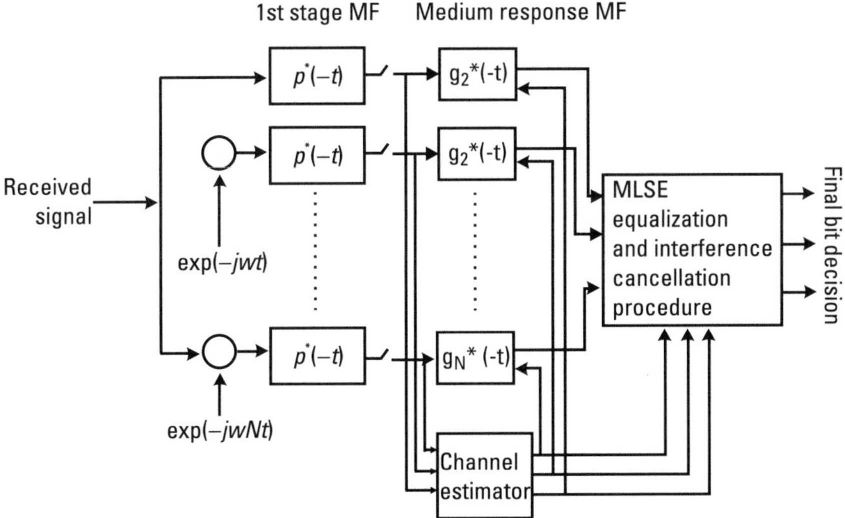

Figure 7.10 Successive cancellation (method 2). (After: [21].)

method, FDMA channelization and channel estimation are performed only once.

7.9 Adaptive Multistage PIC

PIC is attractive for its simplicity and the fact that it can lead to considerable capacity increase without service deterioration. Ideally, when the MAI signal is known a priori, a single stage PIC is equivalent to the optimum detector in a maximum-likelihood (ML) sense [26]. In practical applications, the MAI estimates are used due to the lack of an exact knowledge of MAI. By introducing a multistage architecture [27], MAI estimation can be improved in an iterative way. However, this is not always true for a conventional multistage PIC, especially, when the BER in the previous stage is sufficiently high. A wrong estimation used in MAI cancellation will largely increase the interference power, thus introducing further degradation.

Partial cancellation of MAI at each stage to reduce the cost of wrong MAI estimation has been suggested in [26]. The amount of interference to be canceled is decided by a weighting factor at each stage for all users. This method can ensure a performance improvement after partial interference cancellation. Because the bit decisions become more reliable when more MAI is canceled, an increase of the weighting factors for each successive stage results in an improvement manifested as a capacity increase.

For the approach in [26], a constant weight is used for all users at each stage throughout the cancellation. For a CDMA system operating in a multipath fading channel, the MAI varies from one user to another and from bit to bit according to the PN cross-correlation and the power level of each user at a particular time instant. Hence, adaptive weights that reflect the reliability of data estimation can offer a better solution. Motivated by this thought, a new cost function, which takes the weighting factors into account, has been proposed [28]. The objective is to minimize the mean-square error between the received signal and the weighted sum of the signal estimates of all users' during a bit interval with respect to the weights. The optimum weights can be obtained through an adaptive LMS algorithm.

7.9.1 PIC

Without loss of generality, let us focus on the first user. For the multistage PIC [15] operating in multipath environment, the MAI as can be shown [28] is estimated at the kth stage as follows

$$\hat{I}_1^{(k)} = \sum_{i=2}^{K} \sum_{l=1}^{L} \sqrt{E_{bi}}\, \alpha_{il} c_i(t)\, \hat{a}_i^{(k-1)} \tag{7.48}$$

where

$\alpha_{il}(t)$ represents the time-variant complex channel parameter, which includes the attenuation and phase shift;

$c_i(t)$ is a complex form of PN sequence;

$\hat{a}$ is a binary data sequence decision at the RAKE receiver.

At the kth stage, the estimated MAI is completely removed from the received signal in the conventional multistage PIC. This can be written as

$$r_{c1}^{(k)} = r(t) - \hat{I}_1^{(k)} \tag{7.49}$$

RAKE combining and bit decisions can be carried out in the same way as for single user RAKE receiver. The only difference is that the received signal $r(t)$ should be replaced by $r_{c1}^{(k)}$ for conventional PIC.

In a multipath fading channel, the procedure of interference cancellation can be described as follows:

$$\bar{r}_{p1}^{(k)} = p^{(k)}\lfloor r(t) - \hat{I}_1^{(k)}\rfloor + [1 - p^{(k)}]\bar{r}_{p1}^{(k-1)} \tag{7.50}$$

$$\bar{r}_{p1}^{(0)} = \sum_{l=1}^{L} y_{il}$$

where $p^{(k)}$ is the weighting factor for interference cancellation at the kth stage. RAKE diversity is then carried out based on the interference partially removed signal $\bar{r}_{p1}^{(k)}$.

7.9.2 Adaptive Multistage PIC

In a partial cancellation scheme [26], the weight for each stage remains constant. Intuitively, it is more reasonable to have a set of weights that can reflect the reliability of the bit estimations from previous stages. In this section, an adaptive multistage PIC approach is described, where the weights are updated by an LMS algorithm. In order to incorporate the adaptive algorithm, the received signal must be sampled. Because of that, the received signal $r(t)$ is sampled once per chip, the discrete form received signal is denoted by $r(m)$, where

$$r(m) = \sum_{i=1}^{K} \sum_{l=0}^{L-1} \alpha_{il} s_i(m - l) + n(m) \tag{7.51}$$

where

$s_i(m - l)$ are samples of transmitted signal;

$n(m)$ is additive Gaussian noise.

and

$$s_i^{k}(m) \equiv c_i(m)\, \hat{\alpha}_i^{(k-1)}$$

We try to estimate $r(m)$ at the kth stage from the PN sequence $c_i(m)$, the bit estimate from the previous stage $\hat{\alpha}_i^{(k-1)}$, and the weight $\{\lambda_{il}(m), l = 0, \ldots, L - 1\}$. The estimation is carried out as follows:

$$\hat{r}^{(k)}(m) = \sum_{i=1}^{K} \sum_{l=0}^{L-1} \hat{s}_i(m - l)\, \lambda_{il}(m) \tag{7.52}$$

where $\hat{s}_i(m)$ is defined as

$$\hat{s}_i^{(k)}(m) = c_i(m)\, \hat{\alpha}_i^{(k-1)} \tag{7.53}$$

The objective is to minimize the MSE between the received signal $r(m)$ and its estimate $\hat{r}(m)$ with respect to the weights. The cost function can be expressed as

$$E\left[\,|r(m) - \hat{r}^{(k)}(m)|^2\,\right] \qquad 0 \le m \le N - 1 \tag{7.54}$$

A normalized LMS algorithm is used to search for the optimum weights during each bit interval and on a chip basis. The weights update is given by [26]:

$$\lambda^{(k)}(m + 1) = \lambda^{(k)}(m) + \frac{\mu}{\|\hat{s}^{(k)}(m)\|^2}\, \hat{s}^{(k)}(m)\, [e^{(k)}(m)]^* \tag{7.55}$$

where

μ is a step size and $e^{(k)}(m)$ represents the error between the desired response and the output of the LMS filter of the kth stage:

$$e^{(k)}(m) = r(m) - \hat{r}^{(k)}(m) \tag{7.56}$$

The dimension of vector λ is $L \times K$. The block diagram of the weight estimation via an LMS algorithm is depicted in Figure 7.11. The same concept can be used to develop an adaptive multistage parallel interference cancellation structure for applications in an AWGN environment.

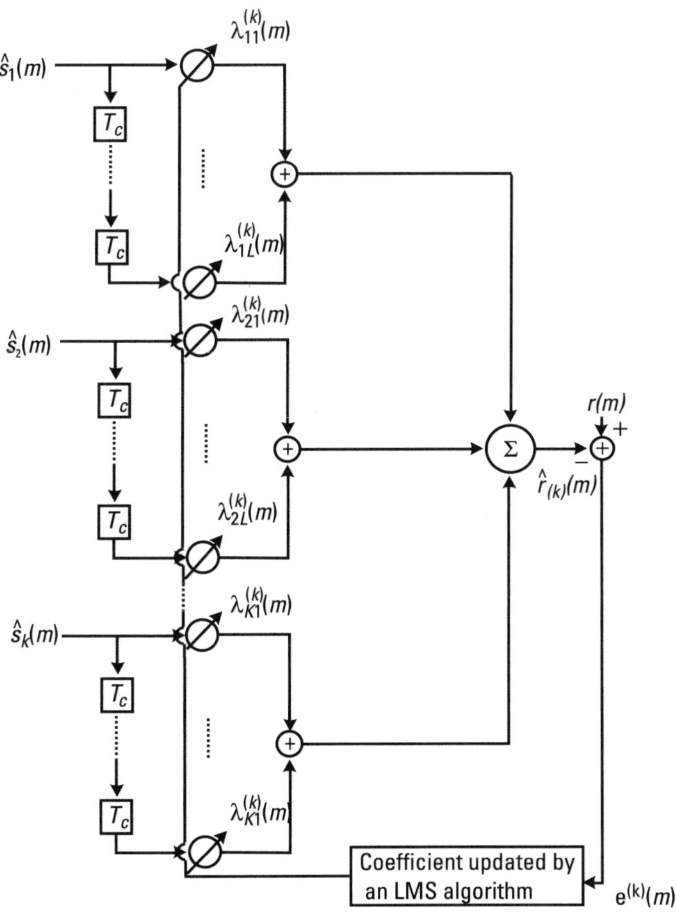

Figure 7.11 Adaptive PIC using an LMS algorithm in multipath fading. (After: [28].)

With the weights provided by the LMS algorithm, the interference cancellation at the kth stage for the ith user is realized, as shown here:

$$y_{ci}^{(k)}(m) = r(m) - \sum_{\substack{j=1 \\ j \neq i}}^{K} \sum_{l=0}^{L-1} \lambda_{il}^{(k)}(N-1)\hat{s}_j^{(k)}(m-l) \qquad (7.57)$$

RAKE diversity is then carried out based on the less interfered signal $y_{ci}^{(k)}(m)$, that is

$$Y_i^{(k)} = \mathrm{Re}\left\{ \frac{1}{N} \sum_{l=0}^{L-1} \sum_{m=0}^{N-1} y_{ci}^{(k)}(m)\, c_i^*(m-l)\, \alpha_{il}^* \right\} \qquad (7.58)$$

A more reliable decision is made as

$$a_i^{(k)} = \mathrm{sgn}\left[Y_i^{(k)} \right] \qquad (7.59)$$

where Y_i is as defined as

$$y_i = \mathrm{Re}\left\{ \sum_{l=0}^{L-1} \alpha_{il}^*\, y_{il} \right\}, \qquad Y_i = \mathrm{Re}\{y_i\} \qquad (7.60)$$

Either exact channel parameters or their estimates can be used as the initial value of the tap coefficients of the LMS filters at each stage. Even if certain MAI estimates are wrong, it is possible for the LMS algorithm to reverse the sign of their corresponding weights, ensuring removal of the interference to some extent.

The step size μ plays an important role in the LMS algorithm. For the normalized LMS algorithm deployed in our approach, μ must satisfy $0 < \mu < 2$ in order to ensure convergence [29]. Generally, a large step size leads to a faster convergence rate; however, it will also cause a greater gradient noise. It is shown therefore that by using an LMS algorithm to search for a set of optimum coefficients, which minimize the MSE between the received signal and its estimate, we then can use these coefficients as weights in parallel interference cancellations. Simulations results [28] show that this method outperforms the conventional PIC and partial PIC [26].

References

[1] Yoshino, H., and H. Suzuki, "Interference Canceling Characteristics of DFE Transversal-Combining Diversity in Mobile Radio Environment-Comparisons with Metric Combining Schemes," *Transaction IEICE Japan,* Vol. J76-B-II, No. 7, 1993, pp. 584–595.

[2] Winter, J. H., "Optimum Combining in Digital Mobile Radio with Co-Channel Interference," *IEEE Journal of Select. Areas Commun.,* Vol. SAC-2, No. 4, 1984, pp. 538–539.

[3] Suzuki, H., "Signal Transmission Characteristics of Diversity Reception with Least-Squares Combining Relationship Between Desired Signal Combining and Interference Canceling," *Transaction IEICE Japan,* Vol. J75-B-II, No. 8, 1992, pp. 524–534.

[4] Van Etten, W., "Maximum Likelihood Receiver for Multiple Channel Transmission Systems," *IEEE Trans. Commun.,* Vol. COM-24, No. 2, 1976, pp. 276–283.

[5] Fukawa, K., and H. Suzuki, "Adaptive Equalization with RLS-MLSE for Fast Fading Mobile Radio Channels," *Proc. IEEE GLOBECOM'91 Conf. Rec.,* Dec. 1991.

[6] Yoshino, H. K. Fukawa, and H. Suzuki, "Interference Canceling Equalizer for Mobile Radio Communication," *IEEE Trans. On Vehic. Techn.,* Vol. 46, No. 4, November 1997.

[7] Suzuki, H., and K. Fukawa, "A Linear Interference Canceller with a Blind Algorithm Form CDMA Mobile Communication Systems," *IEEE VTC'97,* Phoenix, AZ, May 4–7, 1997.

[8] Berangi, R., P. Leung, and M. Faulkner, "Signal Space Representation of Indirect Co-Channel Interference Canceller," *IEEE VTC'97,* Phoenix, AZ, May 4–7, 1997.

[9] Takinami, K., H. Murata, and Susamu Yoshita, "Simple Adaptive Interference Canceller Suitable for DS-CDMA Mobile Radio," *IEEE VTC'97,* Phoenix, AZ, May 4–7, 1997.

[10] Doi, Y., T. Ohgane, and E. Ogawa, "Characteristics of ISI and CCI Adaptive Canceller Combined of Adaptive Array Antennas and Maximum-Likelihood Sequence Estimator in Quasi-Static Rayleigh Fading Channel," *IEICE Technical Report,* RCS95-46, June 1995, pp. 19–24.

[11] Fukasawa, A., et al., "Configuration and Characteristics of an Interference Cancellation System Using a Pilot Signal for Radio Channel Estimation," *Trans. of the Inst. of Elect., Info. and Commun. Engineers of Japan (translated),* Part I, Vol. 79, No. 2, Feb. 1996.

[12] Friedman R., and Y. Bar-Ness, "Combines Channel-Modified Adaptive Array MMSE Canceller and Viterbi Equalizer," *IEEE VTS 53rd,* May 6–9, 2001.

[13] Malik R., V. K. Dubey, and B. McGuffin, "A Hybrid Inreteference Canceller for CDMA Systems in Rayleigh Fading Channels," *IEEE VTS 53rd,* May 6–9, 2001.

[14] Omaya, T., et al., "Performance Comparison of Multi-Stage SIC and Limited Tree-Search Detection in CDMA," *Proc. IEEE Veh. Tech. Conf '98,* pp. 1854–1858.

[15] Varanasi, M. K., and B. Aazhang, "Multistage Detection in Asynchronous Code Division Multiple-Access Communications," *IEEE Trans. Commun.,* Vol. COM-38, 1990, pp. 505–519.

[16]   Kok. L., et al., "Performance of Hybrid Interference Canceller with Zero-Delay Channel Estimation for CDMA," *IEEE GLOBECOM '98,* Sydney, Australia, November 8-12, 1998.

[17]   Sun, S., et al., "A Hybrid Interference Canceller in CDMA," *Proc. IEEE Int. Symp. Spread Spectrum Techs. & Applications (ISSSTA),* 1998.

[18]   Sugimoto, H., et al., "Mapping Functions for Successive Interference Cancellation in CDMA," *Proc. IEEE Veh. Tech. Conf '98,* pp. 1854–1858.

[19]   Moshavi, S., "Multi-User Detection for DS-CDMA Communications," *IEEE Commun. Mag.,* Vol. 34, No. 10, Oct. 1996, pp. 124–136.

[20]   Rydbeck, N., et al., "Mobile-Satellite Systems: A Perspective on Technology Trends," *IEEE 46th Veh. Technol. Conf.,* Atlanta, GA, Apr. 28–May 1, 1996.

[21]   Arslan, H., et al., "Successive Cancellation of Adjacent Channel Signals in FDMA/ TDMA Digital Mobile Radio Systems," *IEEE 48th Annual Intern. VTC'98,* Ottawa, Canada, May 18–21, 1998.

[22]   Varma, V. K., and S. C. Gupta, "Performance of Partial Response CPM in the Presence of ACI and Gaussian Noise," *IEEE Trans. on Comm.,* Vol. 34, Nov. 1986, pp. 1123–1131.

[23]   Sampei, S., and M. Yokoyama, "Rejection Method of ACI for Digital Land Mobile Communications," *Trans. IECE,* Vol. E 69, May 1986, pp. 578–580.

[24]   Bottomley, G. E., and S. Chennakeshu, "Adaptive MLSE Equalization Forms for Wireless Communications," *Virginia Tech's Fifth Symp. Wireless Personal Commun.,* May 31–June 2, 1995.

[25]   Ungerboeck, G., "Adaptive Maximum-Likelihood Receiver for Carrier-Modulated Data-Transmission Systems," *IEEE Trans. Commun.,* Vol. 22, May 1974, pp. 624–636.

[26]   Divsalar, D., M. K. Simon, and D. Raphaeli, "Improved Parallel Interference Cancellation for CDMA," *IEEE Trans. Commun.,* Vol. 46, Feb. 1998, pp. 258–268.

[27]   Varanasi, M. K., and B. Aazhang, "Multistage Detection in Asynchronous Code-Division Multiple-Access Communications," *IEEE Trans. Commun.,* Vol. 38, April 1990, pp. 509–519.

[28]   Xue, G., et al., "Adaptive Multistage Parallel Interference Cancellation for CDMA over Multipath Fading Channels," *IEEE, Int. VTC '99,* Houston, TX, May 1999.

[29]   Haykin, S., *Adaptive Filter Theory,* Englewood Cliffs, NJ: Prentice Hall, 3rd ed., 1996.

Appendix A:
Signal and Spectra in Wireless Communications

In communication systems, the received waveform is usually categorized into the desired part containing the information and the extraneous or undesired part. The desired part is called the signal, and the undesired part is called noise.

This appendix introduces mathematical tools that are used to describe signals and noise from a deterministic and stochastic waveform point of view. The waveforms will be represented by direct mathematical expressions or by the use of orthogonal series representations such as the Fourier series or a continuous frequency spectrum as expressed by the Fourier transform. Measures for characterising these waveforms such as dc value, rms value, normalized power, magnitude spectrum, phase spectrum, power spectral density or energy spectral density, and bandwidth are the main quantitative characteristics.

The waveform of interest may be the voltage as a function of time, $u(t)$, or the current as a function of time, $i(t)$. Often the same mathematical techniques can be used when working with either type of waveform. Thus, for generality, waveforms will be denoted simply as $s(t)$ when the analysis applies to either case.

A.1 Physically Realizable Waveforms

Practical waveforms that are physically realizable (i.e., measurable in a laboratory) satisfy several conditions:

1. The waveform has significant nonzero values over a composite time interval that is finite.

2. The spectrum of the waveform has significant values over a composite frequency interval that is finite.

3. The waveform is a continuous function of time.

4. The waveform has a finite peak value.

5. The waveform has only real values. That is, at any time, it cannot have a complex value $(a + jb)$ where b is nonzero.

The first condition is necessary because systems (and their waveforms) appear to exist for a finite amount of time. Physical signals also produce only a finite amount of energy. The second condition is necessary because any transmission medium—such as wires, coaxial cable, waveguides, or fiber-optic cable—has a restricted bandwidth. The third condition is a consequence of the second—it usually becomes clear from spectral analysis, as we will discuss later. The fourth condition is necessary because physical devices are destroyed if voltage or current of infinite value is present within the device. The fifth condition follows from the fact that only real waveforms can be observed in the real world, although properties of waveforms, such as spectra, may be complex.

Mathematical models that violate some or all of the conditions listed previously are often used, and for one main reason—to simplify the mathematical analysis. In fact, we often have to use a model that violates some of these conditions in order to calculate any type of answer. However, if we are careful with the mathematical model, the correct result can be obtained when the answer is properly interpreted. For example, consider the digital waveforms that are modeled by functions with discontinuities at the switching times [1, 2]. This situation violates the third condition—the physical waveform is continuous.

The physical waveform is of finite duration (decays to zero before $t = \pm\infty$), but the duration of the mathematical waveform extends to infinity.

In other words, this mathematical model assumes that the physical waveform existed in its steady-state condition for all time. Spectral analysis of the model will approximate the correct results, except for the extremely high-frequency components. The average power that is calculated from the model will give the correct value for the average power of the physical signal that is measured over an appropriate time interval. The total energy of the mathematical model's signal will be infinity because it extends to infinite time, whereas that of the physical signal will be finite. Consequently, this model will not give the correct value for the total energy of the physical signal without using some additional information. However, the model can be used to evaluate the energy of the physical signal over some finite time interval of the physical signal. This mathematical model is said to be a power signal because it has the property of finite power (and infinite energy), whereas the physical waveform is said to be an energy signal because it has finite energy. All physical signals are energy signals, although we generally use power signal mathematical models to simplify the analysis.

In summary, waveforms may often be classified as signals or noise, digital or analog, deterministic or stochastic, physically realizable or nonphysically realizable, and belonging to the power or energy type. Concepts such as spectral expansion of signals as well as their representation will be briefly discussed next.

On many occasions, especially in wireless communications, we have to deal with random signals. A random signal can be viewed or defined in two different ways. One way to view such a signal $s(t)$ is to consider that it is a collection of time functions corresponding to various outcomes of a random experiment. Alternatively, we may view the random signal at t_1, t_2, ... as a collection of random variables $s(t_1)$, $s(t_2)$, ...

A complete statistical description of a random signal $s(t)$ is known if for any integer n, and any choice of t_1, t_2, ..., t_n the joint PDF of $s(t_1)$, $s(t_2)$, ... $s(t_n)$ is given by $f_{s(t_1),s(t_2),...,s(t_N)}^{(s_1,s_2,...,s_n)}$. The mean, or expectation, or ensemble, or statistical average of the random process $\bar{s}(t)$ is a deterministic function of time $\bar{s}(t)$ defined by

$$\bar{s}(t) = \int_{-\infty}^{\infty} s f_{s(t)}(s)\, ds \tag{A.1}$$

The autocorrelation function of the random process $s(t)$ denoted as $R_{ss}(t_1, t_2)$ is defined by

$$R_{ss}(t_1, t_2) = E[s(t_1)s(t_2)] = \int_{-\infty}^{\infty} \int_{-\infty}^{\infty} s_1, s_2\, f_{s(t_1)s(t_2)}^{(s_1,s_2)} ds_1\, ds_2 \tag{A.2}$$

When the mean of a random signal is independent of time, and the autocorrelation function depends only on the difference $\tau = t_1 - t_2$, the random signal is called wide-sense stationary. When the time average and the statistical average of any function of a random process are equal, the random signal is called ergodic, whereas the time average is defined

$$\lim_{T \to \infty} \frac{1}{T} \int_{-T/2}^{T/2} g(s(t, \omega)) \, dt \qquad \text{(A.3)}$$

and $g(s(t, \omega))$ is a realization of the random process $g(s(t))$ [3].

The energy and power of each sample function by extension of deterministic signals are defined as

$$E_i = \int_{-\infty}^{\infty} s^2(t, \omega_i) \, dt \qquad \text{(A.4)}$$

and

$$P_i = \lim_{T \to \infty} \frac{1}{T} \int_{-T/2}^{T/2} s^2(t, \omega_i) \, dt \qquad \text{(A.5)}$$

We observe that the energy E_i and power P_i of a random signal are random variables whose expected values are

$$E = \mathrm{E}\left[\int_{-\infty}^{\infty} s^2(t) \, dt \right] \qquad \text{(A.6)}$$

We observe that from A.2 we obtain

$$E = \int_{-\infty}^{\infty} \mathrm{E}[s^2(t)] \, dt \qquad \text{(A.7)}$$

$$= \int_{-\infty}^{\infty} R_s(t, t) \, dt$$

whereas

$$P = \mathrm{E}\left\{ \lim_{T \to \infty} \frac{1}{T} \int_{-T/2}^{T/2} s^2(t)\, dt \right\} \tag{A.8}$$

$$= \lim_{T \to \infty} \frac{1}{T} \int_{-T/2}^{T/2} R_s(t,\, t)\, dt$$

If the process is stationary, $R_s(t,\, t) = R_s(0)$. Hence,

$$E = \int_{-\infty}^{\infty} R_s(0)\, dt \tag{A.9}$$

$$P = R_s(0) \tag{A.10}$$

A.1.1 Energy and Power Waveform

The waveform $s(t)$ is a power waveform if and only if the normalized average power, P, is finite and nonzero (i.e., $0 < P < \infty$). The total normalized energy is given by

$$E = \lim_{T \to \infty} \int_{-T/2}^{T/2} s^2(t)\, dt \tag{A.11}$$

The waveform $s(t)$ is an energy waveform if and only if the total normalized energy is finite and nonzero (i.e., $0 < E < \infty$).

From these definitions, it is seen that if a waveform is classified as either one of these types, it cannot be of the other type. That is, if $s(t)$ has finite energy, the power averaged over infinite time is zero, and if the power (averaged over infinite time) is finite, the energy is infinite. Moreover, mathematical functions can be found that have both infinite energy and infinite power and, consequently, cannot be classified into either of these two categories. Physically realizable waveforms are of the energy type, but we will often model them by infinite-duration waveforms of the power type. Laboratory instruments that measure average quantities—such as dc value, rms value, and average power—are based on a finite time interval. Thus, nonzero average quantities for finite energy (physical) signals can be obtained.

Hence, the average quantities calculated from a power-type mathematical model (averaged over infinite time) will give the results that are measured in the laboratory (averaged over finite time).

A.2 Orthogonal Series Representation of Signals and Noise

An orthogonal series representation of signals and noise such as the Fourier series, sampling function series, and representation of digital signals, has many significant applications in communication problems. Because these specific cases are so important, for a better understanding of the more advanced material in this book, they will be studied in some detail in the sections that follow.

A.2.1 Orthogonal Functions

Before the orthogonal series is studied, a definition for orthogonal functions is needed.

Functions $\varphi_n(t)$ and $\varphi_m(t)$ are said to be orthogonal with respect to each other over the interval $a < t < b$ if they satisfy the condition

$$\int_a^b \varphi_n(t) \, \varphi_m^*(t) \; dt = \left\{ \begin{array}{l} 0, \; n \neq m \\ K_n, \; n = m \end{array} \right\} = K_n \, \delta_{nm} \qquad (A.12)$$

where

$$\delta_{nm} \triangleq \left\{ \begin{array}{l} 0, \; n \neq m \\ 1, \; n = m \end{array} \right\} \qquad (A.13)$$

δ_{nm} is called the Kronecker delta function. If the constants K_n are all equal to one, the $\varphi_n(t)$ are said to be orthonormal functions.

In other words, (A.12) is used to test pairs of functions to determine if they are orthogonal. They are orthogonal over the interval (a,b) if the integral of their product is zero. The zero result implies that these functions are "independent" or in "disagreement." If the result is not zero, they are not orthogonal, and consequently, the two functions have some "dependence" or "likeness" to each other. In a similar manner, we can show that the set of the complex exponential functions $e^{jn\omega_0 t}$ are orthogonal.

A.2.2 Orthogonal Series

Let us assume that $s(t)$ represents some practical waveform (signal, noise, or signal-noise combination) that we wish to represent over the interval $a < t < b$. Then we can obtain an equivalent orthogonal series representation by taking each old $\varphi_n(t)$ and dividing it by $\sqrt{K_n}$ to form the normalized $\varphi_n(t)$.

A waveform $s(t)$ can be represented over the interval (a, b) by the series

$$s(t) = \sum_n a_n \varphi_n(t) \tag{A.14}$$

where the orthogonal coefficients are given by

$$a_n = \frac{1}{K_n} \int_a^b s(t)\, \varphi_n^*(t)\, dt \tag{A.15}$$

and the range of n is over the integer values that correspond to the subscripts that were used to denote the orthogonal functions in the complete orthogonal set.

For (A.14) to be a valid representation for any physical signal (i.e., one with finite energy), the orthogonal set has to be complete. This implies that the set $\{\varphi_n(t)\}$ can be used to represent any function with an arbitrarily small error. In practice, it is usually difficult to prove that a given set of functions is complete. It can be shown that the complex exponential set and the harmonic sinusoidal sets that are used for the Fourier series are complete [4]. Many other useful sets are also complete, such as the Bessel functions, Legendre polynomials, and the $(\sin x)/x$-type sets, expressions of which we shall see in Section A.2.4

Let us try to prove that the set $\{\varphi_n(t)\}$ is sufficient to represent the waveform. Then in order for (A.14) to be correct, we only need to show that we can evaluate the a_n. Using (A.14), we operate on both sides of this equation with the integral operator

$$\int_a^b [\cdot]\, \varphi_m^*(t)\, dt \tag{A.16}$$

obtaining

$$\int_a^b s(t)\,\varphi_m^*(t)\,dt = \int_a^b \left[\sum_n a_n\,\varphi_n(t) \right] \varphi_m^*(t)\,dt$$

$$= \sum_n a_n \int_a^b \varphi_n(t)\,\varphi_m^*(t)\,dt = \sum_n a_n K_n\,\delta_{nm} \quad \text{(A.17)}$$

$$= a_m K_m$$

Thus (A.15) follows.

The orthogonal series is very useful in representing a signal, noise, or a signal-noise combination. The orthogonal functions $\varphi_j(t)$ are deterministic. Furthermore, if the waveform $s(t)$ is deterministic, the constants $\{a_j\}$ are also deterministic and may be evaluated using (A.15). Moreover, if $s(t)$ is stochastic (e.g., in a noisy environment), the $\{a_j\}$ are a set of random variables that give the desired random process $s(t)$.

A.2.3 Fourier Series

The Fourier series is a particular type of orthogonal series representation that is very useful in solving engineering problems, especially communication problems. The orthogonal functions that are used are either sinusoids, or, equivalently, complex exponential functions.

A.2.3.1 Complex Fourier Series

The complex Fourier series uses the orthogonal exponential functions

$$\varphi_n(t) = e^{jn\omega_0 t} \quad \text{(A.18)}$$

where n ranges over all possible integer values, negative, positive, and zero; $\omega_0 = 2\pi/T_0$, where $T_0 = (b - a)$ is the length of the interval over which the series, (A.14), is valid; and from (A.15) $K_n = T_0$. Using (A.14), the Fourier series theorem follows.

A physical waveform (i.e., finite energy) may be represented over the interval $a < t < a + T_0$ by the complex exponential Fourier series

$$s(t) = \sum_{n=-\infty}^{n=\infty} c_n e^{jn\omega_0 t} \quad \text{(A.19)}$$

where the complex Fourier coefficients c_n are given by

$$c_n = \frac{1}{T_0} \int_{a}^{a+T_0} s(t)\, e^{-jn\omega_0 t}\, dt \qquad \text{(A.20)}$$

and $\omega_0 = 2\pi f_0 = 2\pi/T_0$.

If the waveform $s(t)$ is periodic with period T_0, this Fourier series representation is valid over all time (i.e., over the interval $-\infty < t < +\infty$) because the $\varphi_n(t)$ are periodic functions that have a common fundamental period T_0. For this case of periodic waveforms, the choice of a value for the parameter a is arbitrary, and it is usually taken to be $a = 0$ or $a = -T_0/2$ for mathematical convenience. The frequency $f_0 = 1/T_0$ is said to be the fundamental frequency, and the frequency nf_0 is said to be the nth harmonic frequency, when $n > 1$. The Fourier coefficient c_0 is equivalent to the dc value of the waveform $s(t)$, because the integral is identical to that of (A.20) when $n = 0$.

A.2.3.2 Quadrature Fourier Series

The quadrature form of the Fourier series representing any physical waveform $s(t)$ over the interval $a < t < a + T_0$ is

$$s(t) = \sum_{n=0}^{\infty} a_n \cos n\omega_0 t + \sum_{n=1}^{\infty} b_n \sin n\omega_0 t \qquad \text{(A.21)}$$

where the orthogonal functions are $\cos n\omega_0 t$ and $\sin n\omega_0 t$. Using (A.12), we find that these Fourier coefficients are given by

$$a_n = \begin{cases} \dfrac{1}{T_0} \displaystyle\int_{a}^{a+T_0} s(t)\, dt, & n = 0 \\[2em] \dfrac{2}{T_0} \displaystyle\int_{a}^{a+T_0} s(t)\, \cos n\omega_0 t\, dt, & n \geq 1 \end{cases} \qquad \text{(A.22)}$$

and

$$b_n = \frac{2}{T_0} \int_a^{a+T_0} s(t)\, \sin\, n\omega_0 t\, dt, \quad n > 0 \tag{A.23}$$

Once again, because these sinusoidal orthogonal functions are periodic, this series is periodic with the fundamental period T_0, and if $s(t)$ is periodic with period T_0, the series will represent $s(t)$ over the whole real line (i.e., $-\infty < t < \infty$).

A.2.4 Line Spectrum for Periodic Waveforms

For periodic waveforms, with period T_0, the Fourier series representations are valid over all time (i.e., $-\infty < t < \infty$). Consequently, the (two-sided) spectrum, which depends on the waveshape from $t = -\infty$ to $t = \infty$, may be evaluated in terms of the Fourier coefficients c_n as given by (A.20)

$$S(f) = \sum_{n=-\infty}^{n=\infty} c_n \delta(f - nf_0) \tag{A.24}$$

where $f_0 = 1/T_0$.

Equation (A.24) indicates that a periodic function always has a line (delta function) spectrum with the lines being at $f = nf_0$ and having weights given by the c_n values.

Another form of expansion, which is used many times for the approximation of common signal waveforms, is the so-called Bessel-Fourier expansion. For this expansion as basis functions are used, the Bessel functions defined by

$$J_n(t) = \frac{1}{\pi} \int_0^{\pi} \cos\,(n\theta - t\, \sin\, \theta)\, d\theta \tag{A.25}$$

where these functions are also the solutions of the Bessel equation

$$x^2 J_n''(x) + x J_n'(x) + (x^2 - n^2) J_n(x) = 0 \tag{A.26}$$

where

$$J_n'(x) = \frac{d(J_n(x))}{dx}$$

It can be shown [5] that if we change variables, and set $x = \alpha_{nm} \frac{\rho}{a}$ where α_{nm} is the mth zero of $J_n(\alpha_{nm}) = 0$, a is the upper limit of the new variable ρ, then the Bessel function can be expressed as $J_n\left(\alpha_{nm} \frac{\rho}{a}\right)$. This set of functions is orthogonal in the sense

$$\int_0^a J_n\left(\alpha_{nm} \frac{\rho}{a}\right) J_n\left(\alpha_{nm} \frac{\rho}{a}\right) \rho d\rho = 0 \tag{A.27}$$

and the orthogonality is valid in the interval $[0, a]$. Using this orthogonal set, we can expand any well-behaved but otherwise arbitrary signal waveform as shown next.

$$s(t) = \sum_{M=1}^{\infty} c_{nm} J_n\left(\alpha_{nm} \frac{t}{a}\right) \tag{A.28}$$

where $0 \leq t \leq a$ and $n > -1$, and the coefficients α_{nm} can be determined using

$$c_{nm} = \frac{2}{a^2 J_{n+1}(\alpha_{nm})^2} \int_0^a s(\rho) J_n\left(\alpha_{nm} \frac{\rho}{a}\right) \rho d\rho \tag{A.29}$$

Still another series approximation of any signal waveform is the Laguerre series expansion, using as a basis the Laguerre functions. The Laguerre functions $L_n(x)$ are given by the solutions of the equation

$$xL_n''(x) + (1 - x)L_n'(x) + nL_n(x) = 0 \tag{A.30}$$

where it is shown [5] that

$$\int_0^a e^{-x} L_m(x) L_n(x)\, dx = \delta_{nm} \tag{A.31}$$

Similarly converting the Laguerre function into an orthogonal set, $\varphi_n(x)$, given by

$$\varphi_n(x) = e^{-(x/2)} L_n(x) \tag{A.32}$$

the functions $\varphi_n(x)$ satisfies the differential equation

$$x\varphi_n''(x) + \varphi_n'(x) + \left(n + \frac{1}{2} - \frac{x}{4}\right)\varphi_n(x) = 0 \tag{A.33}$$

We can then expand similarly to the expansion of (A.14) to approximate any well-behaved signal waveforms in the interval $0 \leq t \leq a$.

The main purpose of this appendix is to review some approximating techniques and introduce the notion of spectrum, in order to show the techniques by which interference signals in communications systems can be quantitatively modeled and approximated. The accuracy of such an approximation either in the frequency or time domain will determine the accuracy of the mitigation techniques as seen, from the methodology we adapted, in Chapter 4 and onward.

A.3 Fourier Transform and Spectra

In electrical engineering problems, the signal, the noise, or the combined signal plus noise usually consists of a voltage or current waveform that is a function of time. Let $s(t)$ denote the waveform of interest. Theoretically, to evaluate the frequencies that are present, one needs to view the waveform over all time (i.e., $-\infty < t < \infty$) to be sure that the measurement is accurate and to guarantee that none of the frequency components is neglected. The relative level of one frequency as compared to another is given by the spectrum of the waveform. This is obtained by taking the Fourier transform of the signal waveform.

The Fourier transform (FT) of a waveform $s(t)$ is

$$S(f) = F(s(t)) = \int_{-\infty}^{+\infty} (s(t)) e^{-j2\pi ft} \, dt \tag{A.34}$$

where $F(.)$ denotes the Fourier transform operator and f is the frequency parameter with units of hertz.

This is also called a two-sided spectrum of $s(t)$ because both positive and negative frequency components are obtained from (A.34).

In general, because $e^{-j2\pi ft}$ is complex, $S(f)$ is a complex function of frequency. $S(f)$ may be decomposed into two real functions $X(f)$ and $Y(f)$ such that

$$S(f) = X(f) + jY(f) \tag{A.35}$$

This is identical to writing a complex number in terms of pairs of real numbers that can be plotted in a two-dimensional Cartesian coordinate system. For this reason, (A.35) is sometimes called the quadrature form or Cartesian form. Similarly, (A.34) can be written equivalently in terms of a polar coordinate system, where the pair of real functions denotes the magnitude and phase

$$S(f) = |S(f)| e^{j\vartheta(f)} \tag{A.36}$$

where

$$|S(f)| = \sqrt{X^2(f) + Y^2(f)} \text{ and } \theta(f) = \tan^{-1}\left(\frac{Y(f)}{X(f)}\right) \tag{A.37}$$

This is called the magnitude-phase form or polar form. To determine if certain frequency components are present, one would examine the magnitude spectrum $|S(f)|$, and sometimes engineers loosely call this just the spectrum.

It should be clear that the spectrum of a signal waveform is obtained by a mathematical calculation, and that it does not appear physically in an actual circuit. For example, the frequency $f = 10$ Hz is present in the waveform $s(t)$ if and only if $|S(10)| \neq 0$. From (A.34) it is realized that an exact spectral value can be obtained only if the waveform is observed over the infinite time interval $(-\infty, \infty)$. However, a special instrument called a spectrum analyzer may be used to obtain an approximation (i.e., finite time integral) for the magnitude spectrum $|S(f)|$.

The time waveform may be calculated from the spectrum by using the inverse Fourier transform

$$s(t) = \int_{-\infty}^{\infty} S(f) e^{j2\pi ft} \, df \tag{A.38}$$

The functions $s(t)$ and $S(f)$ are said to constitute a Fourier transform pair, where $s(t)$ is the time-domain description and $S(f)$ is the frequency-domain description. Usually, the time-domain function is denoted by a lowercase letter and the frequency-domain function is denoted by an uppercase letter. Shorthand notation for the pairing between the two domains will be denoted by a double arrow: $s(t) \leftrightarrow S(f)$.

The waveform $s(t)$ is Fourier transformable (i.e., sufficient conditions) if it satisfies both Dirichlet conditions:

- Over any time interval of finite width, the function $s(t)$ is single valued with a finite number of maxima and minima and the number of discontinuities (if any) is finite.

- $s(t)$ is absolutely integrable. That is,

$$\int_{-\infty}^{\infty} |s(t)|\, dt < \infty \tag{A.39}$$

Although these conditions are sufficient, they are not necessary. In fact, signal waveforms may not satisfy the Dirichlet conditions and yet their Fourier transform can be found.

A weaker sufficient condition for the existence of the Fourier transform is

$$E = \int_{-\infty}^{\infty} |s(t)|^2\, dt < \infty \tag{A.40}$$

where E is the normalized energy. This is the finite energy condition that is satisfied by all physically realizable waveforms. Thus all physical waveforms encountered in engineering practice are Fourier transformable.

A.3.1　Sampling Theorem

Finally, if the signal $s(t)$ is bandlimited with bandwidth w (i.e., $S(f) = 0$ for $f \geq w$) and the time representation $s(t)$ of the signal is sampled at sampling intervals T_s where $T_s \leq \dfrac{1}{2w}$ obtaining the samples $x(nT_s)$, it can be shown [6] the signal $s(t)$ can be reconstructed from its samples by the formula

$$s(t) = \sum_{n=-\infty}^{\infty} 2w's(nT_s) \text{ sinc } 2w'(t - nT_s)$$

where w' is any arbitrary number that satisfies $w \le w' \le \dfrac{1}{T_s} - w$.

In the special case when $T_s = \dfrac{1}{2w}$, the reconstruction is given by

$$s(t) = \sum_{n=-\infty}^{\infty} x(nT_s) \text{ sinc}\left(\frac{1}{T_s} - n\right)$$

where sinc $(x) \equiv \dfrac{\sin (x)}{x}$.

This is the famous sampling theorem, which allows us to reconstruct signals from their samples and play a major role in the migration from the analog to the digital world. When $T_s = \dfrac{1}{2w}$, this sampling rate is called Nyquist sampling rate.

A.3.2 Parseval's Theorem and Energy Spectral Density

Parseval's theorem gives an alternative method for evaluating the energy by using the frequency-domain description instead of the time-domain definition.

$$\int_{-\infty}^{\infty} s_1(t)s_2^*(t) \, dt = \int_{-\infty}^{\infty} S_1(f)S_2^*(f) \, df$$

If $s_1(t) = s_2(t) = s(t)$, this reduces to

$$E = \int_{-\infty}^{\infty} |s(t)|^2 \, dt = \int_{-\infty}^{\infty} |S(f)|^2 \, df \tag{A.41}$$

which is also known as Rayleigh's energy theorem.

This last equation leads to the concept of the energy spectral density (ESD) function, which is defined for energy waveforms by

$$E(f) = |S(f)|^2 \tag{A.42}$$

where $s(t) \leftrightarrow S(f)$. $E(f)$ has units of joules per hertz.

By using Parseval's theorem, we see that the total normalized energy is given by the area under the ESD function:

$$E = \int_{-\infty}^{\infty} E(f)\, df \tag{A.43}$$

For power waveforms, a similar function called the PSD can be defined. It is further analyzed in the next subsection and plays a central role in interference suppression problems.

A.3.3 PSD

The normalized power of a waveform can be related to its frequency-domain description by the use of a function known as the PSD. The PSD is very useful in describing how the power content of signals and noise is affected by filters and other devices in communication systems. In (A.42), the ESD was defined in terms of the magnitude squared version of the Fourier transform of the waveform. The PSD will be defined in a similar way. The PSD is more useful than the ESD because power-type models are generally used in solving communication problems.

First, define the truncated version of the waveform by

$$s_T(t) = \left\{ \begin{array}{ll} s(t) & -T/2 < t < T/2 \\ 0, & t \text{ elsewhere} \end{array} \right\} = s(t)\, \Pi\left(\frac{t}{T}\right)$$

Using (A.5), we obtain the average normalized power

$$P = \lim_{T \to \infty} \frac{1}{T} \int_{-T/2}^{T/2} s^2(t)\, dt = \lim_{T \to \infty} \frac{1}{T} \int_{-\infty}^{\infty} s_T^2(t)\, dt \tag{A.44}$$

By the use of Parseval's theorem, (A.41), this becomes

$$P = \lim_{T \to \infty} \frac{1}{T} \int_{-\infty}^{\infty} |S_T(f)|^2 \, df = \int_{-\infty}^{\infty} \left(\lim_{T \to \infty} \frac{|S_T(f)|^2}{T} \right) df \quad \text{(A.45)}$$

where $S_T(f) = F(s_T(t))$. The integrand of the right-hand integral has units of watts/Hertz (or, equivalently, volts^2/hertz or amperes^2/hertz, as appropriate) and can he defined as the PSD.

The PSD for a deterministic power waveform is

$$P_w(f) = \lim_{T \to \infty} \left(\frac{|S_T(f)|^2}{T} \right) \quad \text{(A.46)}$$

where $s_T(t) \leftrightarrow S_T(f)$ and $P_w(f)$ has units of watts per hertz.

Note that the PSD is always a real nonnegative function of frequency. In addition, the PSD is not sensitive to the phase spectrum of $s(t)$ because that is lost by the absolute value operation used in (A.46). From (A.45), the normalized average power is

$$P = \langle s^2(t) \rangle = \int_{-\infty}^{\infty} P_w(f) \, df$$

That is, the area under the PSD function is the normalized average power.

References

[1] Haykin, Simon, *Communications Systems*, New York: John Wiley, 1978.

[2] Ziemer, R. E., and W. H. Tranter, *Principles of Communications*, Boston, MA: Houghton, Mifflin Company, 1976.

[3] Papoulis, Athanasios, *Probability, Random Variables, and Stochastic Processes*, Third Edition, New York: McGraw-Hill, 1991.

[4] Courant, R., and D. Hilbert, *Methods of Mathematical Physics*, New York: Wiley (Interscience), 1953.

[5] Arfken, G., *Mathematical Methods for Physics*, New York: Academic Press, 1985.

[6] Proakis, J. G., and Masoud, Salehi, *Communications Systems Engineering*, Englewood Cliffs, NJ: Prentice Hall, 1994.

Appendix B:
HMMs—Kalman Filter

B.1 HMMs

HMMs are used for the statistical modeling of nonstationary signal processes, such as speech signals and image sequences, as shown in Figure B.1 [1–4].

An HMM models the time variations (and/or the space variations) of the statistics of a random process with a Markovian chain of state-dependent stationary subprocesses. An HMM is essentially a Bayesian finite state process, and consists of a Markovian prior for modeling the transitions between the states and a set of state PDFs for modeling the random variations of the signal process within each state.

A discrete-time Markov process $x(m)$ with N allowable states may be modeled by a Markov chain of N states. Each state can be associated with one of the N values that $s(m)$ may assume. In a Markov chain, the Markovian property is modeled by a set of state transition probabilities defined as

$$d_{ij}(m, m - 1) = \text{Prob}\,[x(m) = j \,|\, x(m - 1) = i] \qquad \text{(B.1)}$$

where $d_{ij}(m, m - 1)$ is the probability that at time $m - 1$ the process is in the state i and then at time m it moves to state j. In (B.1), the transition probability is expressed in a general time-dependent form.

An HMM is a double-layered finite-state process, with a hidden Markovian process that controls the selection of the states of an observable process.

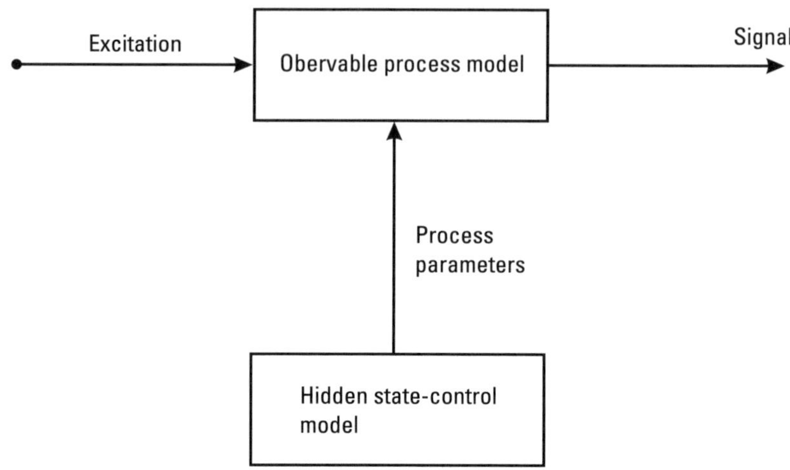

Figure B.1 Two-layered model of a nonstationary process.

In general, an HMM has N states, with each state trained to model a distinct segment of a signal process. An HMM can be used to model a time-carrying random process as a probabilistic Markovian chain of N stationary, or quasi-stationary, elementary subprocesses. A general structure for of a three-state HMM is shown in Figure B.2.

This structure is known as an ergodic HMM. In the context of an HMM, the term *ergodic* implies that there are no structural constraints for connecting any state to any other state.

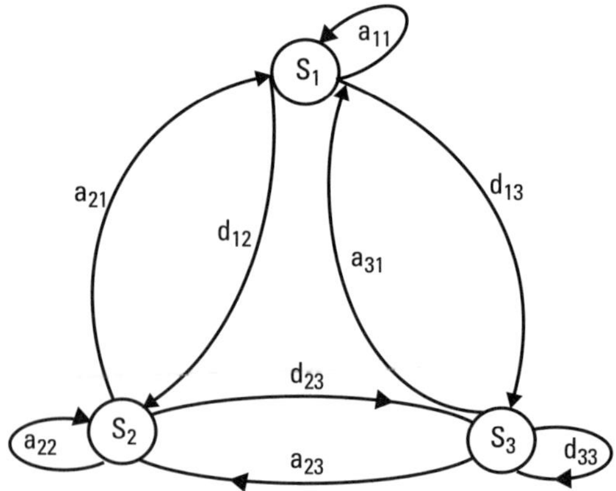

Figure B.2 Three-state HMM.

B.2 Parameters of an HMM

An HMM has the following parameters:

1. *Number of states N.* This is usually set to the total number of distinct, or elementary, stochastic events in a signal process.

2. *State transition-probability matrix $A = \{a_{ij}, \, i, j = 1, \ldots, N\}$.* This provides a Markovian connection network between the states and models the variations in the duration of the signals associated with each state.

3. *State observation vectors $\{\mu_{i1}, \mu_2, \ldots, \mu_{iM}, \, i = 1, \ldots, N\}$.* For each state, a set of M prototype vectors model the centroids of the signal space associated with each state.

4. *State observation vector probability model.* This can be either a discrete model composed of the M prototype vectors and their associated probability mass function (pmf) $P = \{P_{ij}(\cdot), \, i = 1, \ldots, N, j = 1, \ldots, M\}$, or it may be a continuous (usually Gaussian) PDF model $F = \{f_{ij}(\cdot), \, i = 1, \ldots, N, j = 1, \ldots, M\}$.

5. *Initial state probability vector $\pi = [\pi_1, \pi_2, \ldots, \pi_N]$.*

The first step in training the parameters of an HMM is to collect a training database of a sufficiently large number of different examples of the random process to be modeled. The objective is to train the parameters of an HMM to model the statistics of the signals in the training data set.

B.3 HMM—Kalman Filter Algorithm

In this section we shall show how a recursive HMM estimator and a Kalman filter in conjunction with an EM algorithm can be used to estimate narrowband interference, signal, and their parameters. This algorithm is used in Chapter 7 as a narrowband interference suppressor [5, 6].

B.3.1 Problem Formulation

We assume that the received spread-spectrum signal $s(t)$ is sampled at a rate higher than the chip rate of the PN sequence. This yields samples that are correlated in time. Hence, we assume s_k is a finite-state discrete-time homogeneous first order Markov chain. Consequently, the state s_k at time k is one of the finite number of known states $M_q = (q_1, q_2, \ldots, q_M)$.

The transition probability matrix is:

$$A = (a_{mn}) \tag{B.2}$$

where

$$a_{mn} = P\big(s_{t+1} = q_n | s_t = q_m\big);$$

$$m, n \in \{1, \ldots, M\}.$$

Of course, $a_{mn} \geq 0$, $\displaystyle\sum_{n=1}^{N} a_{mn} = 1$ for each m, with π denoting the initial state probability vector: $\pi = (\pi_m)$, $\pi_m = P(s_1 = q_m)$. We assume that the number of states M of the Markov chain is known. Also, for convenience, we assume that $\pi_m = 1/M$, for $m = 1, \ldots, M$.

B.4 Maximum A Posteriori Channel Estimates Based on HMMs

$$\ln f\big(h | y(0), \ldots, y(N-1)\big) = -\sum_{m=0}^{N-1} \ln f(y(m)) - NP \ln (2\pi)$$

$$-\frac{1}{2} \ln \big(|\Sigma_{xx}| \, |\Sigma_{hh}|\big) \tag{B.3}$$

$$-\sum_{m=0}^{N-1} \frac{1}{2} \Big\{ [y(m) - h - \mu_x]^T \Sigma_{xx}^{-1} [y(m) - h - \mu_x]$$

$$+ (h - \mu_h)^T \Sigma_{hh}^{-1} (h - \mu_h) \Big\}$$

The maximum a posteriori (MAP) channel estimate, obtained by setting the derivative of the log posterior function $\ln f_{H|y}\big(h|y\big)$ to zero, is

$$\hat{h}^{MAP} = (\Sigma_{xx} + \Sigma_{hh})^{-1} \Sigma_{hh} (\bar{y} - \mu_x) + (\Sigma_{xx} + \Sigma_{hh})^{-1} \Sigma_{xx} \mu_h \tag{B.4}$$

where

$$\bar{y} = \frac{1}{N} \sum_{m=0}^{N-1} y(m) \tag{B.5}$$

is the time-averaged estimate of the mean of observation vector. Note that for a Gaussian process, the MAP and conditional mean estimate are identical.

The conditional PDF of a channel **h** averaged over all HMMs can be expressed as

$$f_{H|Y}(h|Y) = \sum_{i=1}^{V} \sum_{s} f_{H|Y,S,M}(h|Y, s, M_i) P_{S|M}(s|M_i) P_M(M_i) \tag{B.6}$$

where $P_M(M_i)$ is the prior pmf of the input words. Given a sequence of N P-dimensional observation vectors $Y = [y(0), \ldots, y(N-1)]$, the posterior pdf of the channel **h** along a state sequence s of an HMM M_i is defined as [2].

It can be shown that [1] the MAP estimate along state s, on the left-hand side of (B.3), can be obtained as

$$\hat{h}^{MAP}(Y, s, M_i) = \sum_{m=0}^{N-1} \left[\sum_{k=0}^{N-1} \left(\Sigma_{xx,s(k)}^{-1} + \Sigma_{hh}^{-1} \right) \right]^{-1} \Sigma_{xx,s(m)}^{-1} [y(m) - \mu_{x,s(m)}]$$

$$+ \left[\sum_{k=0}^{N-1} \left(\Sigma_{xx,s(k)}^{-1} + \Sigma_{hh}^{-1} \right)^{-1} \right] \Sigma_{hh}^{-1} \mu_h \tag{B.7}$$

The MAP estimate of the channel over all state sequences of all HMMs can be obtained as

$$\hat{h}(Y) = \sum_{i=1}^{V} \sum_{S} \hat{h}^{MAP}(Y, s, M_i) P_{S|M}(s|M_i) P_M(M_i) \tag{B.8}$$

A MAP differs from maximum likelihood in that MAP includes the prior PDF of a channel. This pdf can be used to confine the channel estimate within a desired subspace of the parameter space. Assuming that the channel input vectors are statistically independent, the posterior PDF of the channel given the observation sequence $Y = y(0) \ldots y(N-1)$ is

$$f_{H|Y}(h \,/\, y(0) \ldots y(N-1)) = \prod_{m=0}^{N-1} \frac{1}{f_y(y(m))} \, f_x(y(m) - h) f_H(h)$$

Assuming also that the channel input $x(m)$ is Gaussian, $f_x(x(m)) = N(x, \mu_x, \Sigma_{xx})$, with mean vector μ_x and covariance matrix Σ_{xx}, and that the channel $\mathbf{h}$ is also Gaussian, $f_N(h) = N(h, \mu_h, \Sigma_{hh})$, with mean vector μ_h and covariance matrix Σ_{hh}, the logarithm of the posterior PDF is given by (B.3).

B.4.1 Notation

Let $D = (d_1, \ldots, d_p)'$. Denote the sequence of observations $(y_1, \ldots, y_T)$ as Y_T. Let $Y_{k_1}^{k_2} = (y_{k_1}, \ldots, y_{k_2})'$. Let $X_T = (i_1, \ldots, i_T)'$ and $S_T = (s_1, \ldots, s_T)'$. Let $S_{k_1}^{k_2} = (s_{k_1}, \ldots, s_{k_2})'$ and $X_{k_1}^{k_2} = (x_{k_1}, \ldots, x_{k_2})'$.

B.4.2 Estimation Objectives

Let $\phi_0 = \left(A, q, D, \sigma_e^2, \sigma_n^2\right)$ denote the true parameter vector that characterizes the narrowband interference—auto regressive (AR) signal—and the spread-spectrum signal (Markov chain).

Given the observations $Y_k = (y_1, \ldots, y_k)$, our aim is twofold.

1. *State estimation.* Compute estimates of the narrowband interference i_k and the spread-spectrum signal s_k.
2. *Parameter estimation.* Derive a recursive estimator $\phi^{(k)}$ for ϕ_0, where $\phi^{(k)} = \left(A^{(k)}, q^{(k)}, D^{(k)}, \sigma_e^{(k)}, \sigma_n^{(k)}\right)$, for $k > 1$, given the observations Y_k.

For maximum generality, the HMM-KF algorithm we present allows for estimation of some or all of the parameters of $\phi^{(k)}$, depending on which parameters are known a priori. In the CDMA signal models q, σ_e, and σ_n are assumed known. For such models, the HMM-KF algorithm provides state estimates and parameter estimates of D.

The HMM-KF algorithm cross couples two recursive EM algorithms, one algorithm for an HMM and the other for a noisy AR model [5, 6].

1. At time k, the KF and recursive estimate maximize (EM) parameter estimator for the narrowband interference yield estimates of the

state of i_k, process noise variance σ_e^2, observation noise variance σ_n^2, and the AR coefficients $d_1, \ldots, d_p$, with p the order of the autoregressive process (narrowband interference).

2. The HMM filter and recursive EM parameter estimator for the spread-spectrum signal gives online estimates of the state of s_k, transition probability matrix A and Markov chain level q.

B.4.3 Spread-Spectrum Signal Estimator Using Recursive HMMs

At time k, the predicted narrowband interference $i_{k|k-1}$ and variance $p_{i_{k|k-1}}$ of the predicted error $w_k \equiv i_k - i_{k|k-1}$ obtained from the KF is available. Therefore, the HMM to be estimated is (HMM signal model)

$$y_k - i_{k|k-1} = s_k + w_k + n_k \tag{B.9}$$

It is assumed that the Kalman predicted error w_k is modeled as a zero-mean white Gaussian process with variance $p_{i_{k|k-1}}$ and is independent of the observation noise n_k.

The recursive HMM estimator recursively updates the state and parameter estimates of the HMM. The recursive HMM parameter vector estimate at k is denoted as

$$\phi_{HMM}^{(k)} \equiv \left(A^{(k)}, q^{(k)}, \sigma_n^{2(k)} \right) \tag{B.10}$$

Given the signal model (B.3), the state and adaptive parameter estimation procedure for the spread-spectrum signal s_k, is presented next.

B.4.3.1 State Estimation
Define the symbol PDF

$$b_m(y_k) \equiv f\left(y_k \mid Y_{k-1}, I_{k|k-1}, s_k = q_m^{(k-1)}, S_{k-1}, \phi_{HMM}^{(k-1)} \right), \tag{B.11}$$

$$m \in \{1, \ldots, M\}$$

$$= \frac{1}{\sqrt{2\pi\left(p_{i_{k|k-1}} + \sigma_n^{2(k-1)}\right)}} \times \exp\left(-\frac{\left(y_k - i_{k|k-1} - q_m^{(k-1)}\right)^2}{2\left(p_{i_{k|k-1}} + \sigma_n^{2(k-1)}\right)} \right)$$

which is obtained directly from (B.2), and the assumptions on the noises w_k, $n_k \cdot q_m^{(k-1)}$, and $\sigma_n^{2(k-1)}$ are the estimates at time $k-1$ of the mth Markov chain level and the observation noise variance, respectively.

Define the nonnormalized filtered Markov state density $\alpha_k(m)$, the filtered conditional mean (CM) state estimate $s_{k|k}^{CM}$, and the filtered MAP state estimate $s_{k|k}^{MAP}$, respectively, as

$$\alpha_k(m) \equiv f\left(s_k = q_m^{(k-1)}, \, Y_k \,|\, I_{k|k-1}, \, \phi_{HMM}^{(k-1)}\right) \tag{B.12}$$

$$s_{k|k}^{CM} \equiv E\left\{s_k \,|\, Y_k, \, I_{k|k-1}, \, \phi_{HMM}^{(k-1)}\right\} \tag{B.13}$$

$$s_{k|k}^{MAP} = q_j^{(k-1)} \tag{B.14}$$

Remark: $s_{k|k}^{MAP}$ is discrete valued, $s_{k|k}^{CM}$ is continuous.

The nonnormalized filtered Markov state density $\alpha_k(m)$ is recursively computed as follows:

$$\alpha_k(n) = b_n(y_k) \sum_{m=1}^{M} \alpha_{mn}^{(k-1)} \, \alpha_{k-1}(m) \tag{B.15}$$

$$\alpha_1(n) = \pi_n^0 b_n(y_1) \tag{B.16}$$

The normalized filtered Markov state density $\gamma_k(m)$ is computed from $\alpha_k(m)$ and given by

$$\gamma_k(m) \equiv f\left(s_k = q_m^{(k-1)} \,|\, Y_k, \, I_{k|k-1}, \, \phi_{HMM}^{(k-1)}\right) = \frac{\alpha_m(m)}{\sum\limits_{n=1}^{N} \alpha_k(n)} \tag{B.17}$$

The filtered CM state estimate $s_{k|k}^{CM}$ and the associated conditional variance $(CV)\, p_{s_{k|k}} \equiv E\left\{\left(s_k - s_{k|k}^{CM}\right)^2 \,|\, Y_k, \, I_{k|k-1}, \, \phi_{HMM}^{(k-1)}\right\}$, which is the expected error in the estimate of $s_{k|k}^{CM}$, are given by

$$s_{k|k}^{CM} = \sum_{m=1}^{M} q_m^{(k-1)} \, \gamma_k(m) \tag{B.18}$$

$$p_{s_{k|k}} = \sum_{m=1}^{M} \left(q_m^{(k-1)}\right)^2 \gamma_k(m) - \left(s_{k|k}^{CM}\right)^2 \tag{B.19}$$

B.4.3.2 Parameter Estimation

The received power levels are time varying, and if asynchronous transmission is used it may be necessary to estimate A and q. To estimate these parameters, the recursive EM algorithm is used. The recursive EM algorithm will be summarized next.

At time k the parameter vector estimate is updated as

$$\phi^{(k)} = \phi^{(k-1)} + (I_{com}(\phi^{(k-1)}))^{-1} S(\phi^{(k-1)}) \qquad \text{(B.20)}$$

where $I_{com}(\phi^{(k-1)})$ and $S(\phi^{(k-1)})$ are the Fisher information matrix (FIM) of the complete data and the incremental score vector at time k, respectively, given by

$$I_{com}(\phi^{(k-1)}) = I_{com}(\phi^{(k-2)}) + V(\phi^{(k-1)}) \qquad \text{(B.21)}$$

$$S(\phi^{(k-1)}) \equiv \left. \frac{\partial L_k(\phi)}{\partial \phi} \right|_{\phi = \phi^{(k-1)}} \qquad \text{(B.22)}$$

where

$$V(\phi^{(k-1)}) \equiv \left. \frac{\partial^2 L_k(\phi)}{\partial \phi^2} \right|_{\phi = \phi^{(k-1)}} \qquad \text{(B.23)}$$

$$L_k(\phi) \equiv E\left\{ \ln f\left(Z_k \mid Z_{k-1}, \phi\right) \mid Z_{k,obs}, \phi^{(k-1)} \right\} \qquad \text{(B.24)}$$

where $Z_k \equiv \left(Z_{k,obs}, Z_{k,mis}\right)$ denotes the complete data and $Z_{k,obs}$ and $Z_{k,mis}$ are the observed and missing data, respectively.

In this case, $Z_{k,obs} = Y_k$ and $Z_{k,mis} = S_k$. Thus, given $I_{k|k-1}$, which are obtained from the recursive KF, we can determine $L_k^{HMM}(\phi)$ from (B.17)

$$L_k^{HMM}(\phi) = E\left\{ \ln f\left(y_k, s_k \mid Y_{k-1}, S_{k-1}, \phi\right) \mid Y_k, \phi_{HMM}^{(k-1)} \right\}$$

$$= -\frac{1}{2} \ln\left(2\pi\left(p_{i_{k|k-1}} + \sigma_n^2\right)\right) - \sum_{m=1}^{M} \gamma_k(m) \times \frac{\left(y_k - i_{k|k-1} - q_m\right)^2}{2\left(p_{i_{k|k-1}} + \sigma_n^2\right)}$$

$$\times \sum_{m=1}^{M} \zeta_k(m, n) \sum_{n=1}^{M} \ln \alpha_{mn} \qquad \text{(B.25)}$$

where $\zeta_k(m, n) \equiv f\left(s_{k-1} = q_m^{(k-1)}, s_k = q_n^{(k-1)} \mid Y_k, I_{k|k-1}, \phi_{HMM}^{(k-1)}\right)$ denotes the normalized filtered joint probability that the Markov chain is in state q_m at $k-1$ time and in state q_n at time k. It is shown in [7] that

$$\zeta_k(m, n) = \frac{b_n(y_k)\, \alpha_{mn}^{(k-1)}\, \alpha_{k-1}(m)}{\displaystyle\sum_{m=1}^{M}\sum_{n=1}^{M} b_n(y_k)\, \alpha_{mn}^{(k-1)}\, \alpha_{k-1}(m)}, \quad m, n \in \{1, \ldots, M\} \tag{B.26}$$

Ignoring the terms $\partial^2 L_k^{HMM}(\phi)/\partial\sigma_n^2 \partial q_m$ for all $m = 1, \ldots, M$, the reestimation equations for $\phi_{HMM}^{(k)}$ are decoupled, then the evaluation of $I_{com}\left(\phi_{HMM}^{(k-1)}\right)$ and $S\left(\phi_{HMM}^{(k-1)}\right)$ in (B.13) yields

$$I_{com}\left(\phi_{HMM}^{(k-1)}\right) = \text{blockdiag}\left(I_{A^{(k-1)}},\, I_{q^{(k-1)}},\, I_{\sigma_n^{2^{(k-1)}}}\right) \tag{B.27}$$

$$S\left(\phi_{HMM}^{(k-1)}\right) = \left(S'_{A^{(k-1)}},\, S'_{q_e^{(k-1)}},\, S'_{\sigma_n^{2^{(k-1)}}}\right) \tag{B.28}$$

Thus, $\phi_{HMM}^{(k)}$ is updated as follows.

B.4.4 Transition Probabilities

The update equation for $\alpha_{mn}^{(k)}$ is somewhat complicated by the two constraints $\alpha_{mn}^{(k)} \geq 0$ and $\displaystyle\sum_{n=1}^{M} \alpha_{mn}^{(k)} = 1$. An elegant way of ensuring both constraints are met is to use the following differential geometric approach.

Let $\alpha_{mn}^{(k)} = (g_{mn}^{(k)})^2$

Then $g_{mn}^{(k)}$ has merely the equality constraint that $\displaystyle\sum_{m=1}^{N} (g_{mn}^{(k)})^2 = 1$. Then computing I_A and S_A by projecting the derivatives to the tangent space yields

$$g_{mn}^{(k)} = g_{mn}^{(k-1)} + I_{g_{mn}^{(k-1)}}^{-1} S_{g_{mn}^{(k-1)}} \tag{B.29}$$

$$\alpha_{mn}^{(k)} = \frac{\left(g_{mn}^{(k)}\right)^2}{\displaystyle\sum_{m=1}^{M}\sum_{n=1}^{M} \left(g_{mn}^{(k)}\right)^2} \tag{B.30}$$

where

$$S_{g_{mn}^{(k-1)}} = 2\left(\frac{\zeta_k(m, n)}{g_{mn}^{(k-1)}} - \gamma_{k-1}(m)g_{mn}^{(k-1)}\right) \tag{B.31}$$

$$I_{g_{mn}^{(k-1)}} = \rho I_{g_{mn}^{(k-2)}} + 2\left(\frac{\zeta_k(m, n)}{\left(g_{mn}^{(k-1)}\right)^2} + \gamma_{k-1}(m)\right) \tag{B.32}$$

B.4.5 Levels of the Markov Chain

The update equation for $q_m^{(k)}$ for $m \in \{1, \ldots, M\}$ is given by

$$q_m^{(k)} = q_m^{(k-1)} + I_{q_m^{(k-1)}}^{-1} S_{q_m^{(k-1)}} \tag{B.33}$$

where

$$S_{q_m^{(k-1)}} = \frac{\left(y_k - i_{k|k-1} - q_m^{(k-1)}\right)\gamma_k(m)}{\sigma_n^{2(k-1)} + p_{i_{k|k-1}}} \tag{B.34}$$

$$I_{q_m^{(k-1)}} = \rho I_{q_m^{(k-2)}} + \frac{\gamma_k(m)}{\sigma_n^{2(k-1)} + p_{i_{k|k-1}}} \tag{B.35}$$

B.4.6 Observation Noise

The update equation for $\sigma_n^{2(k)}$ is given by

$$\sigma_n^{2(k)} = \sigma_n^{2(k-1)} + I_{\sigma_n^{2(k-1)}}^{-1} S_{\sigma_n^{2(k-1)}} \tag{B.36}$$

where

$$S_{\sigma_n^{2(k-1)}} = \frac{\sum_{m=1}^{M}\left(y_k - i_{k|k-1} - q_m^{(k-1)}\right)^2 \gamma_k(m)}{2\left(\sigma_n^{2(k-1)} + p_{i_{k|k-1}}\right)^2} - \frac{1}{2\left(\sigma_n^{2(k-1)} + p_{i_{k|k-1}}\right)} \tag{B.37}$$

$$I_{\sigma_n^{2(k-1)}} = \rho I_{\sigma_n^{2(k-2)}} + \frac{\displaystyle\sum_{m=1}^{M} \left(y_k - i_{k|k-1} - q_m^{(k-1)}\right)^2 \gamma_k(m)}{\left(\sigma_n^{2(k-1)} + p_{i_{k|k-1}}\right)^3}$$

$$- \frac{1}{2\left(\sigma_n^{2(k-1)} + p_{i_{k|k-1}}\right)^2} \tag{B.38}$$

With no forgetting factor ($\rho = 1$), and if we ignore the error in $i_{k|k-1}$ (i.e., $p_{i_{k|k-1}} = 0$ for all k), then update equation for the observation noise is given by

$$\sigma_n^{2(k)} = \sigma_n^{2(k-1)} + \frac{1}{k}\left(\sum_{m=1}^{M} \left(y_k - i_{k|k-1} - q_m^{(k-1)}\right)^2 \gamma_k(m) - \sigma_n^{2(k-1)}\right) \tag{B.39}$$

Conditional mean estimates of x_k are given by KF [4]:

$$\dot{x}_{k|k-1} = F^{(k-1)} x_{k-1|k-1} \tag{B.40}$$

$$P_{k|k-1} = F^{(k-1)} P_{k-1|k-1} F^{(k-1)'} + G\sigma_e^{2(k-1)} G' \tag{B.41}$$

$$u_{k|k-1} = Hx_{k|k-1} \tag{B.42}$$

$$b_k = HP_{k|k-1}H' + p_{s_{k|k}} + \sigma_n^{2(k-1)} \tag{B.43}$$

$$x_{k|k} = x_{k|k-1} + P_{k|k-1}H'(b_k)^{-1}(u_k - u_{k|k-1}) \tag{B.44}$$

$$x_{k|k} = x_{k|k-1} + P_{k|k-1}H'(b_k)^{-1}(u_k - u_{k|k-1}) \tag{B.45}$$

$$P_{k|k} = P_{k|k-1} - P_{k|k-1}H'(b_k)^{-1}HP_{k|k-1} \tag{B.46}$$

where $F^{(k)}$ is the estimate of F in $G = (1 \; 0_{1\times p})'$, $H = (1 \; 0_{1\times p})$ at the kth time instant and

$$x_{k|k-1} = E\left\{x_k \mid Y_{k-1}, S_{k-1|k-1}, \phi_{KF}^{(k-1)}\right\} \tag{B.47}$$

$$x_{k-1|k-1} = E\left\{x_{k-1} \mid Y_{k-1}, S_{k-1|k-1}, \phi_{KF}^{(k-2)}\right\} \tag{B.48}$$

$$P_{k|k-1} = E\left\{\left(x_k - x_{k|k-1}\right)\left(x_k - x_{k|k-1}\right)^T \mid Y_{k-1}, S_{k-1|k-1}, \phi_{KF}^{(k-1)}\right\} \tag{B.49}$$

$$x_{k|k} = E\left\{x_k \mid Y_1, S_{k|k}, \phi_{KF}^{(k-1)}\right\} \tag{B.50}$$

$$P_{k|k} = E\left\{\left(x_k - x_{k|k}\right)\left(x_k - x_{k|k}\right)^T \mid Y_k, S_{k|k}, \phi_{KF}^{(k-1)}\right\} \tag{B.51}$$

The estimate of the narrowband interference $i_{k|k}$ is given by the first element of the vector $x_{k|k}$, while the error covariance $p_{i_{k|k}}$ is given by the element $(1,1)$ of $P_{k|k}$. The parameter estimation procedure given the following subsection requires the evaluation of quantities such as

$$\overline{i_{k-m}\,i_{k-n}}^{(k-1)} \equiv E\left\{i_{k-m}\,i_{k-n} \mid Y_k, S_{k|k}, \phi_{KF}^{(k-1)}\right\} \tag{B.52}$$

References

[1] Vaseghi, Saeed V., *Advanced Digital Signal Processing and Noise Reduction,* Second Edition, New York: John Wiley, 2000.

[2] Rabiner, L. R., and B. H. Juang, "An Introduction of Hidden Markov Models," *IEEE ASSP Magazine,* 1986.

[3] Young, S. J., "HTK: Hidden Markov Model Tool Kit," Cambridge University Engineering Department, 1999.

[4] Einstein, A., "Investigation on the Theory of the Brownian Motion," NY: Dover, 1956.

[5] Chui, C. K., and G. Chen, *Kalman Filtering,* Third Edition, Berlin: Springer, 1999.

[6] Krishnamurthy, V., and A. Logothetis, "Adaptive Nonlinear Filters for Narrowband Interference Suppression in Spread Spectrum CDMA Systems," *IEEE Transactions on Comm.,* Vol. 47, 1999.

[7] Krishnamurthy, V., and J. B. Moore, "Online Estimation of Hidden Markov Parameters Based on the Kullback-Leibler Information Measure," *IEEE Trans. Signal Processing,* Vol. 41, Aug. 1993, pp. 2557–2573.

About the Author

Peter Stavroulakis received his B.S. and Ph.D. from New York University in 1969 and 1973, respectively, and his M.S. from the California Institute of Technology in 1970. He joined Bell Laboratories in 1973 and remained there until 1979, when he joined Oakland University in Rochester, Michigan, as an associate professor of engineering. He worked at Oakland University until 1981, when he joined AT&T International and, subsequently, NYNEX International. In 1990, he joined the Technical University of Crete (TUC), Greece, as a full professor of electrical engineering. His work at Bell Labs and Oakland University resulted in the publication of an IEEE (reprinted) book, *Interference Analysis of Communication Systems,* and the publication of a number of papers in the general area of telecom systems. His book on interference analysis is still referenced in textbooks and relevant international technical journals. He is also the author of four other books—two in distributed parameter systems theory, published by Hutchinson and Ross; one in wireless local loops, published by John Wiley in 2001; and one in third generation mobile telecommunications systems, published by Springer in 2001. He has also served as a guest editor for three special journal issues—one for the *Journal of Franklin Institute on Sensitivity Analysis* and the other two for the *International Journal of Communication Systems on Wireless Local Loops* and the *International Journal of Satellite Systems on Interference Suppression Techniques.*

While at AT&T and NYNEX, Professor Stavroulakis worked as a technical director with the responsibility of leading a team that dealt with technoeconomic studies on various large national and international telephone systems and data networks. When he joined TUC, he led the team for the development of the Technology Park of Chania, Crete, and has had various administrative duties besides his teaching and research responsibilities. Professor Stavroulakis is the founder of the Telecommunication Systems Institute of Crete, a research center for the training of Ph.D. students in telecommunications, associated with and in close collaboration with various research centers and universities in Europe and the United States. He now has a very large research team, the work of which is funded by various public and private sources, including the European Union. He is a member of the editorial board of the *International Journal of Communication Systems* and has been a reviewer for many technical international journals. He has organized more than eight international conferences in the field of communication systems. His current research interests are focused on the application of various heuristic methods on telecommunications, including neural networks, fuzzy systems, and genetic algorithms and also in the development of new modulation techniques applicable to mobile and wireless systems.

Professor Stavroulakis is a member of many technical societies and presently is a senior member of IEEE.

Index

GSM System Engineering, Asha Mehrotra

Handbook of Land-Mobile Radio System Coverage, Garry C. Hess

Handbook of Mobile Radio Networks, Sami Tabbane

High-Speed Wireless ATM and LANs, Benny Bing

Interference Analysis and Reduction for Wireless Systems,
   Peter Stavroulakis

Introduction to 3G Mobile Communications, Juha Korhonen

Introduction to GPS: The Global Positioning System,
   Ahmed El-Rabbany

An Introduction to GSM, Siegmund M. Redl, Matthias K. Weber,
   and Malcolm W. Oliphant

Introduction to Mobile Communications Engineering,
   José M. Hernando and F. Pérez-Fontán

*Introduction to Radio Propagation for Fixed and Mobile
   Communications,* John Doble

*Introduction to Wireless Local Loop, Second Edition:
   Broadband and Narrowband Systems,* William Webb

IS-136 TDMA Technology, Economics, and Services,
   Lawrence Harte, Adrian Smith, and Charles A. Jacobs

Mobile Data Communications Systems, Peter Wong and
   David Britland

Mobile IP Technology for M-Business, Mark Norris

Mobile Satellite Communications, Shingo Ohmori,
   Hiromitsu Wakana, and Seiichiro Kawase

*Mobile Telecommunications Standards: GSM, UMTS, TETRA, and
   ERMES,* Rudi Bekkers

*Mobile Telecommunications: Standards, Regulation, and
   Applications,* Rudi Bekkers and Jan Smits

Multiantenna Digital Radio Transmission,
   Massimiliano "Max" Martone

Understanding GPS: Principles and Applications,
  Elliott D. Kaplan, editor

Understanding WAP: Wireless Applications, Devices, and Services,
  Marcel van der Heijden and Marcus Taylor, editors

Universal Wireless Personal Communications, Ramjee Prasad

WCDMA: Towards IP Mobility and Mobile Internet, Tero Ojanperä
  and Ramjee Prasad, editors

*Wireless Communications in Developing Countries: Cellular and
  Satellite Systems,* Rachael E. Schwartz

Wireless Intelligent Networking, Gerry Christensen,
  Paul G. Florack, and Robert Duncan

Wireless LAN Standards and Applications, Asunción Santamaría
  and Francisco J. López-Hernández, editors

Wireless Technician's Handbook, Andrew Miceli

For further information on these and other Artech House titles,
including previously considered out-of-print books now available
through our In-Print-Forever® (IPF®) program, contact:

Artech House   Artech House
685 Canton Street   46 Gillingham Street
Norwood, MA 02062   London SW1V 1AH UK
Phone: 781-769-9750   Phone: +44 (0)20 7596-8750
Fax: 781-769-6334   Fax: +44 (0)20 7630-0166
e-mail: artech@artechhouse.com   e-mail: artech-uk@artechhouse.com

Find us on the World Wide Web at:
www.artechhouse.com